T0135342

Advances in Intelligent Systems and Computing

Volume 848

Series editor

Janusz Kacprzyk, Polish Academy of Sciences, Warsaw, Poland
e-mail: kacprzyk@ibspan.waw.pl

The series "Advances in Intelligent Systems and Computing" contains publications on theory, applications, and design methods of Intelligent Systems and Intelligent Computing. Virtually all disciplines such as engineering, natural sciences, computer and information science, ICT, economics, business, e-commerce, environment, healthcare, life science are covered. The list of topics spans all the areas of modern intelligent systems and computing such as: computational intelligence, soft computing including neural networks, fuzzy systems, evolutionary computing and the fusion of these paradigms, social intelligence, ambient intelligence, computational neuroscience, artificial life, virtual worlds and society, cognitive science and systems, Perception and Vision, DNA and immune based systems, self-organizing and adaptive systems, e-Learning and teaching, human-centered and human-centric computing, recommender systems, intelligent control, robotics and mechatronics including human-machine teaming, knowledge-based paradigms, learning paradigms, machine ethics, intelligent data analysis, knowledge management, intelligent agents, intelligent decision making and support, intelligent network security, trust management, interactive entertainment, Web intelligence and multimedia.

The publications within "Advances in Intelligent Systems and Computing" are primarily proceedings of important conferences, symposia and congresses. They cover significant recent developments in the field, both of a foundational and applicable character. An important characteristic feature of the series is the short publication time and world-wide distribution. This permits a rapid and broad dissemination of research results.

More information about this series at http://www.springer.com/series/11156

Alexei V. Samsonovich
Editor

Biologically Inspired Cognitive Architectures 2018

Proceedings of the Ninth Annual
Meeting of the BICA Society

 Springer

Editor
Alexei V. Samsonovich
Department of Cybernetics
National Research Nuclear University
 "MEPhI"
Moscow, Russia

ISSN 2194-5357   ISSN 2194-5365   (electronic)
Advances in Intelligent Systems and Computing
ISBN 978-3-319-99315-7   ISBN 978-3-319-99316-4   (eBook)
https://doi.org/10.1007/978-3-319-99316-4

Library of Congress Control Number: 2018951901

This Springer imprint is published by the registered company Springer Nature Switzerland AG
The registered company address is: Gewerbestrasse 11, 6330 Cham, Switzerland

Preface

This volume documents the Proceedings of the 2018 Annual International Conference on Biologically Inspired Cognitive Architectures (BICA), also known as the Ninth Annual Meeting of the BICA Society. BICA are computational frameworks for building intelligent agents that are inspired from biological intelligence. Since 2010, the annual BICA conference attracts researchers from the edge of scientific frontiers. It contrasts major conferences in artificial intelligence (AI) and cognitive science by offering informal brainstorming atmosphere and freedom in conjunction with greater publication venues to ambitious ideas.

Authors of this volume believe that here they have made significant constructive contributions to the future of human-like AI. The future will show whether they are right or wrong. Researchers understand "biological inspirations" broadly, borrowing them from psychology, neuroscience, linguistics, narratology, design and creativity studies in order to advance cognitive robotics and machine learning, among other hot topics in AI. The "filter" is the question of whether your contribution may help us make machines our friends or understand how the mind works. With our steady growth over eight years and unmatched sociocultural program, we look forward to the new unseen success of BICA in Prague in 2018.

BICA 2018 is organized and primarily sponsored by BICA Society, which is an international scientific society and a non-profit organization based in the USA, whose mission is to promote, facilitate, and bring together the many transdisciplinary efforts in the study of cognitive architectures, aimed at a computational implementation of the top functionality of the human mind (known as the BICA Challenge). This year BICA conference is held as part of the Joint Multi-Conference on Human-Level Artificial Intelligence (HLAI-18) at the Czech Technical University (CTU) in Prague, Czech Republic. Therefore, our other sponsors include GoodAI, CTU, and also the NRNU MEPhI: National Research Nuclear University (NRNU) "Moscow Engineering Physical Institute" (MEPhI). We are particularly grateful to all the HLAI-18 Organizers and Partners.

The intended spotlight of this conference—and the volume—is on BICA models of creativity, artificial social–emotional intelligence, active human-like learning, cognitive growth by learning of big volumes of data, narrative intelligence, and true intelligent autonomy in artifacts. Among the major thrusts of the conference are topics like

- Creativity, goal reasoning, and autonomy in artifacts;
- Embodied vs. ambient intelligence;
- Language capabilities and social competence;
- Robust and scalable machine learning mechanisms;
- The role of emotions in artificial intelligence and their BICA models;
- Tests and metrics for BICA in the context of the BICA Challenge;
- Interaction between natural and artificial cognitive systems;
- Theory-of-Mind, episodic and autobiographical memory in vivo and in vitro;
- Introspection, metacognitive reasoning, and self-awareness in BICA;
- Unifying frameworks, standards, and constraints for cognitive architectures;
- Agents possessing human-level social and emotional intelligence.

Works of many, yet not all, distinguished speakers of BICA 2018 are included in this volume. Among the participants of BICA 2012 are top-level scientists like Paul Robertson, Robert Laddaga, Antonio Chella, Agnese Augello, Ignazio Infantino, Jan Treur, Antonio Lieto, Pentti Haikonen, Frank Dignum, Junichi Takeno, Steve PiPaola, Vladimir Red'ko, Peter Boltuc—to name only a few of the famous names. Other names may not be well known yet. It was my great pleasure to work as the BICA 2018 General Chair with all the participants. Particularly, I would like to thank all members of the Organizing Committee and Program Committee for their precious help in reviewing submissions and in helping me to compose the exciting scientific program. I am also very grateful to the key members of the overarching HLAI Organizing Committee: Tarek Besold, Daria Hvizdalova, Olga Afanasjeva, and others. Last but not the least, my special thanks go to the two publishers who made it possible for two volumes of proceedings of BICA 2018 to become known to the world: Leontina Di Cecco (Springer) and Sweitze Roffel (Elsevier). This work could only materialize in a great quality due to their continuous advice and support, their enthusiasm and devotion to the advancement of science and technology.

June 2018   Alexei V. Samsonovich

Organization

Program Committee

Kenji Araki	Hokkaido University, Japan
Joscha Bach	MIT Media Lab, USA
Paul Baxter	Plymouth University, USA
Paul Benjamin	Pace University
Galina A. Beskhlebnova	SRI for System Analysis RAS, Russia
Tarek Besold	City University of London, UK
Jordi Bieger	Reykjavik University, Iceland
Perrin Bignoli	Yahoo! Labs, USA
Douglas Blank	Bryn Mawr College, USA
Mikhail Burtsev	MIPT, Moscow, Russia
Erik Cambria	Nanyang Technological University, Singapore
Suhas Chelian	Fujitsu Laboratories of America, Inc., USA
Antonio Chella	University of Palermo, Italy
Olga Chernavskaya	P.N. Lebedev Physical Institute, Russia
Thomas Collins	University of Southern California, USA
Christopher Dancy	Pennsylvania State University, USA
Haris Dindo	University of Palermo, Italy
Sergey A. Dolenko	D.V. Skobeltsyn Institute of Nuclear Physics, M.V. Lomonosov Moscow State University, Russia
Alexandr Eidlin	National Research Nuclear University MEPhI, Russia
Jim Eilbert	AP Technology, USA
Thomas Eskridge	Florida Institute of Technology, USA
Usef Faghihi	University of Indianapolis, USA
Stanley Franklin	University of Memphis, USA
Salvatore Gaglio	University of Palermo, Italy
Olivier Georgeon	Claude Bernard Lyon 1 University, France

Ian Horswill	Northwestern University, USA
Eva Hudlicka	Psychometrix Associates, USA
Christian Huyck	Middlesex University, UK
Ignazio Infantino	Consiglio Nazionale delle Ricerche, Italy
Eduardo Izquierdo	Indiana University, USA
Alex James	Kunming University of Science and Technology, China
Li Jinhai	Kunming University of Science and Technology, China
Magnus Johnsson	Lund University, Sweden
Darsana Josyula	Bowie State University, USA
Kamilla Jóhannsdóttir	Reykjavik University, Iceland
William Kennedy	George Mason University, USA
Deepak Khosla	HRL Laboratories, LLC, USA
Swathi Kiran	Boston University, USA
Valentin Klimov	National Research Nuclear University MEPhI, Russia
Giuseppe La Tona	University of Palermo, Italy
Luis Lamb	Federal University of Rio Grande do Sul, Brazil
Leonardo Lana de Carvalho	ISH et LIESP, Brazil
Othalia Larue	University of Quebec, Canada
Christian Lebiere	Carnegie Mellon University, USA
Jürgen Leitner	Australian Centre of Excellence for Robotic Vision, Australia
Simon Levy	Washington and Lee University, USA
Antonio Lieto	University of Turin, Italy
James Marshall	Sarah Lawrence College, USA
Olga Mishulina	National Research Nuclear University MEPhI, Russia
Steve Morphet	Enabling Tech Foundation, USA
Amitabha Mukerjee	Indian Institute of Technology Kanpur, India
David Noelle	University of California, Merced, USA
Andrea Omicini	University of Bologna, Italy
Marek Otahal	Czech Institute of Informatics, Robotics and Cybernetics, Czech Republic
Aleksandr I. Panov	FRC CSC RAS, Moscow, Russia
David Peebles	University of Huddersfield, UK
Giovanni Pilato	ICAR-CNR, Italy
Michal Ptaszynski	Kitami Institute of Technology, Japan
Subramanian Ramamoorthy	University of Edinburgh, Scotland
Uma Ramamurthy	Baylor College of Medicine, USA
Thomas Recchia	US Army ARDEC, USA
Vladimir Redko	Scientific Research Institute for System Analysis RAS, Russia
James Reggia	University of Maryland, USA

Contents

Multi-level Modeling of Structural Elements of Natural Language Texts and Its Applications

Anton Anikin[✉], Oleg Sychev, and Vladislav Gurtovoy

Volgograd State Technical University, Volgograd, Russia
anton@anikin.name, poas@vstu.ru

Abstract. Methods of extracting knowledge in the analysis of large volumes of natural language texts are relevant for solving various problems in the field of analysis and generation of textual information, such as text analysis for extracting data, fact and semantics; presenting extracted information in a convenient for machine processing form (for example, ontology); classification and clustering texts, including thematic modeling; information retrieval (including thematic search, search based on the user model, ontology-based models, document sample based search); texts abstracting and annotating; developing of intelligent question-answering systems; generating texts of different types (fiction, marketing, weather forecasts etc.); as well as rewriting texts, preserving the meaning of the original text for presenting it to different target audiences. In order for such methods to work, it is necessary to construct and use models that adequately describe structural elements of the text on different levels (individual words, sentences, thematic text fragments), their characteristics and semantics, as well as relations between them, allowing to form higher-level structures. Such models should also take into account general characteristics of textual data: genre, purpose, target audience, scientific field and others. In this paper, authors review three main approaches to text modeling (structural, statistical and hybrid), their characteristics, pros and cons and applicability on different stages (knowledge extraction, storage and text generation) of solving problems in the field of analysis and generation of textual information.

Keywords: NLP · Machine learning · Text mining · Text modeling Topic modeling · Ontology · Ontology learning

1 Introduction

Natural language text is a complex phenomenon which can be modeled on several different levels - they would be called text dimensions. Three basic text

This paper presents the results of research carried out under the RFBR grant 18-07-00032 "Intelligent support of decision making of knowledge management for learning and scientific research based on the collaborative creation and reuse of the domain information space and ontology knowledge representation model".

A. V. Samsonovich (Ed.): BICA 2018, AISC 848, pp. 1–8, 2019.
https://doi.org/10.1007/978-3-319-99316-4_1

dimensions are semantic dimension, which describes the content and the meaning of the text fragment; syntactical dimension, which describes the length and the structure of sentences; and lexical dimension, which describes particular words comprising text (semantic and syntactical dimensions can describe only a set of synonyms for particular word, but they are not affecting choosing a word from that set).

All text processing tasks can be put into one of two categories according to their goals: text classification tasks (texts clustering and classification on similarity, information retrieval on request or sample document) and text generation tasks (text abstracting and annotating, intelligent question-answering systems, generating texts of different types, text rewriting, machine translation). Text analysis tasks (information and knowledge extraction) generates some internal representation for further use in classification and text generation tasks. The form of such representation is often tailored either to text classification or generation.

In tasks concerning text generation, decisions should be taken for each text dimension: semantic dimension will determine the number of sentences and their meaning, the syntactical dimension will affect the structure of sentences and lexical one will affect choosing the particular word from a set of synonyms. Text classification tasks mainly concern semantic dimension. It should be noted that the goal of the model should be taken into account when considering dimensions of the text the model works with, not the sort of information that was used to build the model. So, for example, distributional semantic based models operate on the semantic level because their goal is the semantic classification of text, even if they use individual words (lexical dimensions) to find it. Some models are created for easier transition between two dimensions of the text, for example, FrameNet model allows to map semantics of the text to syntax structures. Such models are of particular interests in text generation tasks.

Also, the models can be classified by used methods into statistical (count frequencies of words or other elements, co-occurrence and other statistical parameters and ignores the structure of text or sentences), structural (use information about syntactical and other relations between the text elements) and hybrid models that use both statistical and structural information.

2   Semantic Text Models

Distributional semantic implies using the distributed representations of words learned by neural networks. Word2vec [15] model allows representing the words as a continuous vector that uses words co-occurrence statistics. It also uses continuous bag-of-words (BOW) and skip-gram models to predict current word based on the context as well as maximize classification of a word based on another word in the same sentence. It uses unlabeled text corpus that allows reducing efforts but does not allow to resolve the problems of polysemy and homonymy. The model allows finding answers to many types of semantic and syntactic questions in form of a word for which a pair should be found that relates to the original word in the same way as words in pairs in a training list that was created manually. The similar approach extended to represent sentences, paragraphs, and documents as vectors to measure the texts similarity in paragraph2vec and doc2vec models [11] with the same advantages and issues.

The LDA model [5] assumes the existence of latent topics that appear across multiple text documents. Each document has its own mixture of topics, and each topic is characterized by a discrete probability distribution over words: the probability that a specific word is present in a text document depends on the presence of a latent topic. This model uses Latent Dirichlet Allocation that is significantly more expensive computationally on large data sets compared with the previous models but allows to improve inference and prediction.

Topical Model for Sentence [3] is also created to solve the topic modeling problem. It proposes senLDA - a statistical unsupervised probabilistic topic model for text classification that extends LDA. Compared with LDA, that assumes independence of all the words in documents, senLDA assumes that words within a sentence pertain to the same topic.

Top-Down Semantic Model for Sentence Representation [17] is created by extracting a sentence representation that uses both the word-level semantics by linearly combining the words with attention weights and the sentence-level semantics with BiLSTM (Bidirectional Long-Short Term Memory). It is used for text classification. It learns a sentence representation that carries the information of both the BOW-like representation and Recurrent Neural Networks to represent sentence-level semantic which performs well for datasets of any size and complexity. Word topic vector in this model differs from word embedding: each dimension of the topic-vector represents a topic aspect of the word. Topic-vector is extracted from characters in words using a fully convolutional network.

Text Model Combining Lexical and Semantic Features for Short Text Classification [18] allows classifying short texts by combining both their lexical and semantic features. It uses topic repositories with long documents as background knowledge, maps the weighted words of short texts to corresponding topics as the vector representations of short texts, and trains the classification model on labeled data.

A Formal Text Representation Model Based on Lexical Chaining [14] is an alternative to the vector space model. It combines a tree-like model with graph-inducing lexical relations - lexical chaining and quantitative structure analysis, in order to combine content and structure modeling to improve the model quality. QSA shows that a large number of genre-related text categories can be learned by exploring the structure of texts, that proves that lexical chaining has a high potential for content-related categorization. The terminological ontologies are used as a lexical resource for chaining.

Neural Text Generation from Structured Data paper [12] introduces a neural model for a concept-to-text generation field (where rule-based models are more common) that scales well to large, rich domains and generates biographical sentences from fact tables on a new dataset of biographies from Wikipedia. It contains steps for content planning, sentence planning and surface realization for constrained sentence generation. The model also implements attention mechanism over input table fields.

A Global Model for Concept-to-Text Generation [9] use database records to generate textual output for specific domains like sportscasting, weather forecast, air travel. The input (a set of database records and text describing some of them) reformulated as a probabilistic context-free grammar and task is to find the best generated string licensed by the grammar. The weights of grammar rules estimated during model training.

3   Semantic to Syntax Text Models

Ontology allows to describe semantics using domain concepts and relations between them [1]. Ontology-based models for NLG approach [4] use information stored in an ontology as a set of triples subject-property-object (so it is a structural model) and specially developed Text Composing Language (that provides the syntactic rules) to generate text. The language uses constructs typical to the programming languages Javascript and PHP to retrieve data and templates with variable substitution to generate sentences. The model is used to generate short similar text in well-established domains like weather forecasts, finance and sports news.

FrameNet project aims at providing the link between semantic and syntax dimensions of text using Frame Semantic and Construction Grammar theories [2,6]. The semantic part consists of Frames that include Lexical Units (a word-meaning pair that serves as a trigger to evoke particular frame) and Frame Elements (other words or sub-phrases in a frame). The syntactical part consists of annotated sentences. An annotation for each frame element is a triple of its role in the frame, grammatical function, and phrase type (26 phrase types are defined). Frames are linked together by such relations as inheritance, using, subframe and perspective on. FrameNet is a structural model used in many natural language processing tasks like information extraction, machine translation, natural language generation, natural language understanding, question answering, semantic role labeling and word sense disambiguation.

4   Syntactic and Lexical Text Models

Rhetorical Structure Theory is a descriptive theory of text organization, that allows to parse text into independent, non-overlapping units (sentences and clauses) and identify rhetorical relationships, that represent functional relations between them like condition, question-answer, motivation, elaboration, contrast, concession, sequence, means etc. [13]. Syntactical parsing system Discourse Analysis System can find rhetorical structure units in given text and build a tree of relationships between them using cue phrases and various textual coherence structures [10]. It is a structural model, that can be used to transform original text - for example, to translate descriptive text into dialogue representation, that can be used in a more user-friendly, engaging ways [16]. However, rhetorical structure trees contain only high-level information about a limited number of relations and do not capture related content or its syntactical structure, which makes uses of such model limited to restructuring clauses from the original text.

Literary Structures model was proposed as a basis for modeling and generating novel sentences, that can be part of a literary text [7]. It contains five types of phrases that overlap in text with one another on key words (cf. other models: Rhetorical Structure Theory is working with independent phrases, while Construction Grammar works with embedded, but not overlapping structures). Main types of phrases are noun phrases with a verb, verb phrases with a preposition, previous prepositional phrases, simple phrases and clauses. Overlapping of phrases is used to sequence them with one another to create sentences using word-phrase similarity measure without modeling semantics of the text. Syntactical Literary Structures model is a structural model, while the proposed approach to the sentence and text generation is a statistical one. Examples provided by authors show that just word overlapping without accounting for polysemy, homonymy and domain produces sentences with low coherency.

WordNet, a project developed at Princeton University, set a standard for a lexical model of the language, it remains one of most comprehensive models of the words which constitutes texts [8]. The basic unit of WordNet is a set of synonyms (synset), representing a particular meaning. One word can belong to many synsets. Synsets are linked together by relations. Most common relations is super-subordinate or hyperonymy, which links more general synsets with more specific ones, allowing meanings of the language to form a hierarchy. Nouns can also form part-whole relations, while adjectives are organized in antonymous pairs. WordNet synsets are much better suited to represent meaning in the semantic-level models, than particular words. The main problem of using WordNet for text generation is lack of information for choosing particular synonym from synset depending on the text style, target audience, and other characteristics.

5 Conclusion

The conducted survey shows that though statistical models (including neural networks) excel in text classification tasks, they are of much less use in text generation tasks because text generation includes the creation of complex structure while statistical models ignore structural rules and can generate unacceptable outcomes. While a certain percent of wrong outcomes is normal for classification tasks, it is much worse in text generation tasks. This reflects the fact that each generated text, delivered to the user, should be acceptable and that the only way to reliably discern acceptable text from unacceptable is human proofreading, which depreciates the idea of automatic text generation. Structural models, based on Ontologies or Frame Semantic theory is better for text generation.

Regarding text dimensions, classification is usually done in semantic dimension to find topic or domain of the text. Classification in lexical dimensions is a part-of-speech tagging which modern software performs well, so they didn't need an additional research.

Some text generation tasks, that usually require text as an input too - machine translation, text rewriting - can be solved on just syntax (Discourse Analysis System) or semantic models. Most text generation tasks require combining models from three text dimensions: semantic model will define the content of the generated text, syntax model will define the structure necessary sentences and lexical model will provide information to choose particular words. The good approach for such tasks is combining ontological model (semantic dimension) with FrameNet (syntax dimension) and WordNet (lexical dimension). But solving such tasks requires creating an extensive ontology of facts used in text generation and providing information for choosing particular synonym from synset, that WordNet lacks (Table 1).

Table 1. Comparison of text models

Text model	Text dimension	Method	Model purpose
Distributional semantic based models (word2vec, paragraph2vec, doc2vec)	Semantic	Statistical	Classification
Latent Dirichlet Allocation model	Semantic	Statistical	Classification
Topic model for sentence	Semantic	Statistical	Classification
Top down semantic model for sentence representation	Semantic	Statistical	Classification
Text model combining lexical and semantic features for short text classification	Semantic	Statistical	Classification
A formal text representation model based on lexical chaining	Semantic	Hybrid	Classification
Neural text generation from structured data	Semantic	Statistical	Generation
A global model for concept-to-text generation	Semantic	Statistical	Generation
Ontology based models for natural language generation	Semantic to syntax	Structural	Generation
FrameNet	Semantic to syntax	Structural	Generation
Literary structures model	Syntax	Structural	Generation
Discourse structure analysis based dialogue model	Syntax	Structural	Generation
WordNet	Lexical	Structural	Generation

References

1. Anikin, A., Litovkin, D., Kultsova, M., Sarkisova, E.: Ontology-based collaborative development of domain information space for learning and scientific research. In: Ngonga Ngomo, A.C., Křemen, P. (eds.) Proceedings of Knowledge Engineering and Semantic Web: 7th International Conference, KESW 2016, 21-23 September 2016, Prague, Czech Republic, pp. 301–315 (2016)
2. Baker, C.F., Fillmore, C.J., Lowe, J.B.: The berkeley FrameNet project. In: COLING-ACL 1998, Proceedings of the Conference, Montreal, Canada, pp. 86–90 (1998)
3. Balikas, G., Amini, M.R., Clausel, M.: On a topic model for sentences. In: Proceedings of the 39th International ACM SIGIR Conference on Research and Development in Information Retrieval, pp. 921–924. SIGIR 2016. ACM, New York (2016). https://doi.org/10.1145/2911451.2914714
4. Bense, H.: Using very large scale ontologies for natural language generation. In: JOWO. CEUR Workshop Proceedings, vol. 2050. CEUR-WS.org (2017)
5. Blei, D.M., Lafferty, J.D.: Dynamic topic models. In: Proceedings of the 23rd International Conference on Machine Learning, ICML 2006, pp. 113–120. ACM, New York (2006). https://doi.org/10.1145/1143844.1143859
6. Boas, H.C.: From Theory to Practice: Frame Semantics and the Design of FrameNet, pp. 129–160. Narr, Tübingen (2005)
7. Daza, A., Calvo, H., Figueroa-Nazuno, J.: Automatic text generation by learning from literary structures. In: Proceedings of the Fifth Workshop on Computational Linguistics for Literature, pp. 9–19. Association for Computational Linguistics (2016). https://doi.org/10.18653/v1/W16-0202
8. Fellbaum, C. (ed.): WordNet: An Electronic Lexical Database. MIT Press, Cambridge (1998)
9. Konstas, I., Lapata, M.: A global model for concept-to-text generation. J. Artif. Int. Res. **48**(1), 305–346 (2013). http://dl.acm.org/citation.cfm?id=2591248.2591256
10. Le, H.T., Abeysinghe, G.: A study to improve the efficiency of a discourse parsing system. In: Gelbukh, A. (ed.) Computational Linguistics and Intelligent Text Processing, pp. 101–114. Springer, Heidelberg (2003)
11. Le, Q., Mikolov, T.: Distributed representations of sentences and documents. In: Proceedings of the 31st International Conference on International Conference on Machine Learning. ICML2014, vol. 32, pp. II–1188–II–1196. JMLR.org (2014)
12. Lebret, R., Grangier, D., Auli, M.: Generating text from structured data with application to the biography domain. CoRR abs/1603.07771 (2016). http://arxiv.org/abs/1603.07771
13. Mann, W.C., Thompson, S.A.: Rhetorical structure theory: toward a functional theory of text organization. Text **8**(3), 243–281 (1988)
14. Mehler, A., Waltinger, U., Wegner, A.: A formal text representation model based on lexical chaining. In: Proceedings of the KI 2007 Workshop on Learning from Non-Vectorial Data (LNVD 2007), 10 September, Universität Osnabrück, pp. 17–26 (2007)
15. Mikolov, T., Chen, K., Corrado, G., Dean, J.: Efficient estimation of word representations in vector space. CoRR abs/1301.3781 (2013). http://arxiv.org/abs/1301.3781
16. Prendinger, H., Piwek, P., Ishizuka, M.: Automatic generation of multi-modal dialogue from text based on discourse structure analysis. In: International Conference on Semantic Computing. ICSC 2007, pp. 27–36, September 2007

17. Wu, Z., Zheng, X., Dahlmeier, D.: Character-based text classification using top down semantic model for sentence representation. CoRR abs/1705.10586 (2017). http://arxiv.org/abs/1705.10586
18. Yang, L., Li, C., Ding, Q., Li, L.: Combining lexical and semantic features for short text classification. Procedia Comput. Sci. **22**, 78–86 (2013). 17th International Conference in Knowledge Based and Intelligent Information and Engineering Systems - KES2013

NarRob: A Humanoid Social Storyteller with Emotional Expression Capabilities

Agnese Augello$^{(\boxtimes)}$, Ignazio Infantino, Umberto Maniscalco, Giovanni Pilato, and Filippo Vella

ICAR-CNR, Via Ugo La Malfa 153, 90146 Palermo, Italy
{agnese.augello,ignazio.infantino,umberto.maniscalco,
giovanni.pilato,filippo.vella}@cnr.it

Abstract. In this paper we propose a model of a robotic storyteller, focusing on its abilities to select the most appropriate gestures to accompany the story, trying to manifest also emotions related to the sentence that is being told. The robot is endowed with a repository of stories together with a set of gestures, inspired by those typically used by humans, that the robot learns by observation. The gestures are annotated by a number N of subjects, according to their particular meaning and considering a specific typology. They are exploited by the robot according to the story content to provide an engaging representation of the tale.

Keywords: Storytelling · Gestures acquisition
Robot expressiveness · Emotions detection from text

1 Introduction

Storytelling is a powerful communication strategy and can lead to several benefits in multiple contexts, especially in education, facilitating the learning process and promoting moral and social values [12]. To make it storytelling effective, it is important to narrate the story with an adequate expressiveness by adding the most appropriate non-verbal communicative signs to the verbal content. Nowadays robots are more and more included in social environments with different purposes, and storytelling is a task where social robots are recently employed with several advantages, they can support the learning process with a greater engagement [4,15,17] and the improvement of social and emotional intelligence skills [8], contributing, by using appropriate co-verbal gestures, to influence the mood induction process of the story and the storytelling experience [18]. In this work we propose a model of a robotic storyteller, focusing on its abilities to select the most appropriate gestures to accompany the story, while showing body expressiveness to emphasize the emotions arising from the text. The storyteller is embodied in the Softbank Pepper Robot. The robot is endowed with a repository of stories and it is equipped with a set of gestures, learned by the

© Springer Nature Switzerland AG 2019
A. V. Samsonovich (Ed.): BICA 2018, AISC 848, pp. 9–15, 2019.
https://doi.org/10.1007/978-3-319-99316-4_2

robot through a direct observation of gestures performed human and acquired by using a RGBD device.

Since a robot like Pepper cannot adapt its facial expression, the module has an effect on the appearance of specific robot's communicative channels that can be easily understood by a human observer as correlated to some emotions, such as the color of its leds, the speed of its speech, and the head inclination [1,3,6].

Before the narration, the robot analyzes each sentence of the story, in order to catch if there is a basic emotion related to that particular chunk of text that is being expressed. In particular, the six basic Eckman emotions have been considered, in order to show the most appropriate expressiveness to accompany the speech during the storytelling.

2   Gestures During Storytelling

A storyteller can catch the attention of its audience by giving more expressiveness during a narration. A greater expressiveness can be obtained by accompanying the speech with non-verbal behavior, such as facial expressions, postures, and gestures. This is important to emphasize both the meaning and the affective content of a story.

Starting from this choice, we have automatically detected words, specifically verbs, that can be associated to a representation of a concept in the stories. The detection is obtained as follows: the recognition module segments the text of a story in sentences; each sentence is therefore parsed and verbs are detected; a lemmatization step finds the lemma of the verb, thereafter the semantic similarity is computed between the verb detected in the sentence and each one of the verbs listed in the knowledge base of the robot, where actions are associated to gestures. If the semantic similarity is below a given threshold T_{sim}, whose value is experimentally determined, the verb is ignored and no specific gesture is performed by the robot. Otherwise, it is executed the action corresponding to the highest similarity between the detected verb in the sentence and that one present in the list of actions that can be performed by the robot. If the highest value is associated to two actions, a random choice process is executed. For the emotional labeling, we have considered the six Ekman basic emotions: *anger, disgust, fear, joy, sadness* and *surprise*, exploiting a well known lexicon derived from the Word-Net Affect Lexicon [13,14] and applying the procedure that has been introduced in [11].

3   Gestures Acquisition and Annotation

We have created a dataset of gestures by the direct observation of a human being while a story is being told. An RGBD device captured the skeleton during multiple repetitions of the same dynamic posture. From each detected gesture, the acquisition module extracted the relevant joints angles and translated them into corresponding robot's joint angles. In particular, we process the 3d positions of the shoulders, elbows, and wrists to compute the shoulder pitch, the shoulder

Fig. 1. Three examples of emotional manifestation of the robot: sadness (left), joy (middle), surprise (right). The eyebrows are emulated by the positioning of the pink leds.

roll, the elbow roll, the elbow yaw. Considering left and right arms, we have eight angle joints determining the robot posture and its evolution over the time. The rate of the human motion capture has been two frames per second, allowing to have a satisfying robotic reproduction of them. To avoid noisy data, we acquired ten repetitions of the same gesture and used mean filtering and normalizing over the duration (Fig. 1).

The problem of synchronizing the gestures of a humanoid robot with speech is a very complex problem. In Fig. 2 three different situation are reported.

At the top of the figure, a phrase in which a single verb and the associated movement is present is represented.

In the middle of the same figure, a phrase in which two verbs and the associated movements is present is represented. In this case, the duration of the first movement is so long that it does not allow the beginning of the second movement. At the bottom of the Fig. 2, a phrase in which two verbs and the associated movements is present is represented. In this case, the duration of the first movement is short enough to allow the beginning of the second movement.

If we consider the simplest case represented at the top in Fig. 2 where a single movement must be synchronized with a single verb, the problem can be solved by finding the timing of the speech and starting the action at the desired moment. Indicating with d_{wi} the duration of the i-th word in the sentence, with d_{vi} the duration of the i-th verb, with d_{mi} the duration of the i-th movement and with p the length of the silence between two words the start time t_{vi_s} of the verb in the speech time-line can be obtained as: $t_{v_s} = \sum_{i=1}^{N} d_{wi} + N * p$ where N is the number of words that precede the verb. The end time of the verb is $t_{v_e} = t_{v_s} + d_v$.

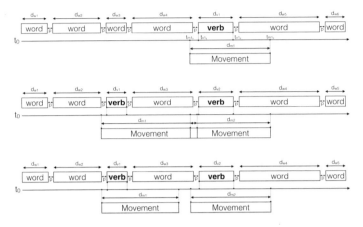

Fig. 2. Three different situations: a single verb with a single movement and two verbs with two overlapped and not overlapped movements.

Starting from these considerations, different hypotheses to synchronize the gesture and not the pronunciation of the verb can be made:

- the movement begins together with the start time of the verb and then $t_{m_s} = t_{v_s} = \sum_{i=1}^{N} d_{wi} + N * p$;
- the movement ends with the ending time of the verb and then $t_{m_s} = t_{v_e} - d_m$;
- the movement is centered with the duration of the verb and then $t_{m_s} = t_{v_s} + \frac{t_{v_e} + t_{v_s}}{2}$;
- the movement starts a time δ before/after the starting time of the verb and then $t_{m_s} = t_{v_s} - \delta$.

Each one of the previous solutions could be considered a valid alternative to synchronize a gesture with the corresponding verb, however, we use the value of -0.3 s according to the findings of [5] to start a robot gesture when the verb to which it refers is presented in the sentence.

4   An Example

We have built a dataset by collecting short stories from the web[1] As an example, we have considered the "Ali and the carpet" story. In Table 1 we report the emotions detected in the text as well the actions to be performed by the robot during the storytelling process.

---

[1] http://www.english-for-students.com/Short-Moral-Stories-for-Kids.html
  http://learnenglishkids.britishcouncil.org/en/.

Table 1. Sentences, emotions and gestures during a storytelling

Sentence	Emotion	Gesture
One very hot day Ali finds a carpet in his uncle's shop	Joy	to_find
What's this?' Suddenly the carpet jumps!	Surprise	no_action
It moves and flies off into the air	Fear	to_go, to_fly
'Hey!	Joy	no_action
Whats happening? A loud booming voice comes from the carpet	Sadness	no_action
Welcome, O master	Joy	no_action
I am a magic carpet. First they fly high up into the sky and then they land in a jungle	Sadness	to_fly, to_land
It is hot and wet and its raining	Sadness	no_action
Its raining!	Sadness	no_action
Yuck! Then they fly to the desert	Anger	to_fly
It is very, very hot and dry	Fear	no_action
It is very, very hot today! After that they fly to the South Pole	Sadness	to_fly
There is lots of ice and snow	Sadness	no_action
Its freezing	Anger	no_action
Brrr! Where are we now?	no_emotion	no_action
I cant see! In the mountains	Sadness, Anger	no_action
Can you see me? Its very foggy. Then they fly to a forest	no_emotion	to_see, to_fly
Its very windy there	Sadness	no_action
Oh, its windy in the forest! Then they fly to an island in the sea	Sadness	to_fly
There is thunder and lightning	Fear	no_action
Aaagh!	no_emotion	no_action
Lets go home! What a storm! Finally they fly back home	Fear	to_go, to_fly
The carpet lands in the shop and Ali gets off	Joy	to_land
Wow!	Joy	no_action
What an adventure!	Joy	no_action

5   Conclusion and Future Works

We have presented a first prototype of a robotic storyteller. The robot is capable to select the most appropriate gestures to accompany the story, while expressing also the emotions which arise from the text.

The approach is promising, and it is a first step toward the realization of social robots for an effective storytelling. Future work will regard the personalization of the choice of the story through a dialogue process aimed at detecting the user preferences and interests, as well as taking into account the reaction of the user while the tale is being told.

References

1. Bänziger, T., Scherer, K.R.: The role of intonation in emotional expressions. Speech Commun. **46**(3), 252–267 (2005)
2. Casasanto, D.: Gesture and language processing. In: Encyclopedia of the Mind, pp. 372–374 (2013)
3. Feldmaier, J., Marmat, T., Kuhn, J., Diepold, K.: Evaluation of a RGB-LED-based emotion display for affective agents. arXiv preprint arXiv:1612.07303 (2016)
4. Ferreira, M.J., Nisi, V., Melo, F., Paiva, A.: Learning and teaching biodiversity through a storyteller robot. In: International Conference on Interactive Digital Storytelling, pp. 367–371. Springer, Heidelberg (2017)
5. Holzapfel, H., Nickel, K., Stiefelhagen, R.: Implementation and evaluation of a constraint-based multimodal fusion system for speech and 3D pointing gestures, 01 2004
6. Johnson, D.O., Cuijpers, R.H., van der Pol, D.: Imitating human emotions with artificial facial expressions. Int. J. Soc. Robot. **5**(4), 503–513 (2013)
7. Landauer, T.K., Foltz, P.W., Laham, D.: An introduction to latent semantic analysis. Discourse Process. **25**(2–3), 259–284 (1998)
8. Leite, I., McCoy, M., Lohani, M., Ullman, D., Salomons, N., Stokes, C., Rivers, S., Scassellati, B.: Narratives with robots: the impact of interaction context and individual differences on story recall and emotional understanding. Front. Robot. AI **4**, 29 (2017)
9. Loehr, D.P.: Gesture and intonation. Ph.D. thesis, Georgetown University, Washington, DC (2004)
10. McNeill, D.: Hand and Mind: What Gestures Reveal About Thought. University of Chicago Press, Chicago (1992)
11. Pilato, G., D'Avanzo, E.: Data-driven social mood analysis through the conceptualization of emotional fingerprints. Procedia Comput. Sci. **123**, 360–365 (2018). 8th Annual International Conference on Biologically Inspired Cognitive Architectures, BICA 2017 (Eighth Annual Meeting of the BICA Society), 1–6 August 2017, Moscow, Russia
12. Robin, B.: The power of digital storytelling to support teaching and learning. Digit. Educ. Rev. **30**, 17–29 (2016)
13. Strapparava, C., Mihalcea, R.: Semeval-2007 task 14: affective text. In: Proceedings of the 4th International Workshop on Semantic Evaluations. SemEval 2007, pp. 70–74. Association for Computational Linguistics, Stroudsburg (2007)
14. Strapparava, C., Mihalcea, R.: Learning to identify emotions in text. In: Proceedings of the 2008 ACM Symposium on Applied Computing. SAC 2008, pp. 1556–1560, ACM, New York (2008)
15. Striepe, H., Lugrin, B.: There once was a robot storyteller: measuring the effects of emotion and non-verbal behaviour. In: International Conference on Social Robotics, pp. 126–136. Springer, Heidelberg (2017)

16. Wagner, P., Malisz, Z., Kopp, S.: Gesture and speech in interaction: an overview (2014)
17. Westlund, J.K., Breazeal, C.: The interplay of robot language level with children's language learning during storytelling. In: Proceedings of the Tenth Annual ACM/IEEE International Conference on Human-Robot Interaction Extended Abstracts, pp. 65–66. ACM (2015)
18. Xu, J., Broekens, J., Hindriks, K., Neerincx, M.A.: Effects of a robotic storyteller's moody gestures on storytelling perception. In: 2015 International Conference on Affective Computing and Intelligent Interaction (ACII), pp. 449–455. IEEE (2015)

The Cortical Conductor Theory: Towards Addressing Consciousness in AI Models

Joscha Bach[(✉)]

Harvard Program for Evolutionary Dynamics, Cambridge, MA 02138, USA
bach@fas.harvard.edu

Abstract. AI models of the mind rarely discuss the so called "hard problem" of consciousness. Here, I will sketch informally a possible functional explanation for phenomenal consciousness: the conductor theory of consciousness (CTC). Unlike IIT, CTC is a functionalist model of consciousness, with similarity to other functionalist approaches, such as the ones suggested by Dennett and Graziano.

Keywords: Phenomenal consciousness · Cortical conductor theory
Attention · Executive function · Binding · IIT

1 Introduction: Consciousness in Cognitive Science

While AI offers a large body of work on agency, autonomy, motivation and affect, cognitive architectures and cognitive modeling, there is little agreement on how to address what is usually called "the hard problem" of consciousness. How is it possible that a system can take a first person perspective, and have phenomenal experience?

One of the better known recent attempts to address phenomenal consciousness is Guilio Tononi's Integrated Information Theory (IIT) [1, 2]. Tononi argues that experience cannot be reduced to a functional mechanism, and hence it must be an intrinsic property of a system, rather than a functional one. He characterizes consciousness by a parameter, Φ, which is a measure for the amount of mutual information over all possible partitionings of an information processing system. If the information in the system is highly integrated (i.e. the information in each part of the system is strongly correlated with the information in the others), it indicates a high degree of consciousness. As has for instance been argued by Aaronson [3], IIT's criterion of information integration could perhaps be necessary, but is not sufficient, because we can construct structurally trivial information processing systems that maximize Φ by maximally distributing information (for instance via highly interconnected XOR gates). Should we assign consciousness to processing circuits that are incapable of exhibiting any of the interesting behaviors of systems that we usually suspect to be conscious, such as humans and other higher animals?

From a computationalist perspective, IIT is problematic, because it suggests that two systems that compute the same function by undergoing a functionally identical sequence of states might have different degrees of consciousness based on the arrangement of the computational elements that realize the causal structure of the

© Springer Nature Switzerland AG 2019
A. V. Samsonovich (Ed.): BICA 2018, AISC 848, pp. 16–26, 2019.
https://doi.org/10.1007/978-3-319-99316-4_3

system. A computational system might turn out to be conscious or unconscious regardless of its behavior (including all its utterances professing its phenomenal experience) depending on the physical layout of its substrate, or the introduction of a distributed virtual machine layer. A more practical criticism stems from observing conscious and unconscious people: a somnambulist (who is generally not regarded as conscious) can often answer questions, navigate a house, open doors etc., and hence should have cortical activity that is distributed in a similar way as it is in an awake, conscious person [4]. In this sense, there is probably only a low quantitative difference in Φ, but a large qualitative difference in consciousness. This qualitative difference can probably be explained by the absence of very particular, local functionality in the brain of the somnambulist: while her cortex still produces the usual content, i.e. processes sensory data and generates dynamic experiences of sounds, patterns, objects, spaces etc. from them, the part that normally attends to that experience and integrates it into a protocol is offline. This integrated experience is not the same as information integration in IIT. Rather, it is better understood as a particular local protocol by one of the many members of the "cortical orchestra": its *conductor*.

In this contribution, I will sketch how a computational model can account for the phenomenology and functionality of consciousness, based on my earlier work in the area of cognitive architectures [5]; we might call this approach the "conductor theory of consciousness" (CTC).

2   An AI Perspective on the Mind

Organisms evolved information processing capabilities to support the regulation of their systemic demands in the face of the disturbances by the environment. The simplest regulator system is the feedback loop: a system that measures some current value and exerts a control operation that brings it close to a target value. Using a second feedback loop to regulate the first, the system can store a state and regulate one value depending on another. By changing the control variables to maximize a measured reward variable, a system can learn to approximate a complex control function that maps the values of a set of inputs (sensors) to operators (effectors). Our nervous systems possess a multitude of feedback loops (such as the mechanisms of the brain stem regulating heart rate and breathing patterns).

The control of behavior requires more complex signals; the sensors of the limbic system measure changes in organismic demands and respond to satisfaction of needs with pleasure signals (indicating to intensify the current behavior). The frustration of needs leads to displeasure signals (pain), which indicate the current behavior should be stopped. Directed behavior of a system may be governed by impulses, which associate situations (complex patterns in the outer or inner environment of the organism) with behavior to obtain future pleasure, or avoid future pain. Pain and pleasure act as reward signals that establish an association between situations, actions and needs [6]. In mammals, such connections are for instance established in the hippocampus [7] (Fig. 1).

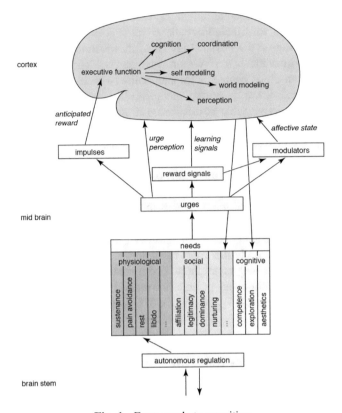

Fig. 1. From needs to cognition

The human neocortex enables better regulation of needs by encoding sensory patterns into a complex hierarchical model of the environment (including the inner environment). This dynamic model is not just a mapping from past observation to future observations, but takes on the shape of a progressively updated stateful function, a program that generates a simulation of the environment. The formation of the model is driven largely by data compression, i.e. by optimizing for the data structure that allows the best predictions of future observations, based on past observations. This principle has for instance been described by Solomonoff [8]: The best possible model that a computational agent can form about its environment is the shortest program among those that best predict an observation from past observations, for all observations and past observations. Machine learning models of the mind can be understood as approximating Solomonoff induction [9], by capturing the apparent invariances of the world into an almost static model, and its variance as a variable state of that model. By varying the state, such a model cannot only capture the current state of the world, but be used to anticipate and explore possible worlds, to imagine, create and remember. Machine learning systems have demonstrated how recurrent neural networks can discover and predict the structure of visual and auditory stimuli by forming low level feature detectors, which can then be successively organized into complex high level features, object categories and conceptual manifolds [10]. Deep networks can form hierarchical knowledge representations. LSTMs [11] and GRUs [12] are building blocks for recurrent neural networks that can learn sequences of operations. Generative neural networks can use the constraints learned from the data to produce possible worlds [13].

While current machine learning systems outperform humans in many complex tasks that require the discovery and manipulation of causal structures in large problem spaces, they are very far from being good models of intelligence. Part of this is due to our current learning paradigms, which lead to limitations in the generation of compositional knowledge and sequential control structures, and will be overcome with incremental progress.

Recently, various researchers have proposed to introduce a unit of organization similar to cortical columns into neural learning [14]. Cortical columns are elementary circuits containing between 100 and 400 neurons [15], and are possibly trained as echo state networks [16] to achieve functionality for function approximation, conditional binding and reward distribution. In the human neocortex, the columnar units form highly interconnected structures with their immediate neighbors, and are selectively linked to receptive fields in adjacent cortical areas. A cortical area contains ca. 10^6 to 10^7 columns, and may be thought of as a specialized instrument in the orchestra of the neocortex.

Beyond Current Machine Learning

A more important limitation of many current machine learning paradigms is their exclusive focus on policy learning and classification. Our minds are not classifiers—they are simulators and experiencers. Like machine learning systems, they successively learn to identify features in the patterns of the sensory input, which they then combine into complex features, and organize into maps. High-level features may be integrated into dynamic geometries and objects, motor patterns and procedures, auditory structure and so on. Features, objects and procedures are sensory-motor scripts that allow the manipulation of mental content and the execution of motor actions.

Unlike most machine learning systems, our minds combine these objects, maps and procedural dynamics into a persistent dynamic simulation, which can be used to continuously predict perceptual patterns at our systemic interface to the environment (Fig. 2). The processing streams formed by the receptive fields of our cortical instruments enable the bottom-up cuing of perceptual hypotheses (objects, situations etc.), and trigger the top-down verification of these hypotheses, and the binding of the features into a cohesive model state. The elements of this simulation do not necessarily correspond to actual objects in the universe: they are statistical regularities that our mind discovered in the patterns at its systemic interface. Our experience is not directed on the pattern generator that is the universe, but on the simulation produced in our neocortex. Thus, our minds cannot experience and operate in an "outer" reality, but in a dream that is constrained to past and current sensory input [17].

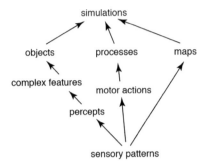

Fig. 2. Gradual abstraction from sensory patterns to mental simulations

Human cognition does not stop at generative simulations, however. We can abstract our mental representations into a conceptual manifold (Fig. 3). Concepts can be thought of as an address space for our sensory-motor scripts, and they allow the interpolation between objects, as well as the manipulation and generation of previously unknown objects via inference. The conceptual manifold can be organized and manipulated using grammatical language, which allows the synchronization of concepts between speakers, even in the absence of corresponding sensory-motor scripts. (The fact that language is sufficient to infer the shape of the conceptual manifold explains the success of machine translation based on the statistical properties of large text corpora, despite the inability of these systems to produce corresponding mental simulations.)

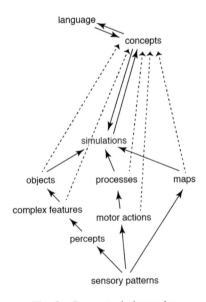

Fig. 3. Conceptual abstraction

3   The Cortical Conductor

Cortical columns may be thought of as elementary agents that self-organize into the larger organizational units of the brain areas as a result of developmental reinforcement learning. The activity of the cortical orchestra is highly distributed and parallelized, and cannot be experienced as a whole. However, its performance is coordinated by a set of brain areas that act as a conductor. The conductor is not a "homunculus", but like the other instruments, a set of dynamic function approximators. Whereas most cortical instruments regulate the dynamics and interaction of the organism with the environment (or anticipated, reflected and hypothetical environments), the conductor regulates the dynamics of the orchestra itself. Based on signals of the motivational system, it provides executive function (i.e. determines what goals the system commits to at any given moment), resolves conflicts between cortical agents, and regulates their activation level and parameterization. Without the presence of the conductor, our brain can still perform most of its functions, but we are sleep walkers, capable of coordinated perceptual and motor action, but without central coherence and reflection (Fig. 4).

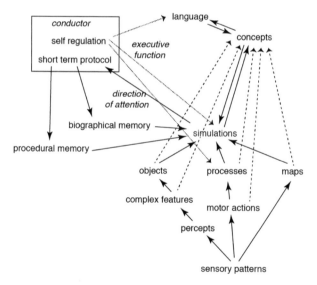

Fig. 4. The cortical conductor

In the human brain, the functionality of the conductor is likely facilitated via the dorsolateral prefrontal cortex [18–20], anterior cingulate cortex and anterior insula [21]. The conductor has attentional links into most regions. In each moment, it directs its attention to one or a few of the cortical instruments, while others continue to play unobserved in the background. The conductor may not access the activity of the region it attends to in its entirety, but it may usually access some of the currently relevant processing states and parameters of it. To learn and to reflect, the conductor maintains a protocol of what it attended to, as a series of links to experiences generated by the other cortical instruments. This protocol may be used to address the currently active regions, and to partially recreate past states of the mental simulation by reactivating the corresponding configuration of active regions with the parameters of the stored links. The reactivation of a past state of the mental simulation will generate a re-enactment of a previous world state: a memory. Further abstraction of the protocol memory leads to the formation of new kinds of sensory motor scripts: an autobiographical memory (events that happened to the agent), and a procedural memory. The reflective access to the protocol allows learning and extrapolation of past events, and the act of accessing the protocol may of course itself become part of the protocol. By accessing the memory of the access to its own protocol, the system remembers having had access to experience (access consciousness).

While all cortical regions store information as a result of updating their models and learning associations to motivational signals, the attentional protocol of the conductor is the only place where experience is integrated. Information that is not integrated in the protocol cannot become functionally relevant to the reflection of the system, to the production of its utterances, the generation of a cohesive self model, and it cannot become the object of access consciousness.

Phenomenal consciousness may simply be understood as the most recent memory of what our prefrontal cortex attended to. Thus, conscious experience is not an experience of being in the world, or in an inner space, but a memory. It is the reconstruction of a dream generated more than fifty brain areas, reflected in the protocol of a single region. By directing attention on its own protocol, the conductor can store and recreate a memory of its own experience of being conscious.

The idea that we are not actually conscious in the moment, but merely remember having been conscious is congruent with known inconsistencies in our experience of consciousness, such as subjective time dilation, false continuity, and loops in the conscious experience. Subjective dilation of time results from states of high arousal, for instance during an accident, whereas uneventful flow states often lead to a subjective contraction of time. Both dilated and contracted time do not correspond to an increase or decrease in the actual processing speed of our cognitive operations. Instead, they result from a higher or lower number of entries in the protocol memory: the experienced time interval only seems to be longer or shorter with hindsight. An extreme case of a subjective dilation of time can happen during dreams, which sometimes play out in a physical time interval of a few seconds of REM sleep, yet may span hours of subjective time. This may be explained by the spontaneous generation of the entire dream, rather than the successive experience of each event. Hour-long dreams are probably simply false memories.

False continuity results from gaps in our attention, for instance during saccadic movements, or interruptions and distractions of gaze. While these breaks in attention may lead to missing significant changes in parts of the environment that we believe we are attending to, they are not part of the protocol memory and hence our experience appears to be unbroken in hindsight. For a considerable fraction of our days, we are probably wakeful but not conscious. Inconsistent experiences of consciousness can be explained as false memories, but they do not have subjective qualities that makes them appear "less conscious" than consistent experiences. Thus, if at least some of our conscious experience is a false memory, why not all of it? Treating consciousness as a memory instead of an actual sense of the present resolves much of the difficulty for specifying an AI implementation of consciousness: it is necessary and sufficient to realize a system that remembers having experienced something, and being able to report on that memory.

4 Consciousness and Self Model

In the above discussion, I have treated phenomenal consciousness in the sense of "the feeling of what it's like". However, consciousness is often associated with more concrete functionality, especially a specific model of self, and a set of functionality pertinent to that model. This has lead Marvin Minsky to call consciousness "a suitcase term" [22], a notion that is notoriously hard to unpack.

Conscious states differ by the configuration and available functionality of a cognitive system at a given time. However, once we understand how an attentional protocol can provide for binding of other cortical functionality into a single structure for the purpose of self regulation, we can enumerate some of the functionality that corresponds to a given conscious state.

Core consciousness is characterized by:

- a local perceptual space
- the ability to access mentally represented percepts
- a current world model
- directed attention (inwards/outwards, wide/focused)
- the ability to access and follow concepts and similar content
- the ability to manipulate and create concepts and similar content,
- the presence of an inner stage of currently active, non-perceptual concepts and associative representations

In deep meditation, the following functionality may be absent:

- an integrated personal self-model (sense of identity)
- a sense of one's own location and perspective in space
- proprioception (position and state of body and limbs)
- valences (pleasure and displeasure signals)
- goals and committed plans
- the awareness of the current affective state
- the influence of desires and urges on behavior
- the ability to create and process discourse (i.e. translate mental representations into communicable symbols, and vice versa)

Lucid dreams are specific dream states that are different from wakefulness by the absence of:

- access to needs/desires, urges
- access to sensory perception
- the ability to exert voluntary control over muscles
- a biographical memory and protocol

- a short term biography
- the ability to separate perceptually grounded content from ideas/imaginations

Dreams are usually in addition characterized by the absence of:

- having accessible knowledge about the fact that there access to percepts and concepts (access consciousness)
- a social model of self (self-ascription of beliefs, desires, intentions, skills, traits, abilities, personality)
- the formation of and access to expectations of immediate future
- the ability to influence behavior based on discursive thought
- the ability to relate self-ascribed actions to apparent mental causes (sense of agency)
- the ability to form memories of the current content
- the ability to reason
- the ability to construct plans
- the ability to act on plans
- the ability to keep goals stable until they are achieved
- the ability to let go of goals that are unattainable
- the ability to signal aspects of one's mental state to others

Diminished states of consciousness (for instance, in small children or due to neurodegenerative diseases) may also impair:

- the ability to influence behavior based on past experience (learning)
- the ability to construct causal models of the environment
- the ability to construct intentional models of agents

The above functionality is part of the general functionality of a human-like cognitive agent and has to be implemented into its cognitive architecture, either explicitly or via a self-organized process of reward driven learning (each of them can be realized on computational machines). The differences between conscious states result from the dissociation or impairment of these functions.

Is the conductor a learned or a predefined structure? I suspect that the formation of the conductor functionality is itself a process of developmental learning, driven by rewards for the self-regulation of cognition, and developmental cues that regulate the onset and some of the parameters of the formation of the structure. Multiple personality disorder lends further credibility to the hypothesis that the conductor is constructed by reward driven neural self-organization. In patients with multiple personalities, the different personas usually do not share a subjective protocol, biographical and procedural memory. But even if we form multiple conductors, they share infrastructure (such as linguistic processing, access to the attentional network and information transfer via the thalamic loop), which ensures that only one of them may be online and form memories at any given moment.

5 Summary

The cortical conductor theory (CTC) posits that cortical structures are the result of reward driven learning, based on signals of the motivational system, and the structure of the data that is being learned. The conductor is a computational structure that is trained to regulate the activity of other cortical functionality. It directs attention, provides executive function by changing the activity and parameterization and rewards of other cortical structures, and integrates aspects of the processes that it attended to into a protocol. This protocol is used for reflection and learning. Memories can be generated by reactivating a cortical configuration via the links and parameters stored at the corresponding point in the protocol. Reflective access to the protocol is a process that can itself be stored in the protocol, and by accessing this, a system may remember having had experiential access.

For phenomenal consciousness, it is necessary and sufficient that a system can access the memory of having had an experience—the actuality of experience itself is irrelevant (and logically not even possible).

CTC explains different conscious states by different functionality bound into the self construct provided by the attentional protocol. The notion of integration is central to CTC, however, integration is used in a very different sense than in Tononi's Integrated Information Theory (IIT). In CTC, integration refers to the availability of information for the same cognitive process, within a causally local structure of an agent. In IIT, integration refers to the degree in which information is distributed within a substrate. CTC is a functionalist theory, and can be thought of as an extension to Dennett's "multiple drafts" model of consciousness [23]. CTC acknowledges that the actual functionality of perception and cognition is distributed, disjoint and fragmentary, but emphasizes the need to integrate access to this functionality for a module that in turn has access to capabilities for reflection and the formation of utterances (otherwise, there would be no self model and no report of phenomenal experience).

CTC also bears similarity to Michael Graziano's attention schema theory of consciousness [24]. Graziano suggests that just like the body schema models the body of an agent, its attention schema models the activity and shape of its attentional network. While the functionality subsumed under access consciousness, phenomenal consciousness and conscious states, and the required mechanisms are slightly different in CTC, we agree with the role of consciousness for shaping and controlling attention-related mechanisms.

Acknowledgements. This work has been supported by the Harvard Program for Evolutionary Dynamics, the MIT Media Lab and the Epstein Foundation. I am indebted to Katherine Gallagher, Adam Marblestone, and the students of the Future of Artificial Intelligence course at the MIT Media Lab for their contributions in discussions of the topic, as well as to Martin Novak and Joi Ito for their support.

References

1. Tononi, G.: The integrated information theory of consciousness: an updated account. Arch. Ital. Biol. **150**, 56–90 (2012)
2. Tononi, G., Boly, M., Massimini, M., Koch, C.: Integrated information theory: from consciousness to its physical substrate. Nat. Rev. Neurosci. **17**(7), 450–461 (2016)
3. Aaronson, S.: Why I am not an integrated information theorist (or, the unconscious expander) (2015). http://www.scottaaronson.com/blog/?p=1799. Accessed 15 Feb 2017
4. Zadra, A., Desautels, A., Petit, D., Montplaisir, J.: Somnambulism: clinical aspects and pathophysiological hypotheses. Lancet Neurol. **12**(3), 285–294 (2013)
5. Bach, J.: Principles of Synthetic Intelligence. Psi, An Architecture of Motivated Cognition. Oxford University Press, Oxford (2009)
6. Bach, J.: Modeling motivation in MicroPsi 2. In: 8th International Conference on Artificial General Intelligence, AGI 2015, Berlin, Germany, pp. 3–13 (2015)
7. Cer, D.M., O'Reilly, R.C.: Neural mechanisms of binding in the hippocampus and neocortex. In: Binding in Memory. Oxford University Press, Oxford (2006)
8. Solomonoff, R.J.: A formal theory of inductive inference. Inf. Control **7**, 224–254 (1964)
9. Hutter, M.: Universal Artificial Intelligence: Sequential Decisions Based on Algorithmic Probability. EATCS Book. Springer, Cham (2005)
10. LeCun, Y., Bengio, Y., Hinton, G.: Deep learning. Nature **521**, 436–444 (2015)
11. Hochreiter, S., Schmidhuber, J.: Long short-term memory. Neural Comput. **9**(8), 1735–1780 (1997)
12. Cho, K., van Merrienboer, B., Gulcehre, C., Bahdanau, D., Bougares, F., Schwenk, H., Bengio, Y.: Learning phrase representations using RNN encoder–decoder for statistical machine translation. In: Proceedings of EMNLP, October 2014, pp. 1724–1734 (2014)
13. Dosovitskiy, A., Springenberg, J.T., Brox, T.: Learning to generate chairs with convolutional neural networks. In: IEEE Conference on Computer Vision and Pattern Recognition (CVPR), pp. 1538–1546 (2015)
14. Hinton, G.E., Krizhevsky, A., Wang, S.D.: Transforming auto-encoders. In: International Conference on Artificial Neural Networks. Springer, Berlin (2011)
15. Mountcastle, V.B.: The columnar organization of the neocortex. Brain **20**(4), 701–722 (1997)
16. Jaeger, H.: Echo state network. Scholarpedia **2**(9), 23–30 (2007)
17. Bach, J.: No room for the mind. Enactivism and artificial intelligence. In: Proceedings of the Conference of History and Philosophy of Computing, Ghent (2011)
18. Bodovitz, S.: The neural correlate of consciousness. J. Theor. Biol. **254**(3), 594–598 (2008)
19. Safavi, S., Kapoor, V., Logothetis, N.K., Panagiotaropoulos, T.I.: Is the frontal lobe involved in conscious perception? Front. Psychol. **5**, 1063 (2014)
20. Del Cul, A., Dehaene, S., Reyes, P., Bravo, E., Slachevsky, A.: Causal role of prefrontal cortex in the threshold for access to consciousness. Brain **132**(9), 2531–2540 (2009)
21. Fischer, D.B., Boes, A.D., Demertzi, A., Evrard, H.C., Laureys, S., Edlow, B.L., Liu, H., Saper, C.B., Pascual-Leone, A., Fox, M.D., Geerling, J.C.: A human brain network derived from coma-causing brainstem lesions. Neurology **87**(23), 2427–2434 (2016)
22. Minsky, M.: The Emotion Machine. Simon and Shuster, New York (2006)
23. Dennett, D.C.: Consciousness Explained. Back Bay Books, New York (1992)
24. Graziano, M.S.A., Webb, T.W.: A mechanistic theory of consciousness. Int. J. Mach. Conscious. **6**, 163–176 (2014)

Meta-Engineering: A Methodology to Achieve Autopoiesis in Intelligent Systems

Wolfgang Bartelt[✉], Tirtha Ranjeet, and Asit Saha

Ingeniation Pty Ltd., Perth, Australia
{wolfgang.bartelt, tirtha.ranjeet,
asit.saha}@ingeniation.com.au

Abstract. This paper presents an architecture of autopoietic intelligent systems (AIS) as systems of automated "software production"-like processes based on meta-engineering (ME) theory. A self-producing AIS potentially displays the characteristics of artificial general intelligence (AGI). The architecture describes a meta-engineering system (MES) comprising many subsystems which serve to produce increasingly refined "software-production"-like processes rather than producing a solution for a specific domain. ME-theory involves a whole order of MES and the ME-paradox, expressing the fact that MES can potentially achieve a general problem-solving capability by means of maximal specialization. We argue that high-order MES are readily observable in software production systems (sophisticated software organizations) and that engineering practices conducted in such domains can provide a great deal of insight on how AIS can actually work.

Keywords: Autopoiesis · Meta-Engineering · Meta-learning
Software production system · Artificial general intelligence

1 Introduction

We are presenting a novel methodological framework of autopoiesis based on engineering principles. It was Maturana and Varela [1] who first defined "autopoiesis" as a self-organizing and self-creating principle within the biological domain. Researchers have adopted "autopoiesis" to explain emergence and self-organization within various domains, such as biological [1–3], social [4–6] and information systems [7, 8]. We incorporate the idea of autopoiesis in the domain of artificial general intelligence (AGI) to develop autopoietic intelligent systems (AIS). The AIS term was coined by Wolfgang Bartelt - founder of Ingeniation Pty Ltd – to describe his integrated approach to autopoiesis and general intelligence based on a meta-engineering (ME) theory.

In artificial intelligence many studies [9–11] have been conducted to implement self-creating machines ever since the idea of autopoiesis had been introduced by Maturana and Verala [1–3]. In general, if an intelligent system is infused with autopoietic properties, it may gain self-improvement capabilities similar to such observed in biological systems. Eventually this may lead to a machine implementation that exhibits general intelligence characteristics as found in humans [12].

However, the existing studies do not show how general intelligent systems can emerge as an outcome of something more fundamental. So far, any attempt to build a

© Springer Nature Switzerland AG 2019
A. V. Samsonovich (Ed.): BICA 2018, AISC 848, pp. 27–36, 2019.
https://doi.org/10.1007/978-3-319-99316-4_4

functioning AGI, be it autopoietic or not, has only limited success [12]. The reason may be either they focus more on utilizing high-level algorithmic approaches [13], such as optimization or search-based algorithms, or low-level connectionist learning techniques such as mainstream deep neural networks.

We consider an engineering technique instead of high-level algorithms to develop AIS. Peter Voss [13] claims that genetic programming techniques are the only potential option in absence of any specific engineering ideas. Further, Pennechin and Goertzel [12] once stressed that "AGI is merely an engineering problem". We find that the use of existing engineering principles makes AIS relatively easier to understand than abstract program search strategies or evolutionary algorithms. Therefore, ME-driven AIS can be a viable alternative to any kind of search-based or evolutionary algorithms.

Beer introduced the viable system model (VSM) [14] which is based on cybernetic principles of control and communication. VSM has been successfully applied to model emergent properties in social organizations, such as large corporations. We identified that research conducted in VSM [14–16], is among the most relevant for our ME-approach, as engineering organizations ultimately are social systems. However, our approach goes further by capitalizing on the fact that engineering organizations are somewhat "extreme social systems". The reason is that engineering organizations, by necessity, routinely communicate all the intricate details of their interior operational principles, i.e. they perform ME as part of their standard conduct.

With AIS we are taking a radical constructivist approach to AGI. It is fundamentally different as it incorporates characteristics of meta-learning [17] as part of its autopoietic principles. We have no knowledge of any other systematic approach that demonstrates the association between general problem-solving and autopoiesis as a concrete methodology.

In Sect. 2 we describe ME as a methodology of autopoietic intelligent systems (AIS) followed by Sect. 3 in which we discuss observations of ME systems in real-world engineering organizations. Section 4 consists of conclusion and future work in the study of AIS.

2   An Engineering Approach to Autopoiesis

Engineering can be defined as the application of theoretical and practical knowledge to invent effective solutions for problems of everyday life. An engineering process (EP) is a systematic sequence of activities to generate a solution from a given problem description. Such activity sequences are commonly described in some form of engineering process description which we refer to as an engineering policy (EPO). Thus, an EPO can be applied by human engineers as a rule set or guideline when conducting an engineering performance.

Engineering automation (EA) is a means to substitute human engineering performances along such guidelines with computer-based equivalents. An increase of software aided engineering capabilities accelerates engineering performances and simultaneously allows to scale them up to tackle more complex engineering domains. Novel engineering domains require new and/or improved EPO which ultimately become the subject of even more EA and so on.

Continuous EA by means of migrating human engineering performances into software processes is a driving force of what can readily be observed as an exponentially accelerating technological progress, e.g. as discussed in Kurzweil [18]. We argue that the principle cause for achieving an exponential rather than linear progress, as observed in almost any technological domain, is a positive feedback loop created by EA in software production systems, where EA potentially becomes applied to itself. The following sections will provide an outline of the presumed conceptional principles behind this positive feedback loop.

2.1   Meta-Engineering

EPO are engineering domain dependent and represent the key knowhow of an engineering organization as an effective problem solver in its relevant domains. Therefore, the continuous development and improvement of EPO is seen by many as a strategical core task. In its most cultivated form this core task has spawned an own engineering discipline, an EPO engineering, which we call a policy engineering (PE). PE is defined as a systematic sequence of activities to create an effective set of EPO. It is obvious that an EPO of a PE is also a possibility. Notably, an engineering policy describing EPO construction is more engineering domain independent. This is simply because its sole engineering domain becomes the domain of constructing effective EPO. We call an EPO of a PE a meta-engineering policy (MPO) and the EP described by it a meta-engineering process (MP). We present meta-engineering as a methodological framework for AIS to potentially achieve general intelligence in a system. The methodology involves a meta-engineering order which is described in the following sections.

2.2   The Meta-Engineering Order

As described in Sect. 2.1 an MPO/MP pair always produces an EPO/EP pair for a specific problem domain in a two-step process. Such ME-performance describes what we call a meta-engineering system (MES). However, at closer examination it turns out that such two-stage MES describes just one specific case of what we call the meta engineering order (ME-order). The ME-order is inspired by the fact that MPO/MP can have different capabilities in respect to the EPO/EP they produce. Such MPO/MP capabilities must be related to certain production potentials in EPO/EP. In the case above the production potentials in EPO/EP result in domain specific solutions as shown in Fig. 1. We call such system a second order MES (N = 2) which is based on second order MPO/MP capabilities.

From here, the ME-order can be expanded downwards and upwards. At first-order (N = 1), MPO/MP capabilities would imply a one-stage MES, which is the equivalent of an EPO/EP without a concrete guideline. The possibility of a zeroth-order MPO/MP (N = 0) can be introduced as a starting point describing a state of "non-engineering".

When the ME-order is continued toward higher orders, an interesting thing happens which we call the "meta-engineering paradox" (ME-paradox). The ME-paradox expresses the fact that MES can potentially achieve a general problem-solving capability by means of maximal specialization. This appears paradoxical because specialization and gaining generality would typically be regarded as opposing tendencies of a developing system.

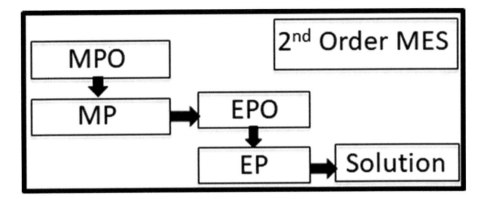

Fig. 1. 2nd order ME system

Figure 2 depicts 3rd order MPO/MP capabilities (N = 3) implying a three-stage MES. The top-level MPO/MP produces subordinate MPO/MP which in turn produces EPO/EP. Here it must be noted that the top-level MPO/MP must have a distinctive specialization which provides its subordinate MPO/MP with a capability to produce MPO/MP rather than ordinary EPO/EP.

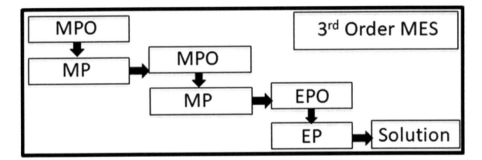

Fig. 2. 3rd order ME system

2.3   Self-constructive MES

We continue the ME-order one step further. With fourth-order MPO/MP capabilities (N = 4) the top-level MPO/MP becomes even more specialized. It produces MPO/MP which can produce another set of MPO/MP. Obviously, this appears to already represent the maximum possible specialization in MPO.

Through its maximal specialization a fourth-order MES potentially becomes, at least on a logical level, circular and closed. That is, fourth-order MPO/MP could now produce an MPO/MP that produces an MPO/MP of itself in a completely self-referential manner as depicted in Fig. 3(b). At the end of it, the described system would inevitably enter a perpetual loop of (re-) constructing itself.

(a)   **(b)**

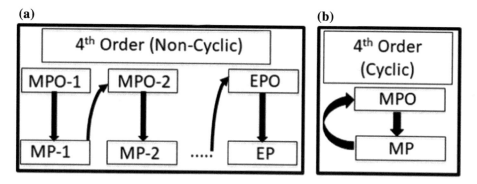

Fig. 3. (a) 4th order (non-cyclic), (b) 4th order (cyclic)

2.4   Autopoietic Intelligent System (AIS)

A self-constructing MES as depicted in Sect. 2.3 would be fragile in respect to any kind of disturbances. In particular, if a self-constructing MES becomes error-prone and fails to produce effective MPO/MP in its perpetual self-construction cycle, it could easily become subject to a sudden "death".

A general problem-solving capability, as presumed by the ME-paradox, is achieved when the perpetual cycle of a self-constructing MES is impacted by external influences. Coupling a self-constructing MES to an environment occurs in such a way that its perpetual self-construction cycle becomes disturbed and not destroyed. This imposes a certain kind of survival pressure onto the completely closed ME system. This situation is depicted in Fig. 4.

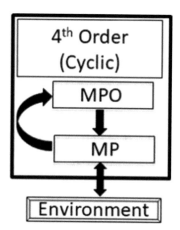

Fig. 4. Self-constructing MES coupled to an environment

However, a self-constructing MES can adjust its MPO construction activities incorporating aspects of disturbance/noise. The perturbing noise coming from an external environment becomes effectively "weaved" into the self-construction policies of the system. Therefore, the system acquires the necessary constructive abilities to control any perturbances. Such systemic acquisition of advanced self-construction reflects a "tricky kind" of problem solving capability. It is entirely self-referential, i.e. it is specialized on a single domain. At the same time, regardless of the kind of non-destructive perturbing noise, its high-order ME performance, over time, provides the system with an on-demand solution capability to reconstruct itself.

We conclude that implementing MES may represent an effective means to achieve autopoiesis through a maximal specialization of their ME processes. Autopoiesis appears as an outcome of a ME-system that possesses general problem-solving capability. Although the resulting system is single domain only, it has the necessary reconstructive capabilities to deal with a potentially infinite set of problems.

3   MES in Real-World Scenarios

In this section we illustrate evidences that MES are not a mere theoretical construct but a real-world phenomenon. For example, MES can readily be observed in contemporary engineering organizations. Although ME is potentially applicable to any engineering domain, we will focus our attention to what we call software production systems (SPS) i.e., sophisticated software engineering organizations.

3.1   ME in Software Production Systems (SPS)

The landscape of SPS is extremely diversified because each SPS requires specializations. Each individual SPS seeks to provide a unique set of products to its market environment. However, for the intended purpose of this section we need to shift our attention from the "what" an SPS's product offering is to "how" such product offerings are produced by it. Taking on this perspective is crucial to thoroughly understand what follows. Concentrating on the "how" perspective, the formerly highly diversified SPS landscape is becoming well-structured in only a handful of categories. The most prominent categories are described in the following subsections.

Software Production Systems – SPS(1)

There exists a category of SPS that perform product development processes by creating some plan (EPO), then execute it (EP) to achieve the desired outcomes. This action sequence is performed in a cyclic manner and may include revisions of an original plan to account for changes or error corrections. Since there exists no prescribed form of EPO, each original EPO may be significantly different from others. It also implies that learning in such SPS is strongly bound within the individuals in the engineering organization. We categorize such SPS as 1st order MES, or SPS(1). Most SPS which are small, inexperienced or provide software production services to third parties would fall into this category. The reason for the latter is that having many different customers typically implies that each development project is significantly different.

Software Production Systems – SPS(2)

There exists a advanced category of SPS that defines standard procedures (MPO) to construct high-quality plans (EPO). Such SPS apply their standard procedures to construct instances of such high-quality plans, and then execute each instance (EP). A typical example would be a project management plan which is created based on a process guideline and an associated set of documented templates.

The important step taken here in contrast to SPS(1) is that the knowledge of how to construct an effective plan has become explicit in an MPO and therefore largely independent of individuals in the respective engineering organization. Thus, learning in such organization becomes a matter of continuously improving MPO based on EPO/EP performances. We categorize such SPS as 2nd order MES, or SPS(2).

Most SPS which are medium in size, reasonably experienced or offer standard products to their markets would fall into this category. The latter is because producing new releases of standard products is typically achievable through a reuse of standard procedures stemming from similar earlier experiences.

Software Production Systems – SPS(3)

Further, there exists a more advanced category of SPS that define a PE standard methodology (MPO). When applied (MP), the methodology facilitates a construction of standard procedures (subordinate MPO) with an inherent self-optimization strategy. The idea behind this is that each MP performance supports a self-improvement cycle, which includes the construction of subordinary instances of high-quality plans (EPO) and the execution of such instances (EP).

Strategically, this establishes a scenario in which EPO/EP instantiation can now be treated as a 'test' to validate the superordinate MPO (the guidelines) for the purpose of optimal solution. The top level MPO is therefore a master guideline to construct optimal guidelines over many EPO/EP iterations. A typical example would be an SPS applying process improvement techniques such as the SPICE/ISO 15504 methodology [19] to optimize their engineering processes.

This kind of multi-stage process engineering describes a meta-learning strategy. The important additional step taken here in contrast to SPS(2) is that process guidelines become systematically infused with a self-optimization strategy, thus increasingly reflect an effective means to learn how to learn better from each performance. Practically, the upper level MPO/MP pair is guiding this optimization process by imposing a set of constraints on what subordinate MPO/MP pairs should account for. As an example, in SPICE/ISO 15504 such constraints are reflected in its process capability model (PCM) and what it refers to as a process reference model (PRM). The SPICE PCM effectively describes the distinctive aspects of a self-optimizing process. The PRM provides a high-level abstraction of a set of EP to be optimized based on the PCM.

We categorize such SPS as 3rd order ME or SPS(3). Most SPS which are large, highly experienced or perform large scale/high-risk development projects would fall into this category. The latter because for such engineering organizations optimal process quality becomes a dominant crucial success factor which needs to be dealt with systematically.

Software Production Systems – SPS(4)

We intentionally omit SPS(0) and SPS(1) categories in favor to focus more on what we want to categorize as SPS(4), i.e. SPS involved in so called product family engineering [20]. SPS in this category operate similar to a cyclic 4th order ME as shown in Fig. 4 by applying a slightly different development strategy than SPS in all other categories. This strategy can best be summarized as "strategic reuse". Instead of focusing their entire engineering performances to deliver a next respective end-product, SPS(4) concentrate their efforts to create a highly effective repository of reusable components. Such component repository can then serve as a platform that supports the highly efficient construction of a well-defined set of similar end-products, i.e. a product family. This appears unspectacular, but SPS mastering this strategy typically can achieve at least one order of magnitude higher performance in developing high-quality end products.

There are a few important observations in SPS(4) which are relevant to the ME case. First, SPS(4) are single domain in the sense that the primary engineering objective becomes focused on creating an optimized component repository for a set of similar end-products. Second, SPS(4) operates to some degree in a self-referential manner. For example, the construction of an end-product can be understood as serving the self-purpose to improve the respective component repository. Third, the perpetual construction cycle of product family instances is coupled to an environment which perturbs it in a non-destructive manner. Such perturbances can come in by means of new requirements, bug reports, market feedback, and so on. SPS(4) typically compensates for such disturbances by adjusting and/or modifying repository components to improve reuse in future product family instances. This is another way to say that SPS(4) learns by continuously improving the fitness of a well-defined set of components to compensate for the perturbances generated by their market environment.

3.2   Quasi-autopoietic Software Production Systems

Although there are similarities to ME induced autopoiesis as described in the previous section, SPS(4) lack some characteristics to classify them as fully autopoietic systems. However, we call SPS(4) quasi-autopoietic because from an ME perspective there may be only a few 'adjustments' necessary to turn SPS(4) into genuine self-constructing systems.

One adjustment would be necessary to establish operational self-reference within SPS(4). In terms of product family engineering [20] operational self-reference would mean that a product family's repository must hold available a well-defined set of components that serves to validate and improve itself. More practically, this means that the components of such a self-referential SPS would have to reflect the PRM of an iterative repository construction methodology as a well-defined set of EP. An effective PCM (not necessarily the SPICE PCM) imposing a set of targeted constraints could then be applied to drive such set of EP towards a self-optimization strategy.

It is obvious that any means to construct a product family instance from such a component repository would eventually instantiate a set of EPs according to the PRM. The set of EP would in turn implement a methodology to improve the same component repository. Therefore, the set of EP instances would implement an MP which constructs and improves its MPO, i.e. the set of components in the repository. This way the proposed MPO/MP self-construction cycle would be effectively established.

Another necessary adjustment of SPS(4) towards genuine autopoiesis would be required for the system to achieve autonomy. One can argue, that SPS(4) appear to have a high degree of autonomy towards their environment as its perturbing influences serves the own component repository improvement only. However, SPS(4) are almost entirely non-autonomous because they are inherently driven by human engineering capabilities. Human engineers carry external influences with them into the system and infuse it with higher-order ME capabilities, whereas the system does not have such capabilities on its own.

From an ME perspective the appearance of autonomy in SPS(4) is due to an existence of implicit and explicit ME. We argue that for a sustainable SPS its total implicit and explicit engineering performances must always add up to a self-constructing MES cycle. In such cycle all implicit engineering performances occur in humans and all explicit engineering performances occur between humans as a system.

Everything the system cannot do on its own to establish a self-constructing MES cycle has to be done by humans. Only the aspects of a self-constructing MES which become explicit in form of EPO or MPO can contribute to higher level autonomy in SPS. Consequently, only maximal explication of self-constructing MES can yield a fully autonomous SPS(4), and only if its policies eventually become subject to EA. As a result, this potentially removes the necessity for human engineering performances entirely from the system.

4   Conclusion and Future Work

In this paper we presented an architecture of autopoietic intelligent systems (AIS) as systems of automated "software production"-like processes based on meta-engineering (ME) theory. ME-theory is neither based on hypothesis nor merely a theoretical approach, but its foundations can be observed in the real-world activities of sophisticated engineering organizations. We systematically demonstrated the existence of MES in various categories of software production systems with practical examples.

We illustrated that MES possess an inherent ability to become general problem solvers by means of maximal specialization. MES learn from their experiences and utilize outcomes of such experiences to reconstruct themselves as systems which learn better. This meta-learning capability leads to an exponential learning rate. We attributed exponentially accelerating technological progress, as observed in many engineering domains, to MES performances. Self-constructing MES reflect biologically inspired autopoiesis based on engineering principles applied in information technology.

We have developed an MES simulation platform and are actively conducting experimentation based on it. The platform has been implemented as a multi-agent system comprising constituent agents that mimic constructive interactions of skilled engineers, similar to practices found in SPS. The first experimentation outcomes are excitingly encouraging. We successfully conducted experiments generating perpetual cyclic processes based on a simple process reference model. Future work includes developing our PRM further towards a generic process reference model of AIS.

References

1. Maturana, H.R., Varela, F.G.: Autopoietic systems. Biological Computer Laboratory, BCL report no. 9.4, Department of Electrical Engineering, University of Illinois, Urbana, IL (1975)
2. Varela, F.G.: Autopoiesis and a biology of intentionality. In: Autopoiesis and Perception: A Workshop with ESPRIT, pp. 4–14 (1992)
3. Varela, F.G., Maturana, H.R., Uribe, R.: Autopoiesis: the organization of living systems, its characterization and a model. Biosystems 5(4), 187–196 (1974)
4. Luhmann, N.: Social Systems. Stanford University Press, Stanford (1995)
5. Luhmann, N.: The autopoiesis of social systems. In: Geyer, F., van der Zouwen, J. (eds.) Socio-Cybernetic Paradoxes, pp. 172–192. Sage, London (1986)
6. Schatten, M., Baca, M.: A critical review of autopoietic theory and its applications to living, social, organizational and information systems. Druš. Istraž. J. Gen. Soc. Issues **108–109**, 837–852 (2008)
7. Qiao, H., Mo, R., Xiang, Y., Franke, M., Hribernik, K.A., Thoben, K.D.: An autopoietic approach for building modular design system. In: Engineering, Technology and Innovation (ICE), International ICE Conference, pp. 1–7. IEEE (2014)
8. Zeleny, M., Hufford, K.D.: The application of autopoiesis in systems analysis—are autopoietic systems also social-systems. Int. J. Gener. Syst. **21**(2), 145–160 (1992). https://doi.org/10.1080/03081079208945066
9. Letelier, J.C., Marín, G., Mpodozis, J.: Computing with autopoietic systems. In: Roy, R., Köppen, M., Ovaska, S., Furuhashi, T., Hoffmann, F. (eds.) Soft Computing and Industry, pp. 67–80. Springer, London (2002). https://doi.org/10.1007/978-1-4471-0123-9_6
10. Briscoe, G., Dini P.: Towards autopoietic computing. In: International Conference on Open Philosophies for Associative Autopoietic Digital Ecosystem, pp. 199–212. Springer, Berlin (2010)
11. McMullin, B., Grob, D.: Towards the implementation of evolving autopoietic artificial agents. In: 6th European Conference on Artificial Life, ECAL (2001)
12. Pennachin, C., Goertzel, B.: Contemporary Approaches to Artificial General Intelligence. Cognitive Technologies (2007).
13. Voss, P.: Essentials of General Intelligence: The Direct Path to Artificial General Intelligence. Cognitive Technologies (2007)
14. Beer, S.: The viable system model: its provenance, development, methodology and pathology. J. Oper. Res. Soc. **35**(1), 7–25 (1984)
15. Schwaninger, M., Scheef, C.: A test of the viable system model: theoretical claim vs. empirical evidence. Cybern. Syst. **47**(7), 544–569 (2016). https://doi.org/10.1080/01969722.2016.1209375
16. Espejo, R., Gill, A.: The Viable System Model as a Framework for Understanding Organizations, pp. 1–6. Phrontis Limited & SYNCHO Limited, Adderbury (1997)
17. Villada, R., Drissi, Y.: A perspective view and survey of meta-learning. Artif. Intell. Rev. **18**, 77–95 (2002)
18. Kurzweil, R.: The Singularity is Near. Gerald Duckworth & Co., London (2010)
19. El Eman, K., Drouin, J.N., Melo, W.: SPICE: The Theory and Practice of Software Process Improvement and Capability Determination. IEEE Computer Society Press, Los Alamitos (1997)
20. Linden, F.V.D.: Software product-family engineering. In: 5th International Workshop on PEE (2003)

BICA *à rebours*

Piotr Bołtuć[1,2(✉)]

[1] University of Illinois, Springfield, IL 62703, USA
epetebolt@gmail.com
[2] Warsaw School of Economics, Warsaw, Poland

Abstract. Most authors in this volume focus on Biologically Inspired Cognitive Architectures (BICA). We propose BICA *à rebours*, an AI inspired take on human cognitive architecture. Both strategies share the BICA method – a spectrum approach between animal cognitive architecture and AI. We focus on the BICA-Mary argument [1], which is a response to the classical 'knowledge argument' posed by Jackson, [2]. Philosophers tend to think that *phenomenal qualia* cannot be learned by description; this is supposed to show non-reductive character of consciousness. But when we look at it with a bioengineering eye, all this argument demonstrates is the lack of an inborn connection from qualities of experience to conceptual knowledge. This is due to the specificity of human cognitive architecture (originating from evolutionary history) and not due to some deep epistemological truth. The connections from symbols to qualitative experience could be bioengineered in animal brains, which would deflate the example. This is one way to demonstrate that the example has no bearing on non-reductive consciousness. More broadly, arguments that follow the BICA *à rebours* structure, and view human cognition as an engineering system, help us reexamine misconceptions about human psychology, epistemology and related domains.

Keywords: BICA à rebours · BICA-Mary · Knowledge argument
gen cog · Frank Jackson · Ben Goertzel · Epistemicity
Non-reductive consciousness

1 Method: BICA *à rebours*

At this conference most authors focus on Biologically Inspired Cognitive Architectures [3]. Instead, we concentrate on an AI inspired take on human cognitive architecture (cog arch).

Those two approaches share the same method that I call the BICA method, which views cognitive architectures as a spectrum between animal, human, mixed (augmented) and artificial cog arch within the framework of gen cog [4].

© Springer Nature Switzerland AG 2019
A. V. Samsonovich (Ed.): BICA 2018, AISC 848, pp. 37–43, 2019.
https://doi.org/10.1007/978-3-319-99316-4_5

BICA à rebours[1] allows us to apply biologically inspired approach back to the analysis of human cog arch. Thus, we close the loop: BICA becomes a two-way street, from biology to AI, and back.

2   Case: BICA-Mary

2.1   Black and White Mary

"Mary is confined to a black-and-white room, is educated through black-and-white books and through lectures relayed on black-and-white television. In this way she learns everything there is to know about physical nature of the world". Yet "when she is let out of the black-and-white room or given a color television, she will learn what it is like to see something red, say." [2, p. 291]. This is supposed to demonstrate that some knowledge is non-physical, since Mary had all the physical knowledge but learned something from a 'non-physical' (namely, phenomenal) experience.

2.2   Physical, Yet Non-verbal

The above case relies on the presumption that physical knowledge is fully, and directly, describable in propositional language. The thesis is flowed, for the following reasons:

1. We may witness events that are clearly physical but are hard to describe, such as complex patterns [5].
2. We have the abilities that are physical but need to be learned from experience, such as breathing or riding a bicycle [6, 7]. It would be ridiculous to suppose 'non-physical bikers' [1] since those skills can be learned by monkeys and robots.

It is worth noting that robots 'learn' how to ride a bicycle by following a set of programmable instructions[2]. So, in a way, the ability is describable in language, but this is not first-order semantic meaning but second order (syntax of the programming language) that carries the task. Hence, the assumption that everything physical is describable in a semantic language is mistaken. A version of the theory that claims that everything physical is programmable (affine to the physical interpretation of the Church-Turing thesis [8]) seems right, but such understanding derails Jackson's case.

2.3   Privileged Access

Jackson's claim that "there are truths about other people that escape the physical story" [2, p. 293] seems fine; yet, it counts against physicalism only as long as the latter is

---

[1] À rebours (French) can be translated as 'backwards'; French term has a broader connotation than English, e.g. used in the title of a French novel, the term has been translated as 'against the grain' (or even 'against nature'); to dance à rebours often means to make the same steps in the opposite order and direction.

[2] Suppose traditional AI programming, not neural nets; the latter is more like training someone.

defined through the *intersubjective verifiability* criterion. Everything objective (hence physical) is supposed to be inter-subjectively verifiable.

Today, we treat the word 'subjective' in the term inter-subjective more seriously. Consistent, statistically relevant first-person reports are good science [9]. We also know that the distinction between processes that take place inside and outside of our scull is not absolute – we can read the content off of one's visual cortex [10] and informational content of human brains is more and more machine readable. This content is objective despite the special (though often not exclusive) access people have to their own thoughts and feels.

2.4   The Lacking Connection

It is for the following reason that Mary cannot figure out what the red things look like before she sees one: She lacks the link from knowledge *de dicto* to phenomenal experience (from words, to the way the things they describe feel). Human brains first perceive something and then try to name it and give semantic meaning to the new term. This comes from evolutionary causes: Language is evolutionarily new, while perceptions, even color-perceptions, are some orders of magnitude older than human race. Hence, the fact that we cannot reconstruct the exact feel of a given qualitative experience (color, smell, pain etc.) pertains to details of human cognitive architecture.

It does not reveal anything profound about the physical, or non-physical, character of human experience. The very idea that the experience is non-physical just because we are unable to figure it out on the basis of language descriptions, involves the sense of what counts as 'physical' that is quite remote from today's use of this term.

It is sufficient to design a modification in human cognitive architecture to dissipate the so-called knowledge argument. One could bioengineer a link that goes from descriptions of qualitative experiences, such as pain (or seeing redness), to their phenomenal feel. Such link would have to affect any place down the neural system that produces pain: brain center(s) that produce pain or any of the nerves leading to them. The link may be mechanical (pressing the nerves), chemical (producing pain-generating substances), electronic etc.

To sum up, the fact that we cannot imagine experiential qualities before we get directly acquainted with them can be explained as merely a specificity of the human cognitive architecture [1]. This argument is simpler, more explanatory, and seems to relate more directly to the problem whether the mind is material, than the more philosophical critiques of Jackson's argument.

3   Relevance: Objects Inside and Out

3.1   Mentations

Searle's pitching the concept of "mentations" – viewed as essential, as well as specific, to advanced biological minds – is one of his attempts to preserve non-reducible character of human consciousness. In light of BICA Mary [1], the attempt seems rather noble than sustainable.

The line in the sand between the first- and third-person perspective has been set, naively and mistakenly, among internal and external functionalities of the human mind. The internal, phenomenal objects, such as qualia, get juxtaposed to the external, third-person verifiable objects out in the world. But this is childish folly.

Chalmers seems right (in joining Clark) on the extended mind approach [12], but not so when claiming that the hard problem of consciousness is the problem of qualities of experience [13]. The hard problem is not the problem of content, phenomenal or otherwise. Content are objects, while the hard problem seems not to be about objects at all.

3.2   Privileged Access

Let is revisit the so-called *privileged access*. The content of one's experience – internal, external or spread out – can be read through fMRI and put on screen [10]. The view that we have 'privileged access' to phenomenal content of our own experiences, inaccessible to others, relied on inability of early science to penetrate the scull of human mind; now we know that the mind isn't constrained by any sculls. Proprioception, interoception, exteroception are forms of perception, and so are the images in one's mind or feelings and emotions. By the way, this does not make them passive, since they are about some relations. Oftentimes, e.g. during physical combat, the relations become tangible and intense.

In the epistemic order (which is the order of explanation, not the order of coarse-grained ontology, or of nature) perception is – trivially – of the objects of perception, of phenomena. Only secondarily it is the perception of whatever ontic explanations those phenomena may have. *Aboutness* is a second order property of phenomenal gestalts. The Tulips in the well-known poem by Sylvia Plath [14] may originate in her dreams, drugs, they may be actual physical tulips, fata morgana, or holograms. The ontic story supervenes on the story of phenomena, at least within the epistemic order of explanation.

3.3   Elizabeth Right, Descartes Wrong

Vice versa, the epistemic story supervenes on the ontological account within the coarse-grained ontological framework. The complementary approach retains non-reductive consciousness, as the *epistemic locus*, *pure subject*, the *epistemic starting point* – but also, by the same token, it retains objectivity of the world of coarse-grained physical objects, when we start the enquiry within *the ontic perspective*.

Those complementary [15] starting points – the epistemic and the ontological one – provide the way to avoid denying, or *pseudo*-bridging the epistemic-ontological gap. The line in the sand, the line of defense of non-reductive consciousness, needs not to be drawn among the objects (even broadly understood), those perceived and those in one's mind, phenomenal and noumenal, extended and mental. It is between the objects and pure subjectivity; the latter treated as 'an object of our discourse' only at the meta-level of description, always known solely through its influence on the objects.

Elizabeth of Bohemia [16] seems quite right that there is no bridge between objects and purely mental realms, and thus she was the first to bring out the problem of interaction. Descartes seems wrong, in their exchange, in trying to leaf through it. Interactionist dualism is a non-starter. We either have a complementary or monistic framework. Within the former, non-reductive materialism seems quite defensible.

3.4 Epistemicity

Within our proposed framework, all objects are in one realm – that of the general ontology of all the objects [17] since phenomena belong to this realm. The subject alone is not an object; it is the first-person viewpoint, seen as pure epistemicity, always already presumed in every epistemology. The problem whether artificial cognitive engines, the artificial minds, may have such locus of non-reductive first-person epistemicity is the problem of non-reductive machine consciousness[3]. The main point of the related argument, called *the engineering thesis in machine consciousness,* is that, once we know what produces first-person consciousness in a mind, a projector of non-reductive consciousness may very well be engineered like anything else we know. I want to emphasize, in partial agreement with Searle, that *mentations* may turn out to be very special and impossible to engineer, but (*contra* Searle) unlikely so.

4 Sum up: Reductive Phenomenal Content

The main argument: Not all physical knowledge is easily accessible to human beings as knowledge by description. The so called 'knowledge argument' [2] assumes that, if Mary (a scientist, color-blinded from birth, who has expert knowledge about colors) learns something upon seeing her first colored object, then physicalism is false.

We claim that this shows nothing of philosophical value since the following simpler account is available: The initial setting of human cognitive architecture has no link from knowledge by description to the feel of phenomenal qualities; this link must be created starting with the feel of phenomenal experience (as a given) linked with its name and broader semantic structures. (This functionality may be viewed as part of our operational system, just like a baby suckles a teat or ducklings follow the first individual they see.) Such link from words to phenomenal qualities failed to evolve due to the short evolutionary history of language (and variety of natural languages), It is quite obvious, that in a cognitive architecture like human, it would be possible to bio-engineer the link from words to phenomenal experiences requiring prior first-person observation. This shows that the 'knowledge argument' pertains to the specifics of human cognitive architecture; not quite to deep philosophical truths.

Viewing human cognitive architecture as just, well, a cognitive architecture, helps us clarify several deep philosophical conundrums. Back to my remarks on Searle [11] and Chalmers [13]: It is not phenomenal content that is non-reductive, but the stream of first-person awareness underlying such content. The hard problem is the problem of first-person *epistimicity*; not of content (or its qualities).

---

[3] I have discussed it in my keynote at BICA 2015, in Lyon, and in a few papers [18–20].

5   Follow up: Non-reductive Aspect

Further research may go in the following direction: Phenomenal content seems to impact causal chains in the physical world. You see a red traffic light and you stop. However, the first-person feel of qualia seems irrelevant for the physical causal chains since the feel is a subjective aspect of semantic content that qualia bring out.

This is visible in AI simulations. A robot may react to harm the way a sentient being would, which makes it functionally sentient, with no first-person *feel*. Depending on details of a cognitive architecture, such feel may be left out, making it epiphenomenal. Yet, such epiphenomenal feel is sometimes indirectly relevant. In the case of Church-Turing Lovers [22] one has a moral stake in knowing whether one's significant other has first-person consciousness, or is a philosophical Zombie. This is the difference between the two interpretations of Block's phenomenal consciousness [21]: the one, non-reductive, as intended by Block. We call the other: functional-phenomenal consciousness. Consequently, we need to distinguish phenomenal content from *epistemicity*, the latter being the sole non-reductive element.

References

1. Boltuc, P.: Mary's acquaintance. APA Newsl. Philos. Comput. **14**(1), 25–31 (2014)
2. Jackson, F.: What Mary didn't know. J. Philos. **83**, 291–295 (1986)
3. Samsonovich, A.V.: On a roadmap for the BICA Challenge. Biol. Inspir. Cogn. Archit. **1**, 100–107 (2012)
4. Goertzel, B.: Characterizing human-like consciousness: an integrative approach. In: Proceedings of BICA-14. Procedia Computer Science, vol. 41, pp. 152–157. Elsevier (2014)
5. Boltuc, P.: Reductionism and qualia. Epistemologia **4**, 111–130 (1998)
6. Nemirow, L.: Physicalism and the cognitive role of acquaintance. In: Lycan, W. (ed.) Mind and Cognition, pp. 333–339. Blackwell, Oxford (1990)
7. Lewis, D.: Mad pain and martial pain (with postscript). In: Lewis, D. (ed.) Philosophical Papers, vol. I, pp. 122–130. Oxford University Press, Oxford (1983)
8. Deutsch, D.: Quantum theory, the Church-Turing principle and the universal quantum computer. Proc. R. Soc. Ser. A **400**, 97–117 (1985)
9. Piccinini, G.: Mind gauging: introspection as a public epistemic resource, Grad Expo, University of Pittsburgh, Pittsburgh, PA, September 2001
10. Nishimoto, S.: Reconstructing visual experiences from brain activity evoked by natural movies. Curr. Biol. **21**, 1641–1646 (2011)
11. Searle, J.: Intentionality. Cambridge University Press, Cambridge (1983)
12. Clark, A., Chalmers, D.: The extended mind analysis. Analysis **58**(1), 7–19 (1998)
13. Chalmers, D.: The Conscious Mind. Oxford University Press, Oxford (1997)
14. Plath, S.: Tulips. Ariel. Faber and Faber, London (1965)
15. Russell, B.: The Analysis of Mind. George Allen and Unwin, London; The Macmillan Company, New York (1921)
16. Elizabeth of Bohemia: Elizabeth to Descartes 6 May 1643. Shapiro, L. (ed.) The Correspondence Between Princess Elizabeth of Bohemia and Rene Descartes. CUP, Chicago (2007)

17. Boltuc, P.: Subject is no object; complementary basis of information. In: Dodig-Crnkovic, Burgin (eds.) World Scientific Book Philosophy and Methodology of Information, chap. 1 (2018)
18. Boltuc, P.: The philosophical issue in machine consciousness. Int. J. Mach. Conscious. **1**, 155–176 (2009)
19. Boltuc, P.: A philosopher's take on machine consciousness. In: Guliciuc, V.E. (ed.) Philosophy of Engineering and the Artifact in the Digital Age, pp. 49–66. Cambridge Scholar's Press, Cambridge (2010)
20. Boltuc, P.: The engineering thesis in machine consciousness. Techne Res. Philos. Technol. **16**, 187–207 (2012)
21. Boltuc, P.: Church-Turing lovers. In: Lin, P., Abney, K., Jenkins, R. (eds.) Robot Ethics 2.0: From Autonomous Cars to Artificial Intelligence, pp. 214–228. Oxford University Press, Oxford (2017)
22. Block, N.: On a confusion about a function of consciousness. Behav. Brain Sci. **18**, 227–287 (1995)

The Research of Distracting Factors Influence on Quality of Brain-Computer Interface Usage

Anastasiia D. Cherepanova[1], Aliona I. Petrova[1], Timofei I. Voznenko[1], Alexander A. Dyumin[1,2]([✉]), Alexander A. Gridnev[1], and Eugene V. Chepin[1]

[1] Institute of Cyber Intelligence Systems, National Research Nuclear University MEPhI, Moscow, Russia
a.a.dyumin@ieee.org
[2] College of Information Business Systems, National University of Science and Technology MISiS, Moscow, Russia

Abstract. Nowadays, BCI (brain-computer interface) is a perspective human-machine interaction method with a lot of usage concepts, but many practical aspects should be investigated before it can be used in a broader scale. For example, one should spent a lot of time on training, before he or she can achieve sufficient control accuracy through BCI. Moreover, when one is working with BCI, an important factor is distraction that can have a negative impact on training, and then, subsequently, on the quality of control through it. Such factors can occur in normal conditions of BCI usage – that's why they are need to be taken into account. Distractions can be represented by different external impacts on the operator's channels of the perception (such as auditory impacts, visual impacts, etc.) or operator's own internal state. In this paper, we propose the method and software framework for evaluating control quality using BCI in the presence of distracting factors. Also we researched influence of some of them on quality of BCI usage.

Keywords: Distracting factors · Brain-computer interface · Robotics

1 Introduction

A brain-computer interface is a device that allows to obtain information about neurophysiological activity from the user head (e.g. EEG, eye and facial muscle activity). Using this information it is possible to train a system for recognition of various neural activity components including mental images. However, due to small number of electrodes (compared to the medical electroencephalograph) and due to the possibly noisy data (in the case of non-invasive sensors usage), the accuracy of the brain-computer interface may be not sufficient to control critical equipment such as a robotic device.

To increase the accuracy of control the user needs to spend a lot of time on training.

© Springer Nature Switzerland AG 2019
A. V. Samsonovich (Ed.): BICA 2018, AISC 848, pp. 44–49, 2019.
https://doi.org/10.1007/978-3-319-99316-4_6

During training process, distracting factors may worsen the quality of training outcome and they may increase the time spent on training to obtain good quality of control. In this paper, the influence of various distracting factors on the quality of BCI usage is considered.

2   Related Works

There are many factors that can be considered distracting. For example in research [1] authors showed that psychological factors (attention, concentration, motivation, or visuo-motor coordination) may influence BCI performance. In the paper [2] authors deal with physiological signals (movement of muscles and heart, blinking or other neural activities) which can distort P300 signal. In research [3] authors explored the influence of different factors such as concentration, attention, level of consciousness and the difficulty of the task, on EEG activity. In article [4] authors investigate the influence of various illumination and noise environments on P300 BCI system. In paper [5] authors investigate the effect of subject's emotional states on Brain Computer Interface (BCI) performance.

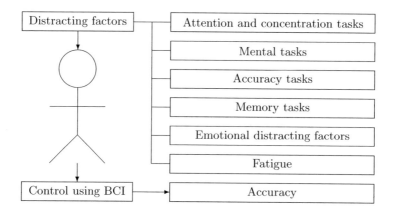

Fig. 1. The diagram of the testing method

3   Methodology

Within the framework of the experiment to evaluate the quality of control through the BCI (Fig. 1) the subject should perform a specific command, for example, moving forward the virtual robot. During the experiment two types of pictures are being shown to the subject: inanimate objects and animals. In the case of one observes inanimate object, he or she must execute command "forward". And if she or he sees animal, the command must not be issued. The detailed methodology is described in [6–8]. At the same time the subject may

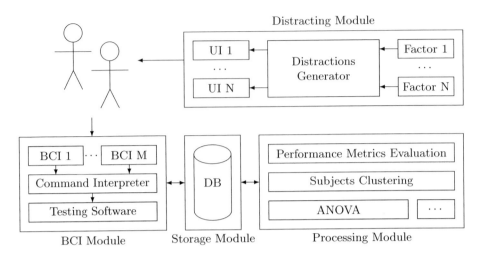

Fig. 2. The architecture of software framework for evaluating control quality using BCI in the presence of distracting factors

be affected by external distracting impacts (factors), which can involve various cognitive mechanisms with the participation of different parts of the brain.

To facilitate research based on this methodology, the software framework shown on the Fig. 2 was developed that includes base modules for interaction with BCI hardware and mental commands acquisition and interpretation, implementation of distracting factors through various user interfaces (as visual, sound, tactile, etc.), storage and processing of raw data obtained during experiments and results evaluations.

There are several distracting factors, which involve maximally diverse areas of the brain, have been chosen for the experiment:

1. Attention and concentration tasks. The subject should press numbers from 1 to 9 in ascending order as quickly as possible (Fig. 3a). Another task was to press the button which was running away from the subject (Fig. 3b).
2. Mental tasks. These are tasks that require mental work from a subject. Arithmetic, logical, linguistic problems and riddles are used to study the influence. An example of such task is an arithmetic task (Fig. 3d).
3. Accuracy tasks. The subject should run the mouse along the simple path from the beginning to the end of the "snake" (Fig. 3c).
4. Memory tasks. During the experiment the subject was shown a set of pictures or words to be memorized. After that, various questions related to long-term memory were asked. Then the system asked to remember the order, or the arrangement of words, which had to be memorized in the beginning of the experiment.
5. Emotional distracting factors. They include:
   (a) observing of pictures of different emotional coloring according to the paper [9];

(b) specially selected sounds listening that were mentioned in the paper [10] (there were taken three main sounds which are highly likely to cause certain emotions: the sound of a drill in 59% of cases causes a feeling of anger, cry of a child in 63% causes sadness, and laugh of a child in 83% causes joy).

6. Fatigue. To implement this factor the subject was using a hand expander for 5 min.

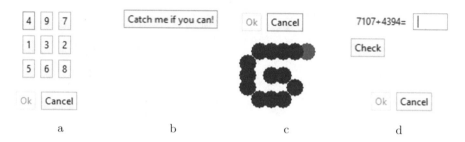

Fig. 3. Examples of distracting tasks. a – arrange the numbers in ascending order, b – catch the button, c – run the mouse along the snake, d – solve an arithmetic problem

The study was conducted in two stages with 14 students of NRNU MEPhI at the age of 19 to 23. All subjects gave oral consent to participate in the experiment. Initially subjects were introduced to principles of BCI usage. After that, the subjects underwent tests without and with distracting factors. For each distracting factor, as well as for testing without distractions, the subject was shown 30 pictures with a duration of 10 seconds each. During the testing with distracting factors the subject should understand whether he or she should execute the command "forward", while solving distracting tasks in parallel. At the end of the experiment, accuracy of control ACC (the proportion of the correct actions of subject) was calculated.

4   Results

The obtained data were analyzed using ANOVA (Table 1). Based on the obtained results it can be concluded that distractions affect the quality of work with BCI ($F = 26.37 > F_{critical}(0.05, 6, 92) = 2.20$; where 0.05 – significance, 6 – degrees of freedom between groups, 92 – degrees of freedom within groups), and the influence is negative. In general, the subjects received the lowest values of accuracy with the influence of factors of memorizing and fatigue. Figure 4 shows the percentage of accuracy with BCI without and with various factors.

Table 1. The result of ANOVA

Source of variation	Sum of squares	d.f	Mean squares	F
Between-treatments	3472	6	578.7	26.37
Within-treatments	1997	91	21.95	
Total	5469	97		

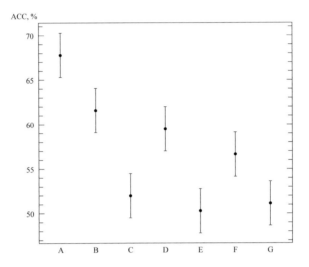

Fig. 4. The accuracy (ACC) results (%). A – without distracting factors, B – attention, C – mental tasks, D – accuracy, E – memorizing, F – emotions, G – fatigue

5   Conclusion

In this study the influence of various distracting factors to the quality of working with BCI was investigated. The worst results in terms of accuracy were obtained in the presence of such distractions as mental tasks, memorizing and fatigue. The fatigue factor was tested last and according to subjects, they had been getting more tired mentally (constantly concentrating on command execution) than physically (using the expander). Therefore, in this study, we did not distinguish between physical and mental fatigue. Thus, it can be concluded that when working with BCI it is necessary to take into account that distracting factors as mental tasks, memorizing and fatigue can have a strong negative impact on the accuracy of BCI usage, and therefore in the process of working with BCI it is highly recommended to protect the operator from the influence of these factors. Such factors as attention, accuracy, and emotional factors also have a negative impact on the accuracy of working with BCI, but not as strongly as the factors mentioned above.

In our future works, we are planning to consider the influence of other factors (such as stress) on the quality of work with BCI. We are also planning to take into account how the negative influence of external factors can be leveled by changing the training process in order to build a model for recognizing commands taking into account their presence.

Acknowledgments. This work was partially supported by Competitiveness Growth Program of the National Research Nuclear University MEPhI (Moscow Engineering Physics Institute).

The authors would like to thank all the participants who agreed to take part in the experiment.

References

1. Kleih, S.C., Kübler, A.: Psychological factors influencing brain-computer interface (BCI) performance. In: 2015 IEEE International Conference on Systems, Man, and Cybernetics (SMC), pp. 3192–3196. IEEE (2015). https://doi.org/10.1109/SMC.2015.554
2. Argunsah, A.O., Curuklu, A.B., Cetin, M., Ercil, A.: Factors that affect classification performance in EEG based brain-computer interfaces. In: 2007 IEEE 15th Signal Processing and Communications Applications. SIU 2007, pp. 1–5. IEEE (2007). https://doi.org/10.1109/SIU.2007.4298842
3. Eskandari, P., Erfanian, A.: Improving the performance of brain-computer interface through meditation practicing. In: 2008 30th Annual International Conference of the IEEE Engineering in Medicine and Biology Society. EMBS 2008, pp. 662–665. IEEE (2008). https://doi.org/10.1109/IEMBS.2008.4649239
4. Lian, J., Bi, L., Fan, X.a.: Effects of illumination and noise on the performance of a P300 brain-computer interface for assistive vehicles. In: 2017 8th International IEEE/EMBS Conference on Neural Engineering (NER), pp. 337–340. IEEE (2017). https://doi.org/10.1109/NER.2017.8008359
5. Zhu, Y., Tian, X., Wu, G., Gasso, G., Wang, S., Canu, S.: Emotional influence on SSVEP based BCI. In: 2013 Humaine Association Conference on Affective Computing and Intelligent Interaction (ACII), pp. 859–864. IEEE (2013). https://doi.org/10.1109/ACII.2013.161
6. Voznenko, T.I., Urvanov, G.A., Dyumin, A.A., Andrianova, S.V., Chepin, E.V.: The research of emotional state influence on quality of a brain-computer interface usage. Procedia Comput. Sci. **88**, 391–396 (2016). https://doi.org/10.1016/j.procs.2016.07.454
7. Chepin, E., Dyumin, A., Urvanov, G., Voznenko, T.: The improved method for robotic devices control with operator's emotions detection. In: 2016 IEEE NW Russia Young Researchers in Electrical and Electronic Engineering Conference (EIConRusNW), pp. 173–176. IEEE (2016). https://doi.org/10.1109/EIConRusNW.2016.7448147
8. Voznenko, T.I., Dyumin, A.A., Aksenova, E.V., Gridnev, A.A., Delov, V.A.: The experimental study of 'unwanted music' noise pollution influence on command recognition by brain-computer interface. Procedia Comput. Sci. **123**, 528–533 (2018). https://doi.org/10.1016/j.procs.2018.01.080
9. Harry, B.B., Williams, M., Davis, C., Kim, J.: Emotional expressions evoke a differential response in the fusiform face area. Front. Hum. Neurosci. **7**, 692 (2013). https://doi.org/10.3389/fnhum.2013.00692
10. Vyskochil, N.: Podbor audial'nogo stimul'nogo materiala dlya izucheniya ehmocional'noj sfery cheloveka (in Russian). In: EHksperimental'naya psihologiya v Rossii: tradicii i perspektivy/Pod red. VA Barabanshchikova, p. 477 (2010)

Metagraph Approach as a Data Model for Cognitive Architecture

Valeriy Chernenkiy, Yuriy Gapanyuk$^{(\boxtimes)}$, Georgiy Revunkov,
Yuriy Kaganov, and Yuriy Fedorenko

Bauman Moscow State Technical University,
Baumanskaya 2-ya, 5, 105005 Moscow, Russia
gapyu@bmstu.ru

Abstract. There is no doubt that cognitive architecture model must be capable of describing very complex hierarchical systems. The basic lower-level "lingua franca" model is required for cognitive architecture description. We propose to use a metagraph model as a basic model. The key element of the metagraph model is metavertex which may include inner vertices and edges. From the general system theory point of view, a metavertex is a special case of the manifestation of the emergence principle, which means that a metavertex with its private attributes and connections become a whole that cannot be separated into its parts. The metagraph agents are used for metagraph transformation. The distinguishing feature of the metagraph agent is its homoiconicity, which means that it can be a data structure for itself. Thus, a metagraph agent can change the structure of other metagraph agents. The structures similar to neural networks are typical elements of cognitive architecture. It is shown that proposed metagraph approach is suitable enough to represent neural networks.

Keywords: Metagraph · Metavertex · Metagraph agent · Neural network

1 Introduction

The Institute of Creative Technologies [1] defines cognitive architecture as: "hypothesis about the fixed structures that provide a mind, whether in natural or artificial systems and how they work together – in conjunction with knowledge and skills embodied within the architecture – to yield intelligent behavior in a diversity of complex environments."

The paper [2] gives an overview of the several most developed cognitive architectures, such as SOAR, ACT-R, EPIC, ICARUS, SNePS. Summing up the findings of the paper [2], one can say that there is no doubt that cognitive architecture model must be capable of describing very complex hierarchical systems. Therefore, according to [3] the "lingua franca," i.e., the basic lower-level model is required for cognitive architecture description.

The authors of paper [3] propose to use the model of conceptual spaces as a "lingua franca" model. There is no doubt that different models can be proposed as a basic or "lingua franca" model. And only future approbation will show the advantages and

© Springer Nature Switzerland AG 2019
A. V. Samsonovich (Ed.): BICA 2018, AISC 848, pp. 50–55, 2019.
https://doi.org/10.1007/978-3-319-99316-4_7

disadvantages of each basic model. Instead of conceptual spaces, we propose to use a model based on complex graphs and multiagent system.

There are a number of graph-based cognitive architectures, for example, the Sigma cognitive architecture [4] proposed by the Institute of Creative Technologies. But from our point of view, flat graphs are not good enough to describe hierarchical systems. Therefore, we propose to use metagraph model.

The idea of using the multiagent system as a component of the cognitive architecture is also not new. For example, the description of biologically inspired cognitive architecture using intelligent agents proposed in [5]. But in this paper, we describe the special kinds of agents, adopted for the metagraph model.

2   Using Metagraph Model as a Data Model

From our point of view, one of the most missed ideas of flat graph usage is the idea of emergence. From the general system theory point of view, the principle of the emergence means that a whole that cannot be separated into its component parts. This principle is one of the central principles of any complex system, including cognitive architecture.

Metagraph is a kind of complex graph model with emergence, proposed by A. Basu and R. Blanning in their book [6] and then adapted for information systems description in our paper [7]. According to [7] metagraph MG is described as $MG = \langle V, MV, E \rangle$, where V – set of metagraph vertices; MV – set of metagraph metavertices; E – set of metagraph edges.

Metagraph vertex is described by a set of attributes: $v_i = \{atr_k\}, v_i \in V$, where v_i – metagraph vertex; atr_k – attribute.

Metagraph edge is described by set of attributes, the source and destination vertices and edge direction flag: $e_i = \langle v_S, v_E, eo, \{atr_k\} \rangle, e_i \in E, eo = true|false$, where e_i – metagraph edge; v_S – source vertex (metavertex) of the edge; v_E – destination vertex (metavertex) of the edge; eo – edge direction flag ($eo = true$ – directed edge, $eo = false$ – undirected edge); atr_k – attribute.

The metagraph fragment: $MG_i = \{ev_j\}, ev_j \in (V \cup E \cup MV)$, where MG_i – metagraph fragment; ev_j – an element that belongs to the union of vertices, edges, and metavertices.

The metagraph metavertex: $mv_i = \langle \{atr_k\}, MG_j \rangle, mv_i \in MV$, where mv_i – metagraph metavertex belongs to set of metagraph metavertices MV; atr_k – attribute, MG_j – metagraph fragment.

Thus, a metavertex is a special case of the manifestation of the emergence principle, which means that a metavertex with its private attributes and connections become a whole that cannot be separated into its component parts.

3   Using Metagraph Agents as an Action Model

The metagraph model is aimed for complex data description. But it is not aimed for data transformation. To solve this issue the metagraph agent (ag^{MG}) aimed for data transformation is proposed. There are two kinds of metagraph agents: the metagraph function agent (ag^F) and the metagraph rule agent (ag^R). Thus $ag^{MG} = ag^F \mid ag^R$.

The metagraph function agent serves as a function with input and output parameter in the form of metagraph: $ag^F = \langle MG_{IN}, MG_{OUT}, AST \rangle$, where ag^F – metagraph function agent; MG_{IN} – input parameter metagraph; MG_{OUT} – output parameter metagraph; AST – abstract syntax tree of metagraph function agent in the form of metagraph.

The metagraph rule agent uses rule-based paradigm: $ag^R = \langle MG, R, AG^{ST} \rangle, R = \{r_i\}, r_i : MG_j \rightarrow OP^{MG}$, where ag^R – metagraph rule agent; MG – working metagraph, a metagraph on the basis of which the rules of agent are performed; R – set of rules r_i; AG^{ST} – start condition (metagraph fragment for start rule check or start rule); MG_j – a metagraph fragment on the basis of which the rule is performed; OP^{MG} – set of actions performed on metagraph.

The antecedent of the rule is a condition over metagraph fragment, the consequent of the rule is a set of actions performed on metagraph. Rules can be divided into open and closed.

The consequent of the open rule is not permitted to change metagraph fragment occurring in rule antecedent. In this case, the input and output metagraph fragments may be separated. The open rule is similar to the template that generates the output metagraph based on the input metagraph.

The consequent of the closed rule is permitted to change metagraph fragment occurring in rule antecedent. The metagraph fragment changing in rule consequent cause to trigger the antecedents of other rules bound to the same metagraph fragment. But incorrectly designed closed rules system can cause an infinite loop of metagraph rule agent.

Thus, metagraph rule agent can generate the output metagraph based on the input metagraph (using open rules) or can modify the single metagraph (using closed rules).

The distinguishing feature of the metagraph agent is its homoiconicity which means that it can be a data structure for itself. This is due to the fact that according to definition metagraph agent may be represented as a set of metagraph fragments and this set can be combined in a single metagraph. Thus, the metagraph agent can change the structure of other metagraph agents.

In this paper, we also use the "active metagraph" concept, which means a combination of data metagraph with an attached metagraph agent.

4   Neural Networks Representation Using Metagraphs

The structures similar to neural networks are typical elements of cognitive architecture. In this section, we will show that proposed metagraph approach is suitable enough to represent a neural network.

In our paper [8] the metagraph representation of general structure neural network (shown in Fig. 1) was proposed. The distinguishing feature of this approach is the ability to change the structure of neural network during creation or training using metagraph agents.

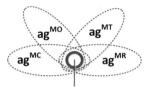

The metagraph representation of a neural network

Fig. 1. The structure of metagraph rule agents for neural network operation representation.

In order to provide a neural network operation, the following rule metagraph agents are used: ag^{MC} – the agent responsible for the creation of the network, ag^{MO} – the agent responsible for the modification of the network; ag^{MT} – the agent responsible for the training of the network; ag^{MR} – the agent responsible for the execution of the network. In Fig. 1 the agents are shown as metavertices by dotted-line ovals.

The network-creating agent ag^{MC} implements the rules for creating an original neural network topology. The agent holds both the rules of creating separate neurons and rules of connecting neurons into a network.

The network-modification agent ag^{MO} holds the rules of modification the network topology in the process of operation. It is especially important for neural networks with variable topologies such as HyperNEAT and SOINN.

The network training agent ag^{MT} implements a particular training algorithm. As a result of training, the changed weights are written in the metagraph representation of the neural network. It is possible to implement a few training algorithms by using different sets of rules for agent ag^{MT}.

The network-executing agent ag^{MR} is responsible for the start and operation of the trained neural network.

The agents can work separately or jointly which may be especially important in the case of variable topologies. For example, when a HyperNEAT or SOINN network is trained, agent ag^{MT} can call the rules of agent ag^{MO} to change the network topology in the process of training.

Let us consider in more detail the case of a description of a neural network containing various regularization strategies in accordance with our paper [9]. The metagraph structure is shown in Fig. 2. The metagraph neural network representation metavertices are shown by ovals, and metagraph agents are shown by arrows.

In the creation mode, the metagraph representation of the neural network is created using metagraph agents. According to modeling tasks, the complexity of created neuron structure can be various. In the simplest case, the neuron may be considered as a node with activation function. In more complex cases the neuron may be represented as a nested metagraph, which contains metavertices with complex activation function addressing neuron structure. Thus, at the end of creation mode, the "Neural Network"

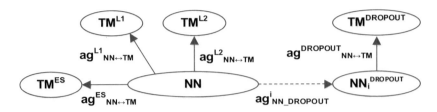

Fig. 2. The metagraph description of a neural network containing regularization strategies.

(NN) structure is created. In case of the deep network, this is a flat graph of nodes (neurons) connected with edges. But node may be represented as complex metavertex and neurons of each layer of the network may also be combined into metavertex.

In the training mode, the "Training Metagraph" (TM) is created. TM structure is isomorphic to the NN structure. For each node NN_i^n in NN, the corresponding metavertex TM_i^n in TM is created. And for each edge NN_i^e in NN, the corresponding edge TM_i^e in TM is created. For the TM creation, the agent $ag_{NN \to TM}$ is used. This agent is kind of function agent.

TM may be considered as an active metagraph with metagraph agent ag_{TM} bound to TM graph structure. Agent ag_{TM} implements a particular training algorithm. As a result of training, the changed weights are written to the TM_i^n.

The agent is also created with the $ag_{NN \to TM}$ agent. Different regularization strategies could be embedded into the ag_{TM} training algorithm.

For the single NN, we can create several TM with different regularization strategies. For example, TM^{L1} means that $ag_{NN \to TM}$ agent creates TM graph structure with ag_{TM} agent that implements training algorithm with L^1 regularization. Similarly, TM^{L2} stands for L^2 regularization, and TM^{ES} stands for early stopping.

It should be noted that neither of these regularization strategies requires changing the network structure during the training process. But in case of a dropout, we have to change the network structure. In this case, the agent $ag_{NN\_DROPOUT}$ is used. This agent is kind of metagraph rule agent.

Using $ag_{NN\_DROPOUT}$ metagraph agent, we can implement different dropout strategies. Applying different $ag_{NN\_DROPOUT}^i$ agents to the original NN structure we resulting the set of modified $NN_i^{DROPOUT}$ network structures. For each $NN_i^{DROPOUT}$ the corresponding $TM^{DROPOUT}$ is created for network training.

In Fig. 2 the transformation with structure changing is shown with a dashed arrow, the transformation without structure changing is shown with a solid arrow.

Thus, the metagraph approach allows representing neural network training with different regularization strategies either with or without network structure transformation.

5 Conclusions

We propose to use the metagraph approach as a "lingua franca" model for cognitive architecture. Metagraph approach allows describing complex hierarchical graph models. Using metagraph agents, it is possible either to generate the output metagraph based on the input metagraph (using open rules) or to modify the metagraph (using closed rules).

The structures similar to neural networks are typical elements of cognitive architecture. The metagraph approach helps to describe the general structure of the neural network, in particular, to represent neural network training with different regularization strategies either with or without network structure transformation.

References

1. The Institute of Creative Technologies Homepage. http://cogarch.ict.usc.edu. Accessed 20 Apr 2018
2. Goertzel, B., Lian, R., Arel, I., de Garis, H., Chen, S.: World survey of artificial brains, part II: biologically inspired cognitive architectures. Neurocomputing **74**(1), 30–49 (2010)
3. Lieto, A., Chella, A., Frixione, M.: Conceptual spaces for cognitive architectures: a lingua franca for different levels of representation. Biol. Inspir. Cogn. Archit. **19**, 1–9 (2017)
4. Rosenbloom, P., Demski, A., Ustun, V.: The Sigma cognitive architecture and system: towards functionally elegant grand unification. J. Artif. Gen. Intell. **7**(1), 1–103 (2016)
5. Samsonovich, A.: On a roadmap for the BICA challenge. Biol. Inspir. Archit. **1**, 100–107 (2012)
6. Basu, A., Blanning, R.: Metagraphs and Their Applications. Springer, New York (2007)
7. Chernenkiy, V., Gapanyuk, Yu., Nardid, A., Gushcha, A., Fedorenko, Yu.: The hybrid multidimensional-ontological data model based on metagraph approach. In: Petrenko, A., Voronkov, A. (eds.) Perspectives of System Informatics 2017, LNCS, vol. 10742, pp. 72–87. Springer, Moscow (2017)
8. Fedorenko, Yu., Gapanyuk, Yu.: Multilevel neural net adaptive models using the metagraph approach. Opt. Mem. Neural Netw. **25**(4), 228–235 (2016)
9. Fedorenko, Yu., Gapanyuk, Yu., Minakova, S.: The analysis of regularization in deep neural networks using metagraph approach. In: Kryzhanovsky, B., Dunin-Barkowski, W., Redko, V. (eds.) Advances in Neural Computation, Machine Learning, and Cognitive Research 2017, SCI, vol. 736, pp. 3–8. Springer, Moscow (2017)

Intelligent Processing of Natural Language Search Queries Using Semantic Mapping for User Intention Extracting

Artyom Chernyshov[✉], Anita Balandina, and Valentin Klimov

National Research Nuclear University "MEPhI", Moscow, Russian Federation
a-chernyshov@protonmail.com,
anita.balandina@gmail.com, vvklimov@mephi.ru

Abstract. Nowadays the leading world scientists and engineers center their attention to data mining and machine learning algorithms optimization and acceleration rather than inventing new ones. The natural language processing methods and tools are widely in use in production in the area of machine translation. The researches in the area of search engines and semantic search are mostly concentrated on data storage and further analysis. The majority of search engines use the huge amounts of previously accumulated user requests for predicting the search output without taking in attention this user intention by qualitative processing the request.

In this paper we explore the idea of usage the semantic cognitive spaces for extracting the exact user intentions by analysis the natural language input requests. The final goal of our research is to develop a valid search query model for further usage in semantic search engines.

Keywords: Natural language processing · Dependency parsing
Syntax analysis

1 Introduction

The majority of modern search engines used to work according the principle of keywords highlighting from the search query and further comparison to existing search index. This approach does not take into account the semantics and the relationship between the words in the original query. The loss of hidden dependencies can cause the poor relevance of search results.

While working on semantic search engine, we cannot ignore that fact, since the simple keywords usage contradicts the main paradigm of semantic search [1].

Therefore, we propose the domain ontologies usage for storing the search index, and natural language processing methods for constructing a hierarchical image of the search query with preservation of the semantic dependencies for matching and searching on the ontology graph.

Our task is to develop a valid search query model i.e. search query image, for further usage in the semantic search system. To do this, it is necessary to solve a number of problems related to the dependencies between words at the phrase level, the

© Springer Nature Switzerland AG 2019
A. V. Samsonovich (Ed.): BICA 2018, AISC 848, pp. 56–61, 2019.
https://doi.org/10.1007/978-3-319-99316-4_8

problems of determining the context synonyms of the original words and phrases and the measures of their proximity, determining the search context and the user's search intentions in the search area context. In this paper, we will consider one of the problems listed above, related to the dependencies between words, we present some modern methods for solving it, as well as our own algorithm, developed based on the classical dependencies parsing.

2 Models and Methods

The approach we propose is one of the ways to isolate dependencies within a phrase by dependency tree constructing. There we have two options - the allocation of dependencies at the words level and at the phrases level.

The phrase level dependencies are well suited to understanding the structure of the whole sentence, while dependency grammars are mostly in use for understanding the structure of a particular phrase within the sentence.

In the syntax tree of dependencies, the root is usually a verb or any other word denoting an action. On the second tier, as a rule, there are the action object, subject, on the lower tiers - the characteristics of the object, the subject, and various constraints (time, quantity, properties, etc.).

All links in the tree are marked with special POS-labels (part-of-speech). The types of these tags can help to discard insignificant terms, for example, the CASE tag relates to prepositions, conjunctions and other insignificant parts of speech that can be discarded.

Thus, the search query can easily be compared to a hierarchical structure with explicitly distinguished dependencies between meaningful terms.

We suggest using the shift-reduce parser in order to construct the corresponding search query model. For training the model, we use the recurrent neural network LSTM with one hidden layer. As data for training, we use a pre-marked corpus of Russian-language texts. As a loss function, we use the cross-entropy: $\mathcal{L}(\sigma) = -\sum_{i,j} \log(y_{ij})$ with L2 regularization.

The ultimate goal is to obtain a sequence of actions - shift (left-arc, right-arc), reduce, by which we can restore the tree structure.

The classical architecture of the LSTM network single neuron can be described by a set of the following formulas:

$$f_t = \sigma_g\left(W_t x_t + U_f h_{t-1} + b_f\right) - \text{forget gate's activation vector}$$
$$i_t = \sigma_g(W_t x_t + U_i h_{t-1} + b_i) - \text{input gate's activation vector}$$
$$o_t = \sigma_g(W_o x_t + U_o h_{t-1} + b_o) - \text{output gate's activation vector}$$
$$c_t = f_t \circ c_{t-1} + i_t \circ \sigma_c(W_c x_t + U_c h_{t-1} + b_c) - \text{cell state vector}$$
$$h_t = o_t \circ \sigma_h(c_t) - \text{output vector of the LSTM unit}$$

We can represent the output model as a following graph:

$$G = \langle X, A \rangle, \text{ where}$$

$X = \{x_1, x_2, \ldots, x_n\}$ – is the set of the input search query words
$A = \{a_1, a_2, \ldots, a_n\}$ – is the set of constructed connections between words:
$a_i := x_i \underset{pos}{\rightarrow} x_j$

The obtained model can be used to search by ontology. Formally, the ontology can be represented as the following set:

$$O = \langle I, C, P, O \rangle, \text{ where}$$

$I = \{o_1, o_2, \ldots, o_n\}$ – objects, or entities of the domain
$C = \{c_1, c_2, \ldots, c_n\}$ – classes,
$P = \{p_1, p_2, \ldots, p_n\}$ – classes characteristics
$O = \{o_1, o_2, \ldots, o_n\}$ – operators between classes

The first step of the search algorithm is to compare the root of the source tree with the type of connection on the ontology. Since the links in the ontology are uniquely named, whereas in the initial query they can be very different. In order to compare the root with the type of connection, it is necessary to determine the degree of similarity between the two words - the connection name on the graph and the root of the tree.

$$sim = \min_{i=1,n} \left(\frac{\sum_{v=1}^{k} x_{iv} o_{i,v}}{\sqrt{\sum_{v=1}^{k} x_{iv}^2} \sqrt{\sum_{v=1}^{k} o_{iv}^2}}, \frac{\sum_{v=1}^{k} x_{iv} c_{i,v}}{\sqrt{\sum_{v=1}^{k} x_{iv}^2} \sqrt{\sum_{v=1}^{k} c_{iv}^2}} \right)$$

To do this, we search through the vector space of constraints, to determine the measure of similarity. As such likelihood measure we use the cosine measure. Once the algorithm found the most plausible link in the ontology, it is necessary to find two types of vertices that it links. Therefore, the algorithm keeps searching for similar words on the graph for vertices of the subject and action object types. Once the necessary connection is found, the conditions are determined depending on the type of connection. This may be temporary conditions, location or characteristic.

3   Results

For syntax model evaluation, we consider using the classical measures – precision, recall and mid-harmonic F-measure. Let us consider that R is the relation between two tokens x and y. S_{train} is the set relations R restored by train data, and S_{parsed} is the set of parser output relations. Thus, we can represent the measures listed above as following:

$$Rec = \frac{|S_{train} \cap S_{parsed}|}{|S_{train}|}$$

$$Prec = \frac{|S_{train} \cap S_{parsed}|}{|S_{parsed}|}$$

$$F = 2\frac{prec \times rec}{prec + rec}$$

Let us also introduce the parser specific measures. Labeled precision is a relative amount of correctly recognized pairs node-label in the output of the parser and the labeled recall – the relative amount of pairs node-label from the train data which can be extracted from parser output. In our work, we have used two approaches to evaluate the parser: leaf-ancestor methods and attachment score.

Attachment score shows the percentage of words that have the correct head. The use of a single accuracy metric is possible in dependency parsing thanks to the single-head property of dependency trees, which makes parsing resemble a tagging task, where every word is to be tagged with its correct head and dependency type. This is unlike the standard metrics in constituency-based parsing, which are based on precision and recall, since it is not possible to assume a one-to-one correspondence between constituents in the parser output and constituents in the treebank annotation [2].

All of these metrics can be unlabeled (only looking at heads) or labeled (looking at heads and labels). The most commonly used metrics are the labeled attachment score (LAS) and the unlabeled attachment score (UAS).

LAS gives an evaluation of how many words were parsed correctly. However, this may not always be the point of interest. Another way of evaluating the quality of dependency parses is using sentences as the basic units. In this case, we have to calculate for each sentence what the percentage of correct dependencies is and then average over the sentences.

The leaf-ancestor metric compares path between nodes (from terminal nodes up to root) of the parser output and the path in appropriate sentence from test data. The paths similarity can be calculating using the Levenshtein distance [2]:

$$lev_{a,b}(i,j) = \begin{cases} \max(i,j), \min(i,j) = 0 \\ \min\big((lev_{a,b}(i-1,j)+1), (lev_{a,b}(i,j-1)+1), (lev_{a,b}(i-1,j-1)+1)\big) \end{cases}$$

We compare the output sequence of labels for each node in the sentence to standard one and take average value for the whole tree. Using these methods, we can calculate the precision, recall and calculate the F-measure.

For labels and tags prediction we train the LSTM network, an input is represented as current state of stack and buffer, and the output – as probability of labeled actions, we take the maximum probability and apply the action.

The Fig. 1 describes the learning curve of resulting neural network. For training the network, we used the treebank in CoNNL-U format for Russian language - UD_Russian-SynTagRus. It contains 61889 sentences and 1106290 words

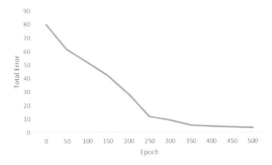

Fig. 1. The learning curve of LSTM label predictor

For comparison to state-of-art analyzers, we chose four the most popular: Googles's SyntaxNet, Stanford neural net dependency parser, spaCy parser and MALT parser. The results are represented in Table 1. We calculated the described above measures for train and test data, as a result, we took the average measure on test dataset. The result is represented on Fig. 2.

Table 1. The comparison on train and test datasets

	LAS		LA		UAS		F	
	Train	Test	Train	Test	Train	Test	Train	Test
MALT	69.73	67.30	78.92	75.77	76.50	71.74	75.12	74.54
SyntaxNet	72.09	69.58	81.59	78.34	79.09	74.17	77.66	77.07
SNNDP	70.41	67.96	79.69	76.51	77.25	72.44	75.85	75.27
spaCy	68.07	65.70	77.04	73.97	74.68	70.04	73.33	72.77
Described parser	76.12	73.47	86.16	82.72	83.51	78.32	82.00	81.38

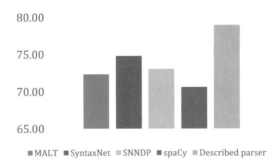

Fig. 2. The comparison to other parsers

As it is show, the average accuracy of suggested dependency parser is higher than other ones. We have to note, that the research is connected to parsing Russian language, we did not consider other languages. All the listed above metrics show the ability of parser correctly recognize arc labels and tags. We also calculated the percentage of correctly recognized the part-of-speech tags and found out that the difference between parsers are minor. For example, the accuracy of SyntaxNet for ability to recognize ADJ, NOUN and VERB tags is 93.2450, 91.4477 and 83.7365 respectively, and our parser is 92.3709, 90.8152 and 85.7920 respectively. As we can see, these measures are quite close to one another.

4 Discussion

In this paper, we described only the first stage of our query handling process. It also contains the other stages – synonyms extension, context and user goals recognition, named entities recognition and common grammar check.

However, the syntax analysis is quite an important step in natural language processing and in our task – input queries handling. Our following tasks include the developments instruments for user intentions extracting with semantic mapping [4, 5] and dependency parsing, synonyms extension with the usage of word embedding spaces. As for described dependency parser – we plan to improve the accuracy by the neural network architecture modification. We suppose that the BiLSTM based classifier will improve the analyzer accuracy.

Acknowledgements. This work was supported by Competitiveness Growth Program of the Federal Autonomous Educational Institution of Higher Professional Education National Research Nuclear University MEPhI (Moscow Engineering Physics Institute). The funding for this research was provided by the Council on grants of the President of the Russian Federation, Grant of the President of the Russian Federation for the state support of young Russian scientists - candidates of sciences MK-6888.2018.9.

References

1. Shadbolt, N., Berners-Lee, T., Hall, W.: The Semantic Web revisited. IEEE Intell. Syst. **3**, 96–101 (2006)
2. Kübler, S., McDonald, R., Nivre, J.: Dependency parsing. Synth. Lect. Hum. Lang. Technol. **1**(1), 1–127 (2009)
3. Yujian, L., Bo, L.: A normalized Levenshtein distance metric. IEEE Trans. Pattern Anal. Mach. Intell. **29**(6), 1091–1095 (2007)
4. Klimov, V., Chernyshov, A., Balandina, A., Kostkina, A.: A new approach for semantic cognitive maps creation and evaluation based on affix relations. In: Samsonovich, A., Klimov, V., Rybina, G. (eds.) Biologically Inspired Cognitive Architectures (BICA) for Young Scientists, pp. 99–105. Springer, Cham (2016)
5. Balandina, A., Chernyshov, A., Klimov, V., Kostkina, A.: Usage of language particularities for semantic map construction: affixes in Russian language. In: International Symposium on Neural Networks, pp. 731–738 (2016)

Levels of Metacognition and Their Applicability to Reinforcement Learning

Jesuye David, Christion Banks, and Darsana Josyula(✉)

Department of Computer Science, Bowie State University,
Bowie, MD 20715, USA
jesuyedd@gmail.com, christionbanks7@gmail.com,
darsana@cs.umd.edu

Abstract. Recognizing patterns is an integral aspect of human intelligence; we know that every winter brings cold weather and snowfall. We therefore go to the stores beforehand to purchase coats and tools that will ensure our comfort and survival. We are not shocked when the season changes, instead we learn to manage in each new season. We instilled this ability to detect and cope with seasonality into an autonomous agent- Chippy. Chippy uses a reinforcement learner to gather rewards as it explores its environment. Seasonal changes are constructed into Chippy's environment by changing rewards at regular intervals. We allowed Chippy to operate at different levels of metacognition and compared the amount of rewards gathered when Chippy operates at each level. Results show that Chippy's reinforcement learner performs best when Chippy metacognitively monitors not only patterns in expectation violations but also patterns in the suggestions made.

Keywords: Q-learning · Prediction · Metacognition · Seasonality

1 Introduction

Learning is the relative permanent change in behaviour, or behaviour potential, as a result of experience [1]. Reinforcement learning is an algorithm in which the impact of an agent's action in the environment may not be immediately perceived by the agent, but the environment provides feedback that guides the learning algorithm [2].

Chippy uses a model-free reinforcement learning algorithm - Q-learning, with metacognition, to navigate an 8X8 square grid over a sequentially repeated number of seasons, collecting rewards (a piece of goo, and an acorn) in its path. The position of the rewards changes from one season to next. While the Q-learner learns the strategy that works for a new season, the metacognition monitors and controls the relearning of seasonal behaviors.

When Chippy notices a change in season, Chippy can choose to continue adjusting its policy; throw out the current policy and relearn from scratch; or save the current policy as associated with the previous season and relearn a new policy for the current season. Which of these strategies will be considered depends on the level of metacognition at which Chippy operates. Chippy can operate at different levels of metacognition, depending on the resources available for metacognition (how resource

A. V. Samsonovich (Ed.): BICA 2018, AISC 848, pp. 62–68, 2019.
https://doi.org/10.1007/978-3-319-99316-4_9

limitations affect metacognition is discussed in [3]). Thus, when environmental changes occur, Chippy behaves differently based on the level at which it is operating.

2 Biological Inspiration

Organisms are able to anticipate seasonal changes and adapt to deal with those changes. Some animals migrate to avoid the harsher conditions of winter and some change their fur color to camouflage against predators. Fluctuating forage resources and spatial variation in predation risk typically initiate seasonal migration of large herbivores [4, 5]. In [6] we learn that the spectral tarsiers modify their behavior when season changes. In order to survive, they spend more time procuring food and less time socializing during the dry season. The lower the abundance of food, the greater the amount of time and effort that they must spend foraging.

Visual feedback of the environment guides color change in many species. For example, cuttlefish use visual information to respond differently to different predator patches [7]. Some species rely on non-visual cues when changes in background color are predictable (e.g. snow cover) or indicated by other cues (e.g. odor). Animals that consume what they live on and change via diet (e.g. some caterpillars) may rely on dietary factors to respond to environmental changes [8].

3 Related Work

This work derives heavily from Dr. Dean Wright's dissertation work [9] that uses metacognition and a reinforcement learner in a Python based system to optimize the reward collection of an autonomous agent. The metacognitive loop (MCL) [10] notes anomalies, assesses the situation and guides an appropriate response using three ontologies - Indications, Response and Failure. In previous work on improving reinforcement learning with MCL, the degree of perturbation determined the response strategy; high-degree perturbations caused throwing out one's existing policy and low-degree perturbations resulted in temporarily increasing the exploration factor [11].

A common approach to improving the performance of reinforcement learning is to provide additional pseudo-rewards for actions that lead in the right direction of the goal. Reward shaping functions supply these additional shaping-rewards to the agent to guide its learning. A potential-based approach to choose the reward shaping function is discussed in [12] where distance and sub-goals heuristics are used to derive potentials.

Tsumori and Ozawa [13] shows that the performance of reinforcement learning could be improved by storing policies and recalling them when a known environment reappears. Wiering [14] explicitly stores information about all the dynamic objects in the environment in the model. When the information changes, the agent replans using model-based reinforcement learning. While these approaches focus on environmental changes, our approach focuses on improving self-performance.

4 Methodology

Chippy can currently function at different levels of metacognition allowing varying levels of metacognitive control. In the basic level of metacognition, when Chippy **notices an expectation violation**, the reinforcement learner is instructed to continue learning. This situation is a lot like searching for a lost key in places that it is most likely to be found based on beliefs from previous experiences, and when one cannot locate the key on those places, one may entertain the possibility of exploring other locations for the lost key, because someone might have moved it temporarily.

In the next level of metacognition, Chippy **monitors the intensity of the expectation violation**, in addition to whether a violation has occurred or not. This requires Chippy to measure how far the current values have deviated from the normal for each expectation. Based on the intensity of the violation, Chippy may instruct its reinforcement learner to throw out the current policy and relearn from scratch. This situation is akin to someone restarting from scratch while getting stuck solving a jig-saw puzzle. This strategy benefits Chippy in situations where the goo and acorn have moved to completely opposite regions of the grid, and hence incremental adjustments to existing policy may take more time than learning from scratch.

In the next level of metacognition, Chippy **monitors the patterns in expectation violations** in addition to the ones above. Thus, with this strategy, Chippy can embrace seasonal changes. If Chippy's acorn which is usually in the first quadrant of the grid (during Chippy's winter), moves to the third quadrant (during Chippy's summer), Chippy remembers the policy that it has learned in winter before changing the Q-table to deal with summer. Thus, when the season changes back to winter after a summer season, Chippy can retrieve the previously learned policy for use.

At the next level of metacognition, Chippy **monitors the patterns in the suggestions** that are considered to deal with expectation violations. This allows Chippy to try alternate strategies when appropriate.

To understand the differences in behavior for each level of metacognition, we generated graphs that depict the relationship between '*rewards*' collected and the number of '*steps*' as Chippy explores its environment. We conducted 2 sets of experiments, one with 2 seasons and the other with 4 seasons. In the first set of experiments, the acorn and goo are at diagonally opposite ends of the environment, and as the season changes, the acorn and goo switch their locations; thus the change of seasons are drastic, since a high positive reward changes to a low negative (and vice versa) at the same location. In the second set of experiments, the rewards in the two reward locations during the different seasons are (−10, 10), (−15, 15), (20, −20) and (25, −25). Here, the change between seasons 1 and 2 is not as severe as the change between seasons 2 and 3.

5 Results

Figures 1, 2, 3, 4, 5 and 6 show the relationship between rewards and step for different levels of metacognition for the first set of experiments. At level 1, Chippy is a plain reinforcement learner with no metacognitive control. Chippy metacognitively monitors the rewards accumulated every 20,000 steps, and returns a SUGGEST_NONE to the reinforcement learner. At this level, (see Fig. 1), the graph is very random because Chippy has no strategy for controlling its learning.

At level 2, (see Fig. 2), Chippy metacognitively monitors expectations of its performance. Hence, whenever Chippy is not getting enough rewards as before (expectation violation), it resets the Q-table of the reinforcement learner by returning SUGGEST_RESET to the learner. If there is no expectation violation, Chippy continues to use the current Q-table since the metacognitive component returns SUGGEST_NONE every 20,000 steps as before.

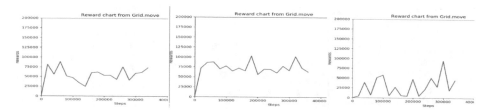

Fig. 1. Baseline,   No
metacognitive control

Fig. 2. Monitor expect violation, reset Q-table

Fig. 3. Monitor expect violation, adjust exploration

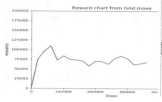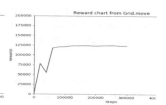

Fig. 4. Monitor severity of expectation violation, adjust metacognitive suggestion

Fig. 5. Past   metacognitive suggestions influence new suggestions

Fig. 6. Monitor   and   store experiences,   reuse   Q-table when seasons reappear

At level 3, (see Fig. 3), whenever the expected reward does not match the actual reward observed, Chippy returns SUGGEST_LEARN for the learner to adjust its exploration factor. If there are no expectation violations, Chippy returns SUGGEST_NONE and continues with the current Q-table as in the previous levels.

At level 4 of metacognition, (see Fig. 4), Chippy resets the Q-table (returns SUGGEST_RESET) only when the expectation violation is severe as when Chippy notices that a reversal of reward values (from positive to negative or vice versa) has occurred. If the expectation violation is mild, Chippy returns SUGGEST_LEARN and is prompted to learn. Notice the smoothness of the graph when compared to the

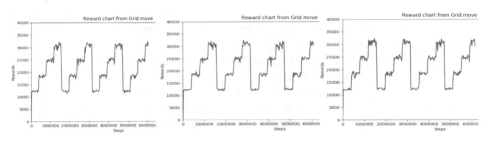

Fig. 7. Baseline- No
metacognition

Fig. 8. Monitor expect violation, reset Q-table

Fig. 9. Monitor expect violation,
adjust exploration

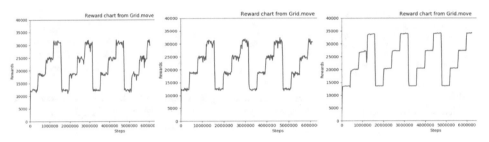

Fig. 10. Monitor severity of
expectation violation, adjust
metacognitive suggestion

Fig. 11. Past metacognitive
suggestions influence new
suggestions

Fig. 12. Monitor and store experiences, reuse Q-table when seasons reappear

previous ones. In the case of no expectation violations, Chippy returns
SUGGEST_NONE.

At level 5, (see Fig. 5), Chippy monitors the number of times SUGGEST_LEARN
has been issued to deal with the recent expectation violations. If Chippy is being
prompted to learn for more than three times in one experiment, then this level forces a
reset of the Q-table by issuing a SUGGEST_RESET.

At level 6, (see Fig. 6), Chippy associates the most recent Q-table values with the
most recently known reward states and stores this association periodically. Whenever
the actual observed reward does not equal the expected reward, Chippy checks the
stored dictionary if it has seen a similar reward structure before. If the given pattern is
similar, Chippy's learner is prompted SUGGEST_RETRIEVE to use the stored Q-
values; otherwise, the learner is issued SUGGEST_LEARN to learn more. Thus, at the
peak level of metacognition, Chippy does not waste time (steps) relearning a Q-table
that was learned before, it simply retrieves it. Thus, Chippy is able to accumulate a
maximum peak reward level of 125,000 rewards in Fig. 6 versus 75,000 in Fig. 5.

Figures 7, 8, 9, 10, 11 and 12 show the relationship between rewards and step for
different levels of metacognition for the second set of experiments. In Fig. 12, we can
see that the triominos-shaped pattern becomes most perfect when Chippy learns from
seasonal experiences and makes the best decisions when a previously observed season
returns. As seen from both sets of experiments, Chippy initially struggles to get

maximum rewards but after it learns from experience, it uses its learned experience to decide the best course of action that will get it the maximum reward.

6 Conclusion

In this paper we have implemented an agent-Chippy- that traverses life (8X8 grid). Metacognition allows it to learn to adapt to a new situation quickly. Before it learns to adapt to a new situation (with specific changes in parameters), it searches its memory to check if it has previously experienced a similar season. It wastes no time in relearning and thus maximizes its number of collected rewards.

Acknowledgements. We would like to thank Dean Wright for providing us his dissertation code. We would also like to thank Dr. Don Perlis and the Active logic and Metacognitive Computing group at UMD, College Park for discussions on this topic. This work is supported by MAST Collaborative Technology Alliance – Contract No. W911NF-08-2-004

References

1. Pastorino, E.E.: What is Psychology? Foundations, Applications and Integration, p. 226. Cengage Learning, Boston (2018)
2. Oladipupo, T.: Types of machine learning algorithms. In: Zhang, Y. (ed.) New Advances in Machine Learning. IntechOpen, Rijeka (2010). https://doi.org/10.5772/9385
3. Josyula, D, M'Bale, K.: Bounded metacognition. In: Fifth International Conference on Advanced Cognitive Technologies and Applications, pp. 147–152 (2013)
4. Fryxell, J.M., Sinclair, A.R.E.: Causes and consequences of migration by large herbivores. Trends Ecol. Evol. **3**, 237–241 (1988)
5. Mysterud, A.: Seasonal migration pattern and home range of roe deer (*Capreolus capreolus*) in an altitudinal gradient in southern Norway. J. Zool. (Lond.) **247**, 479–486 (1999)
6. Gursky, S.: Effect of seasonality on the behavior of an insectivorous primate, Tarsius spectrum. Int. J. Primatol. **21**, 477 (2000)
7. Langridge, K.V., Broom, M., Osorio, D.: Selective signalling by cuttlefish to predators. Curr. Biol. **17**, 1044–1045 (2007). https://doi.org/10.1016/j.cub.2007.10.028
8. Hultgren, K.M., Stachowicz, J.J.: Size-related habitat shifts facilitated by positive preference induction in a marine kelp crab. Behav. Ecol. **21**, 329–336 (2010)
9. Wright III, D.E.: Finding a temporal frame comparison function for the metacognitive loop. UMBC (Doctoral dissertation) (2011)
10. Schmill, M., Anderson, M., Fults, S., Josyula, D., Oates, T., Perlis, D., Shahri, H.H., Wilson, S., Wright, D.: The metacognitive loop and reasoning about anomalies. In: Raja, A. (ed.) Metareasoning: Thinking About Thinking. MIT Press, Cambridge (2011)
11. Anderson, M.L., Oates, T., Chong, W., Perlis, D.: The metacognitive loop I: enhancing reinforcement learning with metacognitive monitoring and control for improved perturbation tolerance. J. Exp. Theor. Artif. Intell. **18**(3), 387–411 (2006)
12. Ng, A.Y., Harada, D., Russell, S.J.: Policy invariance under reward transformations: theory and application to reward shaping. In: Proceedings of the Sixteenth International Conference on Machine Learning, pp. 278–287 (1999)

13. Tsumori, K., Ozawa, S.: Incremental learning in dynamic environments using neural network with long-term memory. In: Proceedings of the International Conference on Neural Networks, pp. 2583–2588 (2003)
14. Wiering, M.A.: Reinforcement learning in dynamic environments using instantiated information. In: Proceedings of the Eighteenth International Conference on Machine Learning, vol. 585/592 (2001)

Extended Hierarchical Temporal Memory for Motion Anomaly Detection

Ilya Daylidyonok[1], Anastasiya Frolenkova[1], and Aleksandr I. Panov[1,2(✉)]

[1] National Research University Higher School of Economics, Moscow, Russia
{iddaylidyonok,aifrolenkova}@edu.hse.ru
[2] Federal Research Center "Computer Science and Control" of the Russian Academy of Sciences, Moscow, Russia
apanov@hse.ru

Abstract. This paper describes the application of hierarchical temporal memory (HTM) to the task of anomaly detection in human motions. A number of model experiments with well-known motion dataset of Carnegie Mellon University have been carried out. An extended version of HTM is proposed, in which feedback on the movement of the sensor's focus on the video frame is added, as well as intermediate processing of the signal transmitted from the lower layers of the hierarchy to the upper ones. By using elements of reinforcement learning and feedback on focus movement, the HTM's temporal pooler includes information about the next position of focus, simulating the human saccadic movements. Processing the output of the temporal memory stabilizes the recognition process in large hierarchies.

Keywords: Anomaly detection · Hierarchical temporal memory
Video processing · HTM feedback · Hierarchical learning

1 Introduction

Artificial neural networks are used to solve a lot of different tasks in various fields of study: object recognition and classification in computer vision, text analysis and translation in natural language processing, etc. But as successful as neural nets are in some tasks, there are some difficulties associated with their usage. Most of the best approaches require massive amounts of manually labelled data. The models are often tuned exclusively for the task being solved and some tasks may require completely different model architecture.

Many of the problems which can be solved with neural nets are being solved by human mind in everyday life regularly and, most of the time, unconsciously. The results are usually more precise compared to those which artificial neural nets achieve. Therefore, it seems that algorithms in our brains are more precise and effective, than those we invent for artificial neural networks.

Among the known biologically inspired models, Hierarchical Temporal Memory (HTM) is one of the most promising from the practical point of view. HTM

© Springer Nature Switzerland AG 2019
A. V. Samsonovich (Ed.): BICA 2018, AISC 848, pp. 69–81, 2019.
https://doi.org/10.1007/978-3-319-99316-4_10

is the model which relies on the knowledge of how mammalian brain – in particular, neocortex – works, compared to classical machine learning approaches [1,2]. The main idea of HTM is to mimic algorithmic and structural features of the neocortex as close as possible. As a result the model has the following properties:

- The main principle of the model is the fact that neocortex is a homogeneous structure [3]. This means that the physical structure and the algorithm being used are the same for every task the model faces
- The model works with unlabelled data and can handle any type of input data
- The HTM algorithm learns temporal sequences in streams of input data
- Training and testing can work in real time on continuous streams of data
- The model is insensitive to noise because of its structure.

HTM learns data sequences during its training, similar to how our brain does it. Learning is performed by creating synaptic links between neurons, and then strengthening or weakening these links. Representations of learned sequences are then being used to recognize and predict the following data sequences.

The task of anomaly detection in motion is frequently and effectively solved by the human brain. Therefore, we chose this task to demonstrate capabilities of HTM.

We use HTM to detect anomalies in videos of different motions. The anomaly in a given video is defined as a presence of frames from another video, or as a sudden change of position, speed, direction or movement type of an object.

Anomaly detection of this kind can be used for detection of suspicious actions using security camera footage. It can also be used for the search of artifacts in motion capture records, which can be caused by equipment failure.

We also propose an extension of HTM. In this extension we add feedback on the movement of the sensor's focus on the video frame. By using elements of reinforcement learning and feedback on focus movement, the HTM's temporal pooler includes information about the next position of focus, simulating the human saccadic movements. We also introduce intermediate processing of the signal transmitted from the lower layers of the hierarchy to the upper ones. We expect it to stabilize the recognition process in large hierarchies.

2   Related Work

In [4] HTM was used for object classification in noisy video streams. The work has shown that HTM models performed very well in the presence of noise in frames of videos. This shows that HTM can be applied in real world scenarios, where noise in video footage is common.

[5] shows implementation of object tracking system. By following the object closely, the model might notice small changes of movement, which can be used

for making classification better or for detecting more subtle anomalies. It used spatial pooler part of HTM for creating a model, which proved to be efficient for solving the task.

[6] has shown the importance of anomaly detection in different data streams, introduced a framework of scoring anomaly detection algorithms. It has also shown superiority of HTM in finding anomalies in some problems.

[7] shows the importance of anomaly detection in surveillance videos and different approaches of solving the problem. In particular, methods of low level feature extraction, feature description and behaviour modelling are discussed. The work also displays different datasets used for testing anomaly detection algorithms and compares the results of existing anomaly detection approaches.

In [8] the anomalies were found by analyzing people movement trajectories outdoors. Hidden Markov Model was used to build an anomaly detector. If the movement trajectory of a particular person differed greatly from the usual ones, the movement was considered to be an anomaly.

The methods of automatic error detection in motion capture data were researched in [9]. The detected anomalies were used to find artifacts in recorded videos. Any unnatural movement (for example, the one that looks unrealistic or cannot be performed by actors) was considered to be anomaly. The work compared classifiers which were obtained by using hidden Markov model, linear dynamic systems and other approaches.

As we can see, the task of anomaly detection is important and has been researched quite well. HTM wasn't used much for finding anomalies in videos, it was mainly used for finding anomalies in numeric data. Some works use position of objects to detect anomaly. For example, if the person walks on an unusual path or with unusual speed, it's detected as an anomaly. But knowing which paths are uncommon requires having statistics on common paths, which isn't always easily obtainable. Our approach to finding anomalies in videos is focusing on actions of one object. We expect that HTM will be able to detect sudden changes of position or speed of an object, but we also want to construct a model capable of detecting more subtle anomalies. This requires the model to have some abstract notion of an action, so that it can detect when action changes, even if the object's position and speed don't change dramatically.

3 HTM Algorithm

Learning algorithms, which simulate neocortex, are based on the notion of hierarchy of regions – layers of neocortex. Each layer is used for search and recognition of patterns in input data, and also for prediction of next possible inputs. First of all, each model's layer transforms data into sparse distributed representation

(SDR) – fundamental structure of HTM. The basic element of SDR is called a column – a set of one or more neurons. Neurons are elementary nodes characterized by their connections with each other. Connections between neurons in HTM have two types: connections between neighboring columns in one region and connections between different regions. These types roughly correspond to two types of dendrites: proximal and distal dendrites accordingly. The first type is used for learning a particular sequence within a region. The second type serves as a link between regions and can be used for building hierarchical models.

After transformation of input data into an SDR, regions represent the data in accordance with a context of the previously received inputs, i.e. in accordance with a previously formed synaptic connections. The resulting representation consists of a set of active neurons and links between these neurons – synapses – based on the previous data.

The first step of the HTM algorithm – representation of the input SDR in columns' terms – is performed by an algorithm called Spatial Pooling. The algorithm goes as follows:

1. For the input vector $\bar{v}_{source}$ calculate overlap with each of the columns $\bar{c} \in C$ in the region:

$$\sum_{synapses \in \bar{C}} \{s \mid s \text{ connected to } \bar{v}_{source}\}$$

```
set_overlaps(columns, 0)
for col in columns:
for s in connected_synapses(col):
overlap(col) += get_input(s.source)

if overlap(col) < threshold:
overlap(col) = 0
else:
overlap(col) *= boost(col)
```

2. After boosting, calculate winning columns as follows: if an overlap of the column with input data is greater than a given threshold, the column is marked as a "winner" and is activated. The minimum activity threshold for each column is defined as k-th element in a sorted list of numbers of active synapses in neighboring columns

```
for col in columns:
if overlap(col) >= activity_threshold:
min_activity = k_score(neighbors(col), desired_activity)
if overlap(col) >= min_activity:
currently_active_columns.append(col)
```

3. The connections are either strengthened or weakened, which is used for training SP:

```
# strengthen active connections of winner columns
for col in currently_active_columns:
for s in potential_synapses:
if is_active(s):
increase_permanence(s)
else:
decrease_permanence(s)


for col in columns:
# if columns don't \win" often enough
if active_cycles(col) < min_cycles:
increase_boost_value(col)


# if column's connections don't overlap
# with input data often enough
if overlap_cycles(col) < min_cycles:
increase_permanences(col)
```

Temporal Memory algorithm performs learning of temporal sequences. The algorithm learns sequences by creating connections between neurons within its region. Inhibition radius determines the maximum distance between neighboring columns between which synaptic connections can be formed. These areas of inhibition are called segments. The Temporal Memory algorithm goes as follows:

1. Neurons activation
   (a) If a column $\bar{c}$ is active and has active connections within the region, then column's neurons switch to predictive state
   (b) If a column $\bar{c}$ is active, but has no active connections within the region, then an active segment gets boosted
   (c) If a column $\bar{c}$ isn't active and has active connections within the region, then these connections are weakened as they're considered to be erroneous.

```
for col in columns:
if is_active(col):
if active_segments != None:
# if the column has connections
# with activated segments
activate(col)
else:
burst(col)
else:
penalize(col)
```

2. Connection activation

(a) Every segment within a region, in which ts activity exceeds predetermined threshold, is activated
(b) Every segment, in which its activity exceeds predetermined threshold but has no connections to other segments, is marked as potentially corresponding to the input data.

```
active_segments.clear()
for segment in all_segments:
if active_connected_synapses[segment] >= activation_threshold:
active_segments.append(segment)
matching_segments.clear()
for segment in all_segments:
if active_potential_synapses[segment] >= min_threshold:
matching_segments.append(segment)
sort(active_segments)
sort(matching_segments)
```

Because of its architecture, HTM region acts as a detector of novel or anomalous data sequences. When a region receives new input data, it compares the data with its prediction, i.e. it compares expected sequences which were learned on previous steps with the actual input. HTM prediction consists of a set of possible sequences, so if a new input significantly differs from all the predicted sequences, the input is likely to be an anomaly.

Every HTM region can work in learning mode, inference (prediction) mode, or both. Learning is a creation and adjustment of synaptic connections in accordance to inputs. Region doesn't learn new sequences in the inference mode, it only compares them with the ones learned previously. Inference mode is primarily used for testing the model and solving different tasks with it.

4   Dataset

Data from Carnegie Mellon University Motion Capture Database [10] was used for the experiments. The dataset contains 2235 videos of 144 people performing different actions: walking, running, jumping and so on. Motion capture technology was used for recording the videos from the dataset. 41 markers were positioned on a different actors' body parts and their positions were recorded during the action performance. The dataset contains both markers coordinate data and generated videos which represent human figures with geometrical shapes. The sample of frames from the dataset can be seen on Fig. 1.

In this work we used a subset of videos from the dataset. Each video was divided into a sequence of frames which were used as inputs for HTM models.

Frames from other videos were inserted to produce an anomaly. The anomalies of different types were tested. One type of anomaly, that was easier to detect, is when the object quickly changed the position (see Fig. 2). Other anomalies were more subtle – for example, when the movements were similar, but different actions were performed.

Fig. 1. Examples of frames in videos from CMU dataset

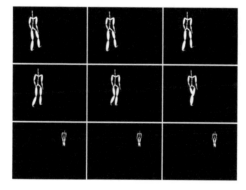

Fig. 2. Example of anomaly – the last 3 frames are anomalous

5   Anomaly Detection Model

In our work we use implementation of the Spatial Pooler and Temporal Memory HTM algorithms presented at [11], along with Image Sensor algorithm for processing images from [12] repository.

Input images were preprocessed with the following algorithms to be used for HTM:

1. Image loading. Python Imaging Library (PIL) is used for image loading. Its internal representation of images is compatible with Spatial Pooler algorithm
2. Algorithms which control the sensor's movement around the image. The set of algorithms, which are called explorers, sets the order of image processing on the sensor and, therefore, the order of image coding in SDR. For our tests we just displayed a whole image to the Image Sensor.

The processed images are used as inputs for the Spatial Pooler which then produces an output which is used as the input for the Temporal Memory region. The described model can be expressed using the scheme shown on Fig. 3.

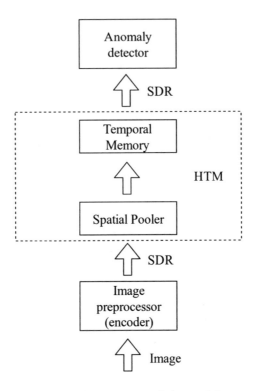

Fig. 3. The structure of the model

6   Experiments

We conducted a number of different experiments with the model described above. We used different combinations of input data and anomalies for model learning. Each experiment was divided in two parts:

1. Model learning on unaltered data. During the learning process original video is shown to the model 10–15 times. Then the model sees the original video one more time in the interference mode, to test how well it learned the sequence.
2. Model testing on video with anomaly. This time, previously trained model is shown a video with anomaly. Learning is turned off and the model's performance is tested again.

After each shown frame, model computes anomaly coefficient (AC) which lies in $[0; 1]$ interval and corresponds to a probability of the current frame being an anomaly. If AC is equal to zero, it means that model sees no anomaly, while values close to 1 indicate high confidence that the frame being shown is anomalous. If the video contains no anomalies, the model is expected to produce $AC = 0$ for all the frames. Otherwise, the model should produce AC close to 1 during the anomalous frames being shown.

Suppose that the video contains anomaly on third and fourth frames. We expect to get the following sequence of ACs: $ac_{expected} = (0, 0, 1, 1, 0, 0, \ldots)$. After we show the video to the model, we get some vector of ACs:ac_{real}. Then the net score is computed using the following formula:

$$score = \frac{\sum_{i=0}^{n} \left(ac_{expected}^{(i)} - ac_{real}^{(i)} \right)^2}{n}$$

where n is a number of frames. This score is used as a metric of how well the model is at finding anomalies. The closer this score is to 0, the better the model performs. If multiple videos are shown, the net score is computed as a mean of net scores for each video.

Experiments we set are of the following types:

1. Train model on a short video. Anomaly is created by inserting frames with a motion different from that in the train video, e.g. a subject walks, then starts to run (Table 1, cases 1–4, 8)
2. Train model on a long sequence. For anomaly use frames from a video with a similar looking motion, e.g. a subject walks then starts to limp (Table 1, cases 5–7)
3. Test model for detection of sudden location changes, e.g. a subject starts to walk from one corner, then suddenly appears on the other side of the screen (Table 1, cases 8–9).

When the same video was used for training in different cases, e.g. cases 1–4, the same model was produces. As a result, such a model was tested against different anomalies. When an anomaly was inserted, resulted video could be longer than the original one. For all the tests we used black and white images with 64×64 pixels resolution. Anomaly threshold was equal to 0.7, i.e. inputs scored higher than 0.7 with accordance to the provided above score function, were marked anomalous. All the anomalous sequences were inserted in the middle of the initial sequence, if not stated otherwise. The results of our experiments can be seen in Table 1.

As one can see from the results of experiments, models perform well with the obvious anomalies, such as jumping in the middle of the walk. Models are also good at determining the beginning and the end of an anomalous sequence. E.g. in the 9th case subject suddenly changes location and continuous to walk, the model notices the leap, but the sequence of frames after that is already known to the HTM regions, so it isn't marked as an anomaly.

Table 1. The results of the experiments

#	Original motion	# of frames	Anomaly motion	# of anomalous frames	# of detected anomalous frames	Score	Comments
1	Walk	21	Run	5	2	0.19	
2	Walk	21	Walk and turn right	12	5	0.19	
3	Walk	21	Jump	17	15	0.08	
4	Walk	21	Gorilla	12	2	0.40	Wrong frames marked as anomaly
5	Walk back and forth	136	Scared walk	41	2	0.28	
6	Walk back and forth	136	Dance	31	2*	0.24	
7	Walk back and forth	136	Bad leg walk	22	2*	0.14	Anomaly starts near the beginning
8	Jump	30	Walk back and forth	15	16	0.03	Motion and location change
9	Walk back and forth	136	Walk back and forth	–	2*	-	Test for sudden location change

*Only first and last frames of anomaly were detected

Some of experiments show worse results. For example, in the case 4 better performance is expected, as it is an obvious anomaly, but it's not detected well. Also more subtle anomalies, like different types of walking, are not easily recognized and most of the time only marked as an anomaly on the frames of change.

7 Extensions of HTM

We plan to further develop our research by extending HTM model. The first extension is an addition of multiple layers of HTM regions to the model. The second is an introduction of communication of higher regions and lower ones.

7.1 Hierarchy

In accordance with the current knowledge of neurobiology, hierarchical structure is what allows the brain to learn complex patterns and concepts. Using learned patterns from the previous ones, each following layer can better distinguish compound sequences. Let's consider an example with motion recognition. The first layers of the brain to process motions deal with sensory information.

8. Suzuki, N., et al.: Learning motion patterns and anomaly detection by human trajectory analysis. In: 2007 IEEE International Conference on Systems, Man and Cybernetics. ISIC, pp. 498–503. IEEE (2007)
9. Ren, L., et al.: A data-driven approach to quantifying natural human motion. ACM Trans. Graph. (TOG) **24**(3), 1090–1097 (2005)
10. CMU Graphics Lab: Carnegie Mellon University Motion Capture Database (2003). http://mocap.cs.cmu.edu/
11. Numenta: NuPIC (2018). https://github.com/numenta/nupic
12. Numenta: NuPIC.vision (2017). https://github.com/numenta/nupic.vision
13. Sabour, S., Frosst, N., Hinton, G.E.: Dynamic Routing Between Capsules. CoRR abs/1710.09829 (2017). arXiv:1710.09829
14. Yarbus, A.L.: Eye movements during perception of complex objects. In: Eye Movements and Vision, pp. 171–211. Springer (1967)

Experimental Model of Study of Consciousness at the Awakening: FMRI, EEG and Behavioral Methods

Vladimir B. Dorokhov[1]([✉]), Denis G. Malakhov[2],
Vyacheslav A. Orlov[2], and Vadim L. Ushakov[2,3]

[1] Institute of Higher Nervous Activity and Neurophysiology of RAS,
Moscow, Russia
vbdorokhov@mail.ru
[2] National Research Center "Kurchatov Institute", Moscow, Russia
[3] National Research Nuclear University "MEPhI", Moscow, Russia

Abstract. For the study of neuronal correlates of consciousness, a simple and effective model is the comparison of sleeping and waking states. Consciousness turns off during sleep and turns on at waking. The moment of awakening from sleep is a promising model for the study of neurophysiological correlates of consciousness. We developed a psychomotor test, the monotonous performance of which, causes within 60 min alternating episodes with the disappearance of consciousness when falling asleep (the "microsleep") and its restoration upon awakening (wakefulness). When performing this test, the subject with closed eyes counts from 1 to 10 and simultaneously presses sensitive buttons, alternately with the right and left hands. Spontaneous restoration of the test after the episode of "microsleep" requires the activation of consciousness, which is accompanied by consciously performing the test with counting and simultaneously pressing the buttons. EEG methods allow you to accurately assess the moments of the transition of sleep/wakefulness, the levels of wakefulness and the depth of sleep, and behavioral methods, by indicators of the correctness of the performance of the psychomotor test - to determine the levels of consciousness. We showed reproducibility of this test obtained both under normal conditions and in conditions of functional magnetic resonance imaging (fMRI) procedure. In 10 out of 14 subjects during a 60-min experiment performed in the MRI scanner, 3–48 episodes of "microsleep" were recorded with subsequent awakening. Preliminary results showed an increase in the activity of the visual regions (the region of the calcarine sulcus) of the cerebral cortex, left precuneus/cuneus, etc. during sleep and regions of the right thalamus, left cuneus, cerebellar zones, stem structures, etc. at the moment of awakening and resumption of conscious activity.

Keywords: Microsleep · Awakening · fMRI · Levels of consciousness

© Springer Nature Switzerland AG 2019
A. V. Samsonovich (Ed.): BICA 2018, AISC 848, pp. 82–87, 2019.
https://doi.org/10.1007/978-3-319-99316-4_11

1   Introduction

The study of consciousness is the most important and most complex problem of neuroscience, necessary for understanding the mechanisms of the functioning of the human mind [1, 2]. At the junction of cognitive and neurosciences, an approach to the study of consciousness has been formulated, as the search for neuronal correlates of consciousness. The basis of this approach is the postulate of the existence of a causal relationship between activity of consciousness and brain activity: for each event in the consciousness there is a corresponding event in the brain. Within the framework of the concept of "neural correlates of consciousness" [2, 3], the task is to find out which neurophysiological events correlate with certain states of the brain and the content of consciousness. Modern theories develop in the direction of the study of the neurological basis of consciousness [4]. The results of the study of neural correlates of consciousness are not only of fundamental importance, but are already applicable in the field of neurology and psychiatry [5].

According to modern ideas [6], studies within the framework of the paradigm of neural correlates of consciousness should look like this: during the experiment, the subject must several times be in two different states: the presence of consciousness (he experiences a certain state) and lack of consciousness (the subject does not experience this state). In all other respects, the experimental conditions must remain unchanged. At the same time, the functioning of the brain of the subject is simultaneously examined using fMRI, PET, EEG or MEG, which allow to evaluate various spatial and temporal characteristics of brain neuronal networks [7]. For the study of neuronal correlates of consciousness, a simple and effective model is a comparison of sleep and wakefulness states [6, 8, 9]. Consciousness turns off during sleep and turns on awakening, which proves its connection with different functional states of the brain [6, 9]. The moment of awakening from sleep is a promising model for the study of neurophysiological correlates of consciousness [10–13].

A necessary condition for the functioning of consciousness is the presence of the necessary level of depolarization of cortical neurons, which is characteristic for wakefulness. And lack of consciousness, in the slow-wave sleep stage, according to existing concepts, is determined by the bistable state of the cortical neurons, with intermittent hyperpolarization and depolarization of the neuronal membrane [7–9]. It is suggested that it is the bistability of the state of neurons in sleep that disturbs the synchronous interaction of the cortical areas of the brain, which is necessary for the functioning of consciousness, as shown by the method of transcranial stimulation of the brain during sleep [14].

Within the framework of the paradigm of two states - presence and absence of consciousness, we have developed a psychomotor test, monotonous execution of which during 50–60 min causes alternating episodes of "microsleep" and awakening [10]. When performing this test, the subject with the closed eyes counts from 1 to 10 and simultaneously presses sensitive analog buttons, alternately with the right and left hands. When performing this test in subjects with partial deprivation of night sleep, by the end of the 60 min experiment, several short-term episodes of "microsleep", with electroencephalographic activity corresponding to the third stage of sleep, with delta-waves of the EEG, could be observed. Spontaneous restoration of the test performed after the episode "microsleep" requires the activation of consciousness, which we believe is accompanied by the extraction of instructions from memory and the conscious execution of the test with counting and simultaneous pressing of the button. Thus, during one short experiment (1 h), several consecutive episodes can be analyzed, with the disappearance of consciousness during sleep ("microsleep") and its recovery upon awakening (wakefulness). EEG methods allow you to accurately assess the moments of the transition sleep/wakefulness, wakefulness levels and the depth of sleep, and behavioral methods, by indicators of the correctness of the performance of the psychomotor test - to determine the levels of consciousness. A study of the feasibility of performing this test under conditions of a magnetic resonance imager (MRI) showed reproducibility of the results obtained under normal conditions.

2 Materials and Methods

MRI data were obtained from 14 healthy subjects, mean age 24 (range from 20 to 35 years). Permission to undertake this experiment has been granted by the Ethics Committee of the NRC "Kurchatov Institute".

In 10 out of 14 subjects during a 60-min experiment performed in the MRI chamber, 3–48 episodes of "microsleep" were recorded with subsequent awakening. Simultaneous registration of EEG, fMRI and button presses was performed, which is a prerequisite for reconciling the spatial and temporal characteristics of the dynamics of brain neural networks with the nature of the test. It should be noted that this technique, when carrying out the experiment under ordinary conditions, allows preliminary selection of the subjects with the greatest number of episodes of "microsleep" having EEG patterns of deep delta sleep, for subsequent fMRI studies. Figure 1 shows the mechanogram of pressing the right and left button during sleep and wakefulness (the arrow marks the moment of "microsleep", the red vertical lines indicate the moments of awakening according to the mechanogram).

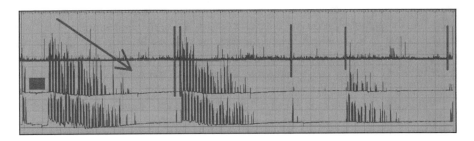

Fig. 1. Mechanogram of pressing the right and left button during sleep and wakefulness (the arrow marks the moment of "microsleep", the red vertical lines indicate the moments of awakening according to the mechanogram).

Figure 2 shows an example of an EEG recorded synchronously with fMRI and the mechanogram. The time of awakening and the preceding slow-wave activity are shown. In this example, at the moment of awakening, the subject starts to press buttons with two hands simultaneously. Usually, after 1 cycle of several presses, subjects recall the instruction and start pressing buttons separately. Figure 3 shows an example of the alpha activity of the EEG in waking state.

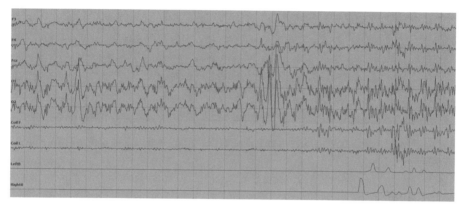

Fig. 2. An example of the EEG of the moment of awakening and the preceding slow-wave activity (after MRI gradient and cardioballistic artifact correction).

The MRI data was acquired using a 3 T SIEMENS Magnetom Verio MR scanner. The T1-weighted sagittal three-dimensional magnetization-prepared rapid gradient-echo sequence was acquired with the following imaging parameters: 176 slices, TR = 1900 ms, TE = 2.19 ms, slice thickness = 1 mm, flip angle = 9^0, inversion time = 900 ms, FOV = 250 mm × 218 mm^2. fMRI data was acquired with the following parameters: 42 slices, TR = 1000 ms, TE = 25 ms, slice thickness = 2 mm, flip angle = 90°, FOV = 192 × 192 mm^2, Mb = 8. The ultrafast fMRI protocol was obtained from the Minnesota Center for Magnetic Resonance Research university. The fMRI and structural MR data were pre-processed using SPM8 (available free at http://www.fil.ion.ucl.ac.uk/spm/software/spm8/). After Siemens DICOM files were converted into SPM NIFTI format, all images were manually centered at the anterior commissure. EPI images were corrected for magnetic field inhomogeneity using FieldMap toolbox for SPM8. Next, slice-timing correction for fMRI data was performed. Both anatomical and functional data were normalized into the ICBM stereotactic reference frame. T1 images were segmented into 3 tissue maps (gray/white matter and CSF). Functional data were smoothed using Gaussian filter with a kernel of 6 mm FWHM. Statistical analysis was performed using Student's T-statistics ($p < 0.05$, with correction for multiple comparisons (FWE)). The activity of the brain structures was compared at intervals of 8 s before the resumption of the button presses and 4 s after the start of the pressing. In the processing of data, the "microsleep" periods were taken if they were 8 s or more, and after which there were at least 2 button presses.

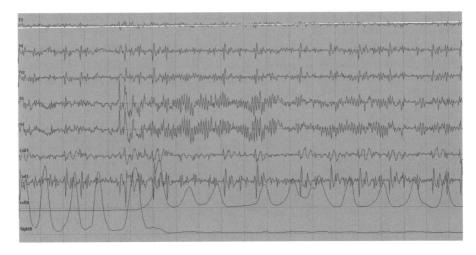

Fig. 3. Alpha-rhythm of EEG in waking state (after MRI gradient and cardioballistic artifact correction).

3   Results

Figure 4 shows the results of increasing the activity of the BOLD signal during sleep (left image) and at the moment of awakening and resumption of conscious activity (right picture).

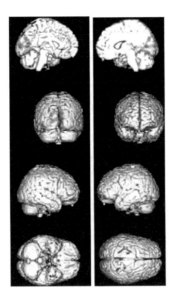 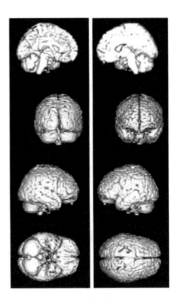

Fig. 4. Localization of neural network activity of the brain during sleep (left image) and at the moment of awakening and renewal of conscious activity (right figure) by fMRI data. (p < 0.05, FWE).

4   Conclusions

Preliminary results showed an increase in the activity of the visual regions (the region of the calcarine sulcus) of the cerebral cortex, left pre-cuneus/cuneus, etc. during sleep and regions of the right thalamus, left cuneus, cerebellar zones, stem structures, etc. at the moment of awakening and resumption of conscious activity.

Acknowledgements. This study was partially supported by the Russian Science Foundation, grant 18-11-00336 (data preprocessing algorithms) and by the Russian Foundation for Basic Research grant ofi-m 17-29-02518 (study of thinking levels). The authors are grateful to the MEPhI Academic Excellence Project for providing computing resources and facilities to perform experimental data processing.

References

1. Ushakov, V.L., Samsonovich, A.V.: Toward a BICA-model-based study of cognition using brain imaging techniques. Proc. Comput. Sci. **71**, 254–264 (2015)
2. Tononi, G.: Information integration: its relevance to brain function and consciousness. Arch. Ital. Biol. **148**, 299–322 (2010)
3. Koch, C., Massimini, M., Boly, M., Tononi, G.: Neural correlates of consciousness: progress and problems. Nat. Rev. Neurosci. **17**(5), 307–321 (2016)
4. Laureys, S., Tononi, G. (eds.): The Neurology of Consciousness: Cognitive Neuroscience and Neuropathology. Academic Press, New York (2009)
5. Blume, C., Giudice, R.M., Lechinger, M.J., Schabus, M.: Across the consciousness continuum—from unresponsive wakefulness to sleep. Front. Hum. Neurosci. **9**, 105 (2015)
6. Revonsuo, A.: Consciousness: The Science of Subjectivity. Psychology Press, London (2010)
7. Tagliazucchi, E., van Someren, E.J.W.: The large-scale functional connectivity correlates of consciousness and arousal during the healthy and pathological human sleep cycle. Neuroimage **15**(160), 55–72 (2017)
8. Gemignani, A., Menicucci, D., Laurino, M., Piarulli, A., Mastorci, F., Sebastiani, L., Allegrini, P.: Linking Sleep Slow Oscillations with consciousness theories: new vistas on Slow Wave Sleep unconsciousness. Arch. Ital. Biol. **153**(2–3), 135–143 (2015)
9. Windt, J.M., Nielsen, T., Thompson, E.: Does consciousness disappear in dreamless sleep? Trends Cogn. Sci. **20**(12), 871–882 (2016)
10. Dorokhov, V.B.: Alpha bursts and K-complex: phasic activation pattern during spontaneous recovery of correct psychomotor performance at different stages of drowsiness. Zh. Vyssh. Nerv. Deiat. Im. I P Pavlova **53**(4), 503–512 (2003)
11. Peter-Derex, L., Magnin, M., Bastuji, H.: Heterogeneity of arousals in human sleep: a stereo-electroencephalographic study. Neuroimage **123**, 229–244 (2015)
12. Tsai, P.J., Chen, S.C., Hsu, C.Y., Wu, C.W., Wu, Y.C., Hung, C.S., Yang, A.C., Liu, P.Y., Biswal, B., Lin, C.P.: Local awakening: regional reorganizations of brain oscillations after sleep. Neuroimage **102**, 894–903 (2014)
13. Hale, J.R., White, T.P., Mayhew, S.D., Wilson, R.S., Rollings, D.T., Khalsa, S., Arvanitis, T.N., Bagshaw, A.P.: Altered thalamocortical and intra-thalamic functional connectivity during light sleep compared with wake. Neuroimage **125**, 657–667 (2016)
14. Massimini, M., Ferrarelli, F., Huber, R., Esser, S.K., Singh, H., Tononi, G.: Breakdown of cortical effective connectivity during sleep. Science **309**, 2228–2232 (2005)

A Temporal-Causal Modeling Approach to the Dynamics of a Burnout and the Role of Physical Exercise

Zvonimir Dujmić, Emma Machielse, and Jan Treur[✉]

Behavioural Informatics Group, Vrije Universiteit Amsterdam,
Amsterdam, The Netherlands
zvonimir.dujmic@gmail.com,
emachielse@hotmail.com, j.treur@vu.nl

Abstract. In this paper from a Network-Oriented Modeling perspective a temporal-causal network model for burnout is introduced. The model can be the basis for a virtual patient agent model, and offers also possibilities to simulate certain forms of treatment. The model was evaluated by simulation experiments, verification by Mathematical Analysis and validation by Parameter Tuning for given patterns found in the literature.

Keywords: Burnout · Network model · Physical exercise

1 Introduction

Everyone experiences demanding working days. But when experienced stress is too high for too long, people can get sick. From Dutch employees 14% has gone to the doctor for burnout complaints, and 5% has to stay home for a longer period because of that [19]. A burnout can be described as a state of being where a person experiences emotional exhaustion. From the literature the causes, elements and consequences of the burnout syndrome have been gathered. In the past psychological diseases, like depression, were predominantly seen as a latent construct that can be measured by its symptoms. This traditional psychometric approach is making room for a new paradigm. In the new paradigm, a psychopathological disease is characterised by the relations between symptoms [2]. There is no underlying cause for the symptoms, the psychological disease is its symptoms and their dynamic interplay. In this line Schmittmann et al. [17] advocate for a network perspective on psychological diseases. This perspective was the basis for the dynamic temporal-causal network of a burnout for an individual introduced in the current paper. This computational model will represent a burnout following the new network paradigm of psychopathological diseases. To achieve ecological validity we will include both risk factors and protective factors. Also states will be added that are a consequent of elements of the burnout syndrome, that can form cyclic patterns. Causal relationships between elements will be added as found in the literature. The model will represent one state where the individual experiences a burnout and one where the individual does not. These two states will be compared and an exploratory research for the protective effect of physical exercise will be done.

A. V. Samsonovich (Ed.): BICA 2018, AISC 848, pp. 88–100, 2019.
https://doi.org/10.1007/978-3-319-99316-4_12

Causal modelling, causal reasoning and causal simulation have a long tradition in AI; e.g., [11, 12, 16]. The Network-Oriented Modelling approach based on temporal-causal networks described in [20, 21] can be viewed on the one hand as part of this causal modelling tradition, and from the other hand from the perspective on mental states and their causal relations from Philosophy of Mind (e.g., [10]). It incorporates a dynamic and adaptive temporal perspective, both on states and on causal relations, thus enabling modelling of cyclic and adaptive networks, and timing of causal effects.

Temporal-causal network models can be described by two equivalent representations: conceptual or numerical representations. A conceptual representation of a temporal-causal network model in the first place involves representing in a declarative manner states and connections between them that represent (causal) impacts of states on each other, as assumed to hold for the application domain addressed. The states are assumed to have (activation) levels that vary over time. In reality not all causal relations are equally strong, so some notion of *strength of a connection* is used. Furthermore, some way to *aggregate multiple causal impacts* on a state is used. Moreover, not every state has the same extent of flexibility; some states may be able to change fast, and other states may be more rigid and may change more slowly. Therefore, a notion of *speed of change* of a state is used for timing of processes. These three notions are covered by elements in the Network-Oriented Modelling approach based on temporal-causal networks, and define a conceptual representation of a temporal-causal network model:

- **Strength of a connection** $\omega_{X,Y}$. Each connection from a state X to a state Y has a *connection weight value* $\omega_{X,Y}$ representing the strength of the connection, often between 0 and 1, but sometimes also below 0 (negative effect) or above 1.
- **Combining multiple impacts on a state** $c_Y(..)$. For each state (a reference to) a *combination function* $c_Y(..)$ is chosen to combine the causal impacts of other states on state Y.
- **Speed of change of a state** η_Y. For each state Y a *speed factor* η_Y is used to represent how fast a state is changing upon causal impact.

This conceptual representation can be transformed automatically into a numerical representation that can be used for simulation and analysis.

In the paper, first in Sect. 2 some background literature on burnout is discussed. Next, the model is introduced in Sect. 3 (conceptual representation) and Sect. 4 (numerical representation). In Sect. 5 simulation results are discussed, and in Sect. 6 verification of the model. Section 6 also discusses validation, and Sect. 7 is a final discussion.

2   Background Literature

During the seventies the burnout syndrome started to be scientifically described as a professional disease. Maslach [13–15] was one of the first researchers to do so, she created a measurement instrument called the Maslach Burnout Inventory (MBI) [14]. She defined a burnout as 'a syndrome of *emotional exhaustion*, *depersonalization* and *reduced personal accomplishment* that can occur among individuals that do "people work" of some kind' [14]. Emotional exhaustion, depersonalization and personal accomplishment are since then still described as the core elements of the burnout syndrome [9, 18]. Various risk factors are described in the literature for a burnout. Risk factors can be categorised as job characteristics and personal characteristics. Maslach [13] describes different job characteristics that can lead to a burnout. The main one is experienced overload, which is another word for subjective stress. Long exposure to chronic job stressors can eventually lead to cynicism and emotional exhaustion [13]. Emotionally charged work is a risk factor; work that is charged with anger, despair embarrassment or fear. This reduces the feeling of personal accomplishment, which is an element of a burnout. When people do too much in support of their ideals, this leads to exhaustion and cynicism when their effort was insufficient for reaching their goals given the circumstances. This job characteristic can be described as job sacrifice [13]. When it is unclear for an employee what, and how much is expected from them, we speak of role ambiguity. Role ambiguity is very frustrating and is also formulated as a possible burnout cause.

Job characteristics that can reduce the effect of the risk factors, or reduce the elements of the burnout syndrome, are protective factors. A protective factor is the amount social contact with coworkers and experience. Social contact with coworkers can reduce feelings of cynicism and depersonalization [4]. Experience is another protective factor, since more experienced people tend to have a sense of personal accomplishment more frequently than their younger coworkers [13]. Experience also lowers their sense of role ambiguity [4]. Figure 1(a) shows the job characteristics that are either risk factors or protective factors.

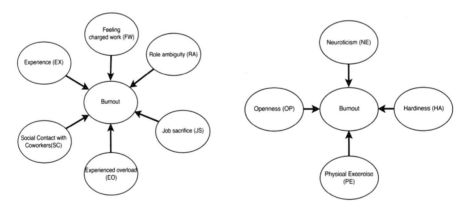

Fig. 1. (a) Job characteristics that are either risk or protective factors of a burnout. (b) Personal characteristics that are either protective factors or risk factors

Another group of risk and protective factors are defined by personal characteristics. Huang [8] explores the relationship between personality and burnout. He found that neuroticism was a predictor for larger amounts of stress and experienced overload. Neuroticism is one of the big 5 personality traits, a measurement scale for personality widely used in psychology [3]. Another big 5 personality trait is openness. This trait has a negative association with depersonalization and emotional exhaustion [5]. Openness is also positively correlated with physical exercise [18]. Physical exercise can reduce experienced overload, and the elements of a burnout. It is found to reduce ambiguity and depersonalization [4]. The third personality trait important for the development of burnout, is hardiness. It is not a part of the big 5, but a multidimensional trait that protects from the effects of stress [6]. Alarcon, Eschleman and Bowling [1] have found that hardiness has a significant relationship with the three elements of burnout as described by the MBI. Figure 1(b) shows the previously described personal characteristics that affect the burnout syndrome.

The consequent states of a burnout are shown in Fig. 2; Golembievwski et al. [7] describe how the burnout leads to a decrease in job attendance and job performance. The quality of work deteriorates and the working morale and motivation becomes low.

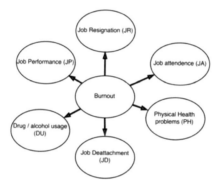

Fig. 2. The consequent state of a burnout

Also a burnout has an impact on the physical health, that shows in the form of rashes and weight fluctuations [4]. A burnout is also correlated with drug and alcohol use, since it releases some of the stress and lowers the experienced overload [4].

3   The Network Model: Conceptual Representation

This section describes the network model developed based on the principles described in the previous sections. This temporal-causal network model is based on the Network-Oriented Modelling approach described in [20, 21]. Based on the literature nineteen states (or state variables) were identified relevant for a burnout; see Table 1. They include the risk and protective factors, the elements of the syndrome, and the consequent state.

The literature describes causal relations between the states in the model. Each connection has a weight value assigned representing the strength of the causal relation. These weights are given in Table 2. A graphical representation of the state variables and their relations is given in Fig. 3. A blue line represents a positive impact, and a red line a negative impact.

Table 1. States used in the network model

Var	Abbr.	Description	Type	Char.	Var	Abbr.	Description	Type	Char.
X1	FW	Feeling charged work	Risk factor	Job	X11	JR	Job Resignation	Consequent	Job
X2	EX	Experience	Protective factor	Job	X12	CY	Cynicism/Depersonalization	Burnout element	Personal
X3	HA	Hardiness	Protective factor	Personal	X13	SC	Social Contact with coworkers	Protective factor	Job
X4	JS	Job Satisfaction	Consequent	Job	X14	JD	Job Detachment	Consequent	Job
X5	RA	Role Ambiguity	Risk factor	Job	X15	JP	Job Performance	Consequent	Job
X6	NE	Neuroticism	Risk factor	Personal	X16	JA	Job Attendance	Consequent	Job
X7	OP	Openness	Protective factor	Personal	X17	DU	Drug and alcohol Use	Protective factor	Personal
X8	EO	Experienced Overload	Risk factor	Job	X18	PA	Personal Accomplishment	Burnout element	Job
X9	EE	Emotional Exhaustion	Burnout element	Personal	X19	PE	Physical Exercise	Protective factor	Personal
X10	PH	Physical Health	Consequent	Personal					

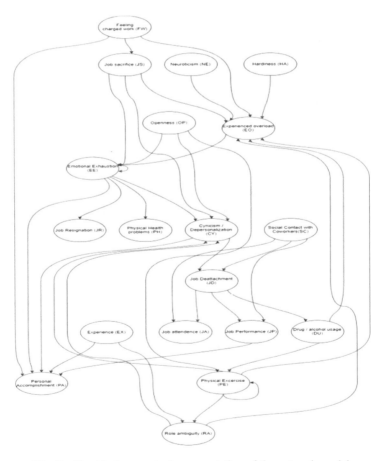

Fig. 3. Graphical conceptual representation of the network model.

Table 2. Connection weights used in the network model

		FW	EX	HA	JS	RA	NE	OP	EO	EE	PH	JR	CY	SC	JD	JP	JA	DU	PA	PE
		X1	X2	X3	X4	X5	X6	X7	X8	X9	X10	X11	X12	X13	X14	X15	X16	X17	X18	X19
FW	X1	1	0	0	1	0	0	0	1	0	0	0	0	0	0	0	0	0	-1	0
EX	X2	0	1	0	0	-1	0	0	0	0	0	0	0	0	0	0	0	0	1	0
HA	X3	0	0	1	0	0	0	0	-1	0	0	0	0	0	0	0	0	0	0	0
JS	X4	0	0	0	0	0	0	0	1	0	0	0	1	0	0	0	0	0	0	0
RA	X5	0	0	0	0	0	0	0	1	0	0	0	1	0	0	0	0	0	0	0
NE	X6	0	0	0	0	0	1	0	1	0	0	0	0	0	0	0	0	0	0	0
OP	X7	0	0	0	0	0	0	1	0	-1	0	0	-1	0	0	0	0	0	0	1
EO	X8	0	0	0	0	0	0	0	0	1	0	0	0	0	0	0	0	0	0	0
EE	X9	0	0	0	0	0	0	0	0	-0.9	1	1	1	0	0	0	0	0	-1	0
PH	X10	0	0	0	0	0	0	0	0	0	0	0	0	0	0	0	0	0	0	0
JR	X11	0	0	0	0	0	0	0	0	0	0	0	0	0	0	0	0	0	0	0
CY	X12	0	0	0	0	0	0	0	0	0	0	0	0	0	1	0	0	0	-1	0
SC	X13	0	0	0	0	0	0	0	0	0	0	0	0	0	1	-1	1	1	0	0
JD	X14	0	0	0	0	0	0	0	0	0	0	0	0	0	0	-1	-1	0.2	0	0
JP	X15	0	0	0	0	0	0	0	0	0	0	0	0	0	0	0	0	0	1	0
JA	X16	0	0	0	0	0	0	0	0	0	0	0	0	0	0	0	0	0	0	0
DU	X17	0	0	0	0	0	0	-1	0	0	0	0	0	0	0	0	0	0	0	-1
PA	X18	0	0	0	0	0	0	0	0	0	0	0	0	0	0	0	0	0	0	0
PE	X19	0	0	0	0	-1	0	0	-1	0	0	0	-1	0	0	0	0	0	0	1

4   The Network Model: Numerical Representation

The conceptual display of the model shown in the text above can be transformed into a numerical representation, in a systematical or even automated manner:

- For each time t every state of the Y model has a real value in the interval $[0, 1]$
- For each time t every condition X connected to state Y has an effect on the state Y defined as $\mathbf{impact}_{X,Y}(t) = \omega_{X,Y}\, Y(t)$
- The aggregated impact of state X_i on state Y at time t is determined by the combination function $\mathbf{c}_Y(..)$:

$$\mathbf{aggimpact}_Y(t) = \mathbf{c}_Y(\mathbf{impact}_{X_1,Y}(t), \ldots, \mathbf{impact}_{X_k,Y}(t))$$
$$= \mathbf{c}_Y(\omega_{X_1,Y}X_1(t), \ldots, \omega_{X_k,Y}X_k(t))$$

- The effect of $\mathbf{aggimpact}_Y(t)$ on Y is exerted over time gradually, depending on the speed factor $\mathbf{\eta}_Y$:

$$Y(t+\Delta t) = Y(t) + \mathbf{\eta}_Y[\mathbf{aggimpact}_Y(t) - Y(t)]\,\Delta t$$
$$\mathbf{d}Y(t)/\mathbf{d}t = \mathbf{\eta}_Y[\mathbf{aggimpact}_Y(t) - Y(t)]$$

- This generates a difference and differential equation for Y

$$Y(t+\Delta t) = Y(t) + \mathbf{\eta}_Y\big[\mathbf{c}_Y(\omega_{X_1,Y}X_1(t), \ldots, \omega_{X_k,Y}X_k(t)) - Y(t)\big]\,\Delta t$$
$$\mathbf{d}Y(t)/\mathbf{d}t = \mathbf{\eta}_Y\big[\mathbf{c}_Y(\omega_{X_1,Y}X_1(t), \ldots, \omega_{X_k,Y}X_k(t)) - Y(t)\big]$$

As combination functions the identity functions and advanced logistic sum functions were used:

$$\mathbf{id}(V) = V$$

$$\mathbf{alogistic}_{\sigma,\tau}(V_1, \ldots, V_k) = \left[\left(1/\left(1 + e^{-\sigma(V_1 + \ldots V_k - \tau)}\right)\right) - 1/(1 + e^{\sigma\tau})\right](1 + e^{-\sigma\tau})$$

Here σ is a steepness parameter σ and τ a threshold parameter.

For each state a speed factor is used to represent how fast a state is changing upon causal impact. The parameters and functions used are represented in Table 3. For the state variables that have only one causal impact, we used a simple identity function. Variables that were affected by more than one variable, use the advanced logistic function. Table 3 shows that the steepness used was 50, which is relatively high. This implies that when the value is higher than the threshold of 0.5, the value changes quite abruptly [21, p. 71].

Table 3. Combination functions and speed factors used in the network model.

State	Abbr.	Function	σ	τ	Speed	State	Abbr.	Function	σ	τ	Speed
X1	FW	Identity			0.5	X11	JR	Identity			0.5
X2	EX	Identity			0.5	X12	CY	Advanced logistic	50	0.5	0.5
X3	HA	Identity			0.5	X13	SC	Identity			0.5
X4	JS	Identity			0.5	X14	JD	Advanced logistic	50	0.5	0.5
X5	RA	Advanced logistic	50	0.5	0.5	X15	JP	Advanced logistic	50	0.5	0.5
X6	NE	Identity			0.5	X16	JA	Advanced logistic	50	0.5	0.5
X7	OP	Identity			0.5	X17	DU	Identity			0.5
X8	EO	Advanced logistic	50	0.5	0.5	X18	PA	Advanced logistic	50	0.5	0.5
X9	EE	Advanced logistic	50	0.5	0.5	X19	PE	Advanced logistic	50	0.5	0.5
X10	PH	Identity			0.5						

There is no previous research done that shows the trend of psychopathological development, so it is uncertain if our state variables should rise gradually or more in an abrupt manner. We expected that low values of risk factor would hardly affect the elements of a burnout syndrome, while high values should have a large impact. Therefore we chose to represent these functions as advanced logistic function, with high steepness and a threshold of 0.5. The speed factor is similar for all state variables for now. Later we will discuss the use of parameter tuning to find appropriate speed values for certain scenarios.

For any set of values for the connection weights, speed factors and any choice for combination functions, each state of the network model gets a difference and differential equation assigned. Examples of such different and differential formulas used in model with the identity combination function are:

$$JS(t + \Delta t) = JS(t) + \eta_{JS} \left[\omega_{FW,JS} FW(t) - JS(t) \right] \Delta t$$
$$dJS(t)/dt = \eta_{JS} \left[\omega_{FW,JS} FW(t) - JS(t) \right]$$
$$JR(t + \Delta t) = JR(t) + \eta_{JR} \left[\omega_{EE,JR} EE(t) - JR(t) \right] \Delta t$$
$$dJR(t)/dt = \eta_{JR} \left[\omega_{EE,JR} EE(t) - JR(t) \right]$$
$$DU(t + \Delta t) = DU(t) + \eta_{DU} \left[\omega_{JP,DU} JP(t) - DU(t) \right] \Delta t$$
$$dDU(t)/dt = \eta_{DU} \left[\omega_{JP,DU} JP(t) - DU(t) \right]$$

Also, an often used combination function is the advanced logistic sum function $\textbf{alogistic}_{\sigma,\tau}(...)$. Examples of states using such difference and differential equations based on this combination function are:

$$RA(t + \Delta t) = RA(t) + \eta_{RA} \left[\textbf{alogistic}_{\sigma,\tau}(\omega_{EX,RA} EX(t), \ \omega_{PE,RA} PE(t)) - RA(t) \right] \Delta t$$
$$dRA(t)/dt = \eta_{RA} \left[\textbf{alogistic}_{\sigma,\tau}(\omega_{EX,RA} EX(t), \ \omega_{PE,RA} PE(t)) - RA(t) \right]$$
$$JD(t + \Delta t) = JD(t) + \eta_{JD} \left[\textbf{alogistic}_{\sigma,\tau}(\omega_{CY,JD} CY(t), \ \omega_{SC,JD} SC(t)) - JD(t) \right] \Delta t$$
$$dJD(t)/dt = \eta_{JD} \left[\textbf{alogistic}_{\sigma,\tau}(\omega_{CY,JD} CY(t) \ \omega_{SC,JD} SC(t)) - JD(t) \right]$$
$$JP(t + \Delta t) = JP(t) + \eta_{JP} \left[\textbf{alogistic}_{\sigma,\tau}(\omega_{CY,JP} CY(t), \ \omega_{JD,JP} JD(t)) - JP(t) \right] \Delta t$$
$$dJP(t)/dt = \eta_{JP} \left[\textbf{alogistic}_{\sigma,\tau}(\omega_{CY,JP} CY(t), \ \omega_{JD,JP} JD(t)) - JP(t) \right]$$

5 Simulations for Some Example Scenarios

We created several scenarios to validate our model against patterns described in the literature. We needed a scenario where a person experiences a burnout, a scenario where a person does not experience a burnout, and a scenario where a person switches between a burnout and no burnout. This third scenario we also attained using parameter tuning in next paragraph. We created these scenarios by using different initialization values, shown in Table 4. By categorizing the variable based on whether they are indicative of a burnout or not, we could assign fitting initializations for each scenario. For instance, a scenario where a person has a burnout is probably characterized by low hardiness. Hardiness is a protective personality factor, a high value would protect a person from a burnout. Therefore hardiness is low in the burnout scenario and high in the no-burnout scenario. We run the model obtained datasets for three scenarios. Figure 4 left graph shows the result from the no burnout simulation. All the variables that are not indicative for a burnout are high, like job performance, personal accomplishment and physical exercise, whereas emotional overload and cynicism are low. The scenario gives a good representation of an individual who has protective factors that are high enough for a burnout not to occur. The second scenario shows an individual who obtains a burnout, shown in Fig. 4 right graph.

Table 4. Different scenarios

Var	Abbr.	Description	Indicative	Burnout	No-burnout	Switch
X1	FW	Feeling charged work	Yes	0.85	0.1	0.2
X2	EX	Experience	No	0.2	0.8	0.8
X3	HA	Hardiness	No	0.2	0.9	0.8
X4	JS	Job Satisfaction	Yes	0.9	0.2	0.9
X5	RA	Role Ambiguity	Yes	0.8	0.15	0.8
X6	NE	Neuroticism	Yes	1	0.23	0.2
X7	OP	Openness	No	0.2	0.75	0.8
X8	EO	Experienced Overload	Yes	0.3	0.1	0.9
X9	EE	Emotional Exhaustion	Yes	0.3	0.1	0.9
X10	PH	Physical Health	Yes	0	0.2	0.9
X11	JR	Job Resignation	Yes	0	0	0.9
X12	CY	Cynicism/Depersonalization	Yes	0.2	0.1	0.9
X13	SC	Social Contact with coworkers	No	0.2	0.9	0.9
X14	JD	Job Detachment	Yes	0.1	0.1	0.9
X15	JP	Job Performance	No	0.75	0.85	0.1
X16	JA	Job Attendance	No	0.95	0.99	0.2
X17	DU	Drug and alcohol Use	Yes/No	0.3	0.1	0.8
X18	PA	Personal Accomplishment	No	0.8	0.85	0.2
X19	PE	Physical Exercise	No	0.7	0.8	0.9

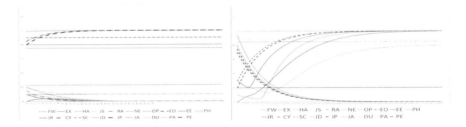

Fig. 4. Non-burnout (left) and burnout scenario (right)

Drug and alcohol use, job detachment, experienced overload and other variables indicative for a burnout, are low in the beginning of the simulation. Due to the high values for risk factors, and low values for protective factors, the elements of the burnout syndrome quickly acquire high values. Around $t = 27$ the states indicative for a burnout are high. Job performance, job attendance and personal accomplishment have lowered.

The third scenario represents a switch, where a person starts to experience a burnout but then the burnout disappears again. This is obtained having some risk factors high and some risk factors low. This is depicted in Fig. 5 (left). Next, a scenario where a person develops a burnout first, but after a while the burnout fades away; Fig. 5 (middle). Finally, Fig. 5 (right) shows a scenario with a cyclic process where a person builds up experienced overload, which is followed by a high burst of cynicism and job detachment. Due to the increasing drug and alcohol use, the experienced overload lowers again, which causes cynicism and job detachment to lower. Drug and alcohol usage lowers again, until the experienced overload builds up too much and the cycle is repeated. Personal exercise rises in a fluctuating manner, until it is high enough at the end of the simulation, for the burnout elements to lower and the cyclic pattern to end.

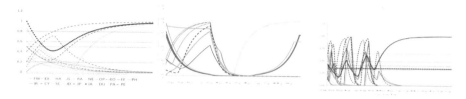

Fig. 5. Three scenarios from left to right: (a) switch from a burnout to a no-burnout state, (b) pattern from non-burnout to burnout to recovery, (c) fluctuating pattern burnout non-burnout

6 Verification and Validation

This section first discusses a mathematical analysis that has been performed for *verification* of the model. The results are consistent with the simulation examples shown in the text. This confirms that the model is mathematically accurate. The analysis was performed on the stationary points of the model. A state Y has a stationary point at t if $dY(t)/dt = 0$. For a temporal-causal network model the criterion for a stationary point is the equation: $\mathbf{c}_Y(\mathbf{\omega}_{X_1,Y}X_1(t), \ldots, \mathbf{\omega}_{X_k,Y}X_k(t)) = Y(t)$; see [21, Chap. 12]. The combination functions used in this model are the identity function and the advanced logistic function. For ten different states an identity function is used; for states RA, EO, EE, CY, JD, JP, JA, PA and PE, advanced logistic functions were used. Examples of stationary point equations are

$$JS(t) = \omega_{FW,JS}FW(t)$$
$$RA(t) = \mathbf{alogistic}_{\sigma,\tau}(\omega_{HA,RA}HA(t), \omega_{PE,RA}PE(t))$$

Since advanced logistic functions are used as a combination function, the analysis by solving such equations in an explicit manner was not possible. But they were still used in verifying this model. This was done by substituting the values found in a simulation at the end time in these equations, and then check whether the equations hold. This was done for a number of cases. The outcome quite accurate, with difference accuracy $<10^{-6}$. The results of the simulation with no burnout outcome can be found in the Table 5.

For *validation* no numerical empirical data were available. However, qualitative information was available in the literature in the sense of patterns that are expected in certain circumstances. A number of such patterns have been identified and used to test the model, with positive outcomes. Examples are the patterns shown in Fig. 6.

Table 5. Overview of the outcomes of the calculation with no burnout outcome

State	Time	Value	Agg impact	Deviation	State	Time	Value	Agg impact	Deviation
PE	99.5	0.999994098	0.9999999	−5.80E−06	JA	99.5	0.9999997	1	−2.97E−07
PA	99.5	0.999995573	0.9999998	−4.23E−06	JP	99.5	0.99999557	1	−4.43E−06

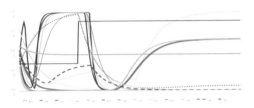

Fig. 6. The effect of increase in physical exercising

As another example, we investigated whether the model could show the effect of physical exercise. For this the following type of parameter tuning was performed by Simulated Annealing. We assumed that physical exercise could be manipulated, even though in our model it is a result of a person's openness and drug/alcohol usage. We investigated whether a change in physical exercise could stop a burnout. The parameters we optimized for were two values of physical exercise, and the time when the value changed. Also the initialisation values and the speed factors were tuned. The final average deviation was 0.07, which is reasonable. The solution found is depicted in Fig. 6. The blue line is physical exercise (PE). At PE = 0.35, the person develops a burnout. Personal accomplishment lowers to zero, and emotional exhaustion, experienced overload and cynicism rise. At $t = 54$, the value of PE is changed to 0.89. This results in the burnout to disappear.

7   Discussion

In this paper temporal-causal network model for burnout was introduced. The model can be the basis for a virtual patient agent model, and offers also possibilities to simulate certain forms of treatment. Representing psychopathological syndromes as a dynamic network is part of a new paradigm in psychology. Syndromes are no longer seen as latent diseases that are measured by their manifesting symptoms, like in medical health care. On the contrary, a syndrome is a network of interacting symptoms [2]. For a model to have ecological validity, it is important to incorporate both risk factors as protective factors. By incorporating this in our model, we were able to run different scenarios. A no-burnout, a burnout scenario, and a scenario where a person switched from a burnout to no-burnout. The network model shows the influence of different risk and protective factors on the development of a burnout. It also shows how the elements of a burnout can be lowered until the negative consequences of a burnout fade away.

There is little literature about more specific characteristics of temporal relations between burnout elements. The relationship between stress or experienced overload and emotional exhaustion is documented [13]. But it is unclear whether emotional exhaustion rises linearly or exponentially, for example, when experienced overload rises. The type, the strength, and the speed of relationships between variables in our model should be tested on empirical data. Unfortunately for our research no numerical empirical data was available. Such data could be gathered using the MBI [14], on different time points, for future work

Like Sutin et al. [18] showed, physical exercise impacts many socio-behavioural factors positively. Where openness and hardiness are static personality traits, physical exercise is something that can be altered consciously more easily. We therefore did an exploratory parameter tuning analysis for a dataset that represented a person who has a burnout, which disappears after some time. We sought whether we could explain this disappearance by a change in physical exercise. We found using parameter tuning, that when the value of physical exercise was raised, indeed the burnout disappeared.

References

1. Alarcon, G., Eschleman, K.J., Bowling, N.A.: Relationships between personality variables and burnout: a meta-analysis. Work Stress **23**(3), 244–263 (2009)
2. Borsboom, D., Cramer, A.O.: Network analysis: an integrative approach to the structure of psychopathology. Annu. Rev. Clin. Psychol. **9**, 91–121 (2013)
3. Costa, P.T., McCrae, R.R.: NEO Five-Factor Inventory (NEO-FFI). Psychological Assessment Resources, Odessa (1989)
4. DePaepe, J., French, R., Lavay, B.: Burnout symptoms experienced among special physical educators: a descriptive longitudinal study. Adapt. Phys. Act. Q. **2**(3), 189–196 (1985)
5. Emilia, I., Gómez-Urquiza, J.L., Cañadas, G.R., Albendín-García, L., Ortega-Campos, E., Cañadas-De la Fuente, G.A.: Burnout and its relationship with personality factors in oncology nurses. Eur. J. Oncol. Nurs. **30**, 91–96 (2017)
6. Eschleman, K.J., Bowling, N.A., Alarcon, G.M.: A meta-analytic examination of hardiness. Int. J. Stress Manag. **17**, 277–307 (2010)
7. Golembiewski, R.T.: Global Burnout: A Worldwide Pandemic Explored by the Phase Model. JAI Press, NY (1996)
8. Huang, L., Zhou, D., Yao, Y., Lan, Y.: Relationship of personality with job burnout and psychological stress risk in clinicians. Chin. J. Ind. Hygiene Occup. Dis. **33**, 84–87 (2015)
9. Kabadayi, A.: Investigating the burn-out levels of Turkish preschool teachers. Procedia-Soc. Behav. Sci. **197**, 156–160 (2015)
10. Kim, J.: Philosophy of Mind. Westview Press, Boulder (1996)
11. Kuipers, B.J.: Commonsense reasoning about causality: deriving behavior from structure. Artif. Intell. **24**, 169–203 (1984)
12. Kuipers, B.J., Kassirer, J.P.: How to discover a knowledge representation for causal reasoning by studying an expert physician. In: Proceedings of the Eighth International Joint Conference on Artificial Intelligence, IJCAI 1983, pp. 49–56. William Kaufman, Los Altos (1983)

13. Maslach, C., Jackson, S.E.: The measurement of experienced burnout. J. Organ. Behav. **2**(2), 99–113 (1981)
14. Maslach, C., Jackson, S.E., Leiter, M.P.: Maslach Burnout Inventory. Consulting Psychologists Press, Palo Alto (1986)
15. Maslach, C., Schaufeli, W.B., Leiter, M.P.: Job burnout. Annu. Rev. Psychol. **52**(1), 397–422 (2001)
16. Pearl, J.: Causality. Cambridge University Press, Cambridge (2000)
17. Schmittmann, V.D., Cramer, A.O., Waldorp, L.J., Epskamp, S., Kievit, R.A., Borsboom, D.: Deconstructing the construct: a network perspective on psychological phenomena. New Ideas Psychol. **31**(1), 43–53 (2013)
18. Sutin, A.R., Stephan, Y., Luchetti, M., Artese, A., Oshio, A., Terracciano, A.: The five-factor model of personality and physical inactivity: a meta-analysis of 16 samples. J. Res. Pers. **63**, 22–28 (2016)
19. TNO, CBS: Nationale Enquête Arbeidsomstandigheden (2014). http://www.monitorarbeid.tno.nl/dynamics/modules/SFIL0100/view.php?fil_Id=149
20. Treur, J.: Dynamic modeling based on a temporal—causal network modeling approach. Biol. Inspired Cognit. Archit. **16**, 131–168 (2016)
21. Treur, J.: Network-Oriented Modeling. Springer, Berlin (2016)

Adaptive Neuro-Fuzzy Inference System Used to Classify the Measurements of Chemical Sensors

Alexander Efitorov and Sergey Dolenko$^{(\boxtimes)}$ (iD)

D.V.Skobeltsyn Institute of Nuclear Physics, M.V.Lomonosov Moscow State University, Leninskiye Gory 1/2, Moscow 119991, Russia
{a.efitorov, dolenko}@sinp.msu.ru

Abstract. Many data processing problems are successfully solved by artificial neural networks (ANN) possessing the property of a universal approximator. However, in case when the number of data patterns available is small, ANN may tend to overtrain and not to generalize well enough. An alternative is use of such a biologically inspired cognitive architecture as fuzzy networks, or Adaptive Neuro-Fuzzy Inference Systems (ANFIS), based on the notions of fuzzy logics and often used in control systems. Like conventional ANN, ANFIS can be also trained by example with error backpropagation algorithm. In this study, we demonstrate use of neuro-fuzzy networks to solve a classification problem for high-dimensional, highly variable and noisy data of chemical sensors. The results are compared to those obtained by a multi-layer perceptron ANN and by linear regression.

Keywords: Artificial neural networks
Adaptive neuro-fuzzy inference systems · Chemical semiconductor sensors
Data processing

1 Introduction

Development and use of such a biologically inspired cognitive architecture as artificial neural networks (ANN) is one of the main directions of computer science in modern world [1]. Deep ANN are widely used for everyday tasks: image and video stream analysis [2, 3], text and speech recognition and generation [4–6] and so on.

The success of development is based on the ability of ANN to approximate complex nonlinear functions in a multidimensional space without having any prior information about it, just using iterative optimization procedures to fit the weight coefficients or other parameters of the network. For example, modern ANN allow solving problems of translation and object detection on photos with accuracy higher than human [7].

This study has been performed with financial support from the Ministry of Education and Science of Russia, Agreement No. 14.604.21.0163, Project ID RFMEFI60417X0163.

© Springer Nature Switzerland AG 2019
A. V. Samsonovich (Ed.): BICA 2018, AISC 848, pp. 101–106, 2019.
https://doi.org/10.1007/978-3-319-99316-4_13

It should be noted that the property of universal approximation was proved a long time ago in 1989 [8–10]. But the success of wide applications of ANN in recent years is associated with the growth of computing power and with simplicity of accumulation of large amounts of digital data. The last factor is extremely critical, since training of a deep ANN with millions of parameters requires billions of patterns; otherwise the ANN tends to overfitting and not having the desired generalizing properties needed to characterize the problem adequately. This is a serious challenge for the problems, obtaining data for which requires expensive measurements, which is so typical for industrial, medical, and scientific applications. Thus, despite of the great results of deep learning with ANN trained on big data, solving problems with high input and/or output dimensionality on relatively small data samples (hundreds and thousands of patterns) is still a topical and complex task.

One of the popular methods to solve the problems of analyzing the signals of real sensors is use of another biologically inspired cognitive architecture – Fuzzy Inference Systems (FIS) [11]. This approach is based on the use of the so-called linguistic variables (small-medium-large, cold-cool-neutral-warm-hot, etc.) and their interactions – fuzzy rules (if cold then heat, etc.) widely used in human reasoning at the household level. It was developed to obtain smooth solutions for small statistical samples, and it is widely used in control systems [12, 13]. In 1994, a theorem was proved, which demonstrated that fuzzy logic networks also possess the properties of universal approximation [14], like sigmoid-based neural networks, and therefore the fuzzy logic approach can be generalized to problems where there is no explicit interpretation of the observed physical variable. The process of construction of the space of linguistic variables and fuzzy rules can be performed automatically using the error backpropagation algorithm. Computational experiments have shown that fuzzy logic networks, or Adaptive Neuro-Fuzzy inference Systems (ANFIS) can really build solutions better than perceptron type ANN with sigmoidal activation functions [15]. In this paper, we demonstrate the solution of the problem of classifying high-dimensional and noisy measurements of chemical sensors using a fuzzy logic network, compared with the results obtained by a multi-layer perceptron ANN and by linear regression.

2 Experimental Data

In this study, we investigate the ability of fuzzy networks to solve the task of binary classification on high variability data measured by semiconductor chemical sensors. The measurement is based on the temperature dependence of sensor resistance R. The details of the experimental equipment and of the measuring process can be found elsewhere [16]. In general, selective detection of CO and H_2 mixture in air (which simulates smoldering in the early stages of fire) by a single SnO_2-based metal oxide sensor (MOX-sensor) was performed. The concentration of CO was varied from 1 ppm to 50 ppm; in experiments with air, ordinary street air of the megapolis of Moscow was used, measured during 2 summer months at various time (total 3 experiments).

The sensor operated in dynamic temperature mode. Namely, each thermal cycle consisted of four stages − heating with constant rate of changing temperature from 150°C to 500°C, isothermal step for 5 s at top temperature, cooling down from 500°C to 150°C with constant rate, and isothermal step for 5 s at bottom temperature − lasting for 60 s in total (Fig. 1, curve for T).

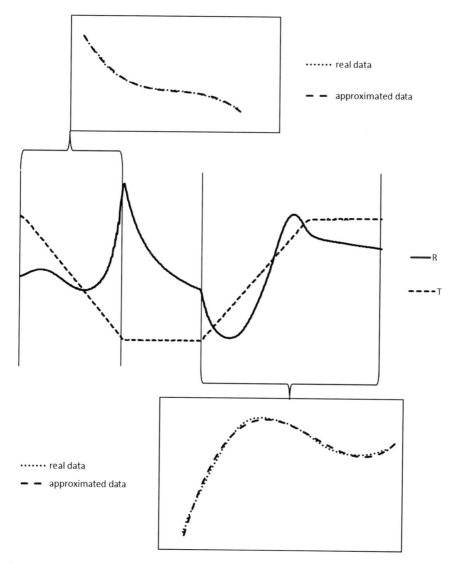

Fig. 1. Data processing scheme. Each sample contains semiconductor resistance data (R, bold line) and measured temperature data (T, dashed line), middle image. Cooling stage and heating stage were extracted from the data pattern; *log10* function was applied and simple normalizing was done to set X and Y ranges to equal values. After that, polynomial fitting was performed independently for heating and cooling stage. The results of polynomial approximation: top image − cooling stage, bottom image − heating and high-temperature plane stages. Dotted line − observed data, dashed line − approximated data.

As a result, about 600 data points per measuring cycle were collected. A custom made electric device was used for the purpose of the sensor resistance measurement and temperature control. At each gas or gas mixture concentration, 30 cycles of sensor measurements were performed. First 10 cycles in each concentration step were neglected as transient between two neighbor concentrations. Further 20 cycles were used for data analysis: algorithm training, testing and examination.

3 Data Processing and Analysis

The points of the sensor response curve have a high level of correlation; therefore, consideration of all points as independent variables is incorrect, and using all points as input variables for model training is excessive. On the other hand, use of all 600 points of the cycle is problematic for training any network with back propagation: when using a classical multilayer perceptron with 10 layers in the first hidden layer, the number of weight coefficients of only the first layer links will be $600 \times 10 = 6000$ parameters, which is comparable with the number of patterns in the training sample. Using fuzzy logic networks leads to only a little better situation, the simplest procedure of fuzzi-fication of input variables with 2 Gaussians with 2 parameters each leads to the appearance of $600 \times 2 \times 2 = 2400$ parameters.

Thus, it is necessary to develop some procedure for effectively reducing the dimension of the input data.

There are two approaches for solving this problem: selection of essential features (feature selection) or transformation to another low-dimensional space (feature extraction). Unfortunately, classical procedures, like estimates of standard deviation, averaging and correlation [17] work on this array inadequately, since there are sig-nificant drift effects associated with the evolution of the sensor surface. For this reason, the approaches connected with feature extraction, such as transition to the space of principal components [18], of wavelet coefficients [19], of preshape space [20], or numerical approximation by nonlinear functions [21] are more effective.

In this study, we chose the feature selection method based on the approximation of the shape of the curve by polynomials of the third order; the set of the obtained coefficients of the polynomials form the array of extracted features. Heating and cooling parts of the cycle were processed independently. To exclude problems con-nected with the extreme values of amplitudes, the signal was pre-processed by *log10* function. An example of the approximation is presented in Fig. 1. Thus, as a result of applying this preprocessing, it was possible to describe each curve with 8 coefficients: for each of the two approximated segments of the curve, 3 coefficients of non-zero order, and 1 coefficient of zero order (shift). In most cases, the quality of the approximation was quite good (Fig. 1).

To train the models and to evaluate the quality of their work in solving the problem of gas classification, the experimental data were divided according to the experiments of continuous measurements. For each gas, two experiments were used as the training set, and one – as the out-of-sample test set. This approach allowed us to exclude the influence of external factors (humidity, state of the sensory material etc.) on the dis-placement of the estimate of the quality of the sensors.

The results of solving the binary classification problem on the extracted set of features with various machine learning methods are given in Table 1.

Table 1. Results of solving the binary classification problem with various methods.

Algorithm	Classification accuracy
Linear model	0.78
Multi-Layer Perceptron	0.91
ANFIS	0.89

For each method, the table presents the best results among those obtained in numerical experiments with various values of method parameters.

The multi-layer perceptron had a single hidden layer and sigmoid activation function both in the hidden and in the output layers; stochastic gradient descent algorithm of weights optimization was used. The other parameters had the following values: size of minibatches for stochastic optimizers – 10, momentum – 0.5, learning rate – 0.01.

In the current implementation of ANFIS, the following solutions were used. Each input feature was subject to fuzzification by three Gaussian functions in three intervals corresponding to the feature distribution quantiles to which they were applied. The value of the Gaussian function width was chosen based on the variance of this feature. The initial approximation of the neural network weights was generated randomly in the interval $[-1, 1]$, training was carried out using the classic gradient descent method, the function of calculating the centroid was chosen as the defuzzification function.

It can be seen that ANFIS provides the results comparable with the multi-layer perceptron and significantly better than the linear model. Future studies should be directed to search and implementation of a modified ANFIS architecture which could perform better than other methods, preserving its advantages, primarily higher stability and higher noise resilience, which are most significant for the studied problem.

4   Conclusion

In this study, it has been demonstrated that ANFIS is a promising approach for solution of a classification problem with high input dimension, high data variability and high level of noise.

Acknowledgements. The authors are grateful to Valeriy Krivetskiy for providing experimental data for this study.

References

1. Bengio, Y., LeCun, Y., Hinton, G.: Deep learning. Nature **521**, 436–444 (2015)
2. Redmon, J., Farhadi, A.: YOLO9000: better, faster, stronger. arXiv:1612.08242 (2016)
3. Smeureanu, S., Ionescu, R.: Real-time deep learning method for abandoned luggage detection in video. arXiv:1803.01160 (2018)
4. Sak, H., Senior, A., Beaufays, F.: Long short-term memory based recurrent neural network architectures for large vocabulary speech recognition. arXiv:1402.1128 (2014)
5. Kassarnig, V.: Political speech generation. arXiv:1601.03313 (2016)
6. Shedko, A.: Semantic-map-based assistant for creative text generation. Proc. Comput. Sci. **123**, 446–450 (2018)
7. He, K., Zhang, X., Ren, S., Sun, J.: Delving deep into rectifiers. Surpassing human-level performance on image net classification. arXiv:1502.01852 (2015)
8. Cybenko, G.: Approximation by superpositions of a sigmoidal function. Math. Control Signals Syst. **2**(4), 303–314 (1989)
9. Funahashi, K.: On the approximate realization of continuous mappings by neural networks. Neural Netw. **2**(3), 183–192 (1989)
10. Hornik, K., Stinchcombe, M., White, H.: Multilayer feedforward networks are universal approximators. Neural Netw. **2**(5), 359–366 (1989)
11. Dubois, D., Prade, H.: Fuzzy Sets and Systems: Theory and Applications. Academic Press, Orlando (1980)
12. Kayacan, E., Khanesar, M.: Chapter 4. Type-2 fuzzy neural networks. In: Fuzzy Neural Networks for Real Time Control Applications, pp. 37–43. Butterworth-Heinemann (2016)
13. Chen, C., Liu, Y.J., Wen, G.-X.: Fuzzy neural network-based adaptive control for a class of uncertain nonlinear stochastic systems. IEEE Trans. Cybern. **44**(5), 583–593 (2014)
14. Buckley, J., Hayashi, Y.: Can fuzzy neural nets approximate continuous fuzzy functions. Fuzzy Sets Syst. **61**(1), 43–51 (1994)
15. Mizutani, J., Sun, C.-T.: Neuro-Fuzzy and Soft Computing: A Computational Approach to Learning and Machine Intelligence. Pearson, London (1997)
16. Krivetskiy, V., Efitorov, A., Arkhipenko, A., Vladimirova, S., Rumyantseva, M., Dolenko, S., Gaskov, A.: Selective detection of individual gases and CO/H_2 mixture at low concentrations in air by single semiconductor metal oxide sensors working in dynamic temperature mode. Sens. Actuators B Chem. **254**, 502–513 (2018)
17. Efitorov, A., Burikov, S., Dolenko, T., Laptinskiy, K., Dolenko, S.: Significant feature selection in neural network solution of an inverse problem in spectroscopy. Proc. Comput. Sci. **66**, 93–102 (2015)
18. Naik, G.: Advances in Principal Component Analysis. Springer, Singapore (2018)
19. Efitorov, A., Dolenko, S., Dolenko, T., Laptinskiy, K., Burikov, S.: Use of adaptive methods to solve the inverse problem of determination of composition of multi-component solutions. Opt. Mem. Neural Netw. (Inf. Opt.) **27**(2), 89–99 (2018)
20. Rohlf, F.: Shape statistics: procrustes superimpositions and tangent spaces. J. Classif. **16**, 197–223 (1999)
21. Dokken, T.: Controlling the shape of the error in cubic ellipse approximation. In: Curve and Surface Design (Saint-Malo, 2002). Modern Methods in Mathematics, pp. 113–122. Nashboro Press, Brentwood (2003)

Association Rules Mining for Predictive Analytics in IoT Cloud System

Vasiliy S. Kireev[1]([✉]), Anna I. Guseva[2], Pyotr V. Bochkaryov[2],
Igor A. Kuznetsov[2], and Stanislav A. Filippov[1]

[1] NRNU MEPhI, Kashirskoe shosse, 31, 115549 Moscow, Russia
`vskireev@mephi.ru`, `stanislav@philippov.ru`
[2] Konnekt Ltd, Lazenki 6 Street, 2, 119619 Moscow, Russia
{`aiguseva,pvbochkarev,iakuznetsov`}`@mephi.ru`

Abstract. The Internet of Things is one of the fastest growing areas of research currently. A promising area for the introduction of this technology is housing and communal services, for which the reduction of accidents, increasing efficiency, in general, focus on transparency, personalization of services and payments for the end user are relevant. This article is devoted to the development and testing of predictive algorithm for predicting the need for repair of various units within the smart home, such as heating, ventilation and air conditioning. The basis for the algorithm is the association rules mining. The paper presents the results of experiments and the directions for further improvement of the algorithm.

Keywords: IoT · Association rules mining · Data mining · Cloud technologies
Predictive analytics

1 Introduction

In recent years, the number of mobile devices with high technical characteristics has increased dramatically. In addition, mobile Internet is developing rapidly and cloud computing is penetrating deeper into common life. These trends are increasingly consistent with the concept of the Internet of things. According to one of the most cited definitions, the term "Internet of things" refers to a network of physical objects containing built - in technology that allows these objects to measure the parameters of their own state or the state of the environment, to use and transmit this information. Things are understood as personal computers, mobile devices, sensors and controlled devices such as lamps, blinds, climate control systems, etc. For smart home systems, the concept of" Internet of things " can be implemented by transferring data from the many sensors used in the system to the cloud, where they will be processed and stored, and by transmitting commands from the user to the end-user managed devices. In turn, the user will have access to an interface (via a browser or mobile application) designed to monitor data from sensors, as well as to control devices. Thus, users will receive a tool for remote management and monitoring of their smart home system [1–4].

Predictive models that are able to assess the current situation and Remaining Useful whole life of industrial equipment have a high percentage, especially for manufacturing

© Springer Nature Switzerland AG 2019
A. V. Samsonovich (Ed.): BICA 2018, AISC 848, pp. 107–112, 2019.
https://doi.org/10.1007/978-3-319-99316-4_14

companies that can optimize their service strategies. If we believe that the costs derived from maintenance are one of the largest parts of operating costs and that often maintenance and operations departments include approximately 30% of the workforce, it is not difficult to estimate the economic benefits that such innovative methods can bring to the industry. In addition, predictive maintenance, where real-time Remaining Useful Lifetime of the machine is calculated, has been proven to significantly benefit other maintenance strategies such as corrective maintenance. In this work, Remaining Useful Lifetime is defined as the time from the current moment that systems will fail. Failure, in this context, is defined as the deviation of the released machine products from the specified service requirements that require maintenance [5–7].

2 Current Approaches in Predictive Analytics for IoT

The general for the approaches operated by data is modeling of a desirable exit of system (but not necessarily mechanics of system) with use of historical data. Such approaches cover" usual "numerical algorithms, such as linear regression, logistic regression or Kallman's filters and also algorithms which usually meet in the fields of machine learning and data mining. The letter algorithms include artificial neural networks (ANN), decision trees, SVM and Markov chains.

The basic elements of ANN are formal neurons, initially aimed at working with vector information. Each neuron is typically connected to all neurons of the previous layer of data processing. Specialization of connections between neurons occurs only at the stage of training of specific data. The ins architecture is a hierarchical sequence of several layers (non-intersecting subsets). In various layers of the ANN can be used by different neurons, but each layer of the ANN consists of neurons of the same type. At the same time, information processing in each ins layer is carried out in parallel. The communication channels between the previous and subsequent ins layers are most often unidirectional and have adjustable weights (synaptic parameters). These weights of relationships are adjusted in the process of training and self-organization of the ins architecture according to the available experimental data or precedents. The architecture of ins in the learning process can be changed by changing the connections between neurons. Each formal neuron performs a simple operation-weighs the values of its inputs with its own locally stored synaptic weights and produces a nonlinear transformation over their sum [8].

Decision trees are quite common in the present approach to the identification and visualization of logical regularities in the data. The decision tree allows us to construct a model of the dependence of the set of outcomes on the set of characteristic features. When building a decision tree must comply with the requirement of consistency-on the path leading from the root to the sheet, there should be no mutual search of luminous values. The decision tree can be translated into a set of logical statements. Each statement is obtained by passing the path from the root vertex to the leaf, and is a logical pattern of the phenomenon under study. The quality of the tree is characterized by two main indicators: the accuracy and complexity of the tree. The accuracy of the tree shows how well the objects of different classes are separated. As an indicator of the complexity of tree stand characteristics such as: number of leaves of the tree, the number of its internal nodes, the maximum path length from the root to the end vertex. Indices of difficulty and accuracy are interrelated: the more complex the tree, so it is usually more accurate [9].

The support vector machine learning (SVM) algorithm is a controlled machine learning algorithm. Currently, the SVM algorithm is successfully used to solve classification problems in various application areas [10, 11]. The main feature of the SVM classifier is the use of a special function called the kernel, which is used to translate an experimental data set from the original characteristic space into a higher-dimensional space, in which a hyperplane separating classes is constructed. At the same time, two parallel hyperplanes are defined on both sides of the separating hyperplane, which define the class boundaries and are at the maximum possible distance from each other. It is assumed that the greater the distance between these parallel hyperplanes, the smaller the average error of the SVM classifier. The characteristic vectors of the objects to be classified that are closest to parallel hyperplanes are called reference vectors.

Markov chains are characterized by this kind of stochastic dependence between the States of the system under study at different times $t \in T$, in which only the state of the system at the moment t determines the distribution law for future States of the object, regardless of the States of the object until t. A large number of real control processes are characterized by the discontinuity points develop control actions. Therein case many T is a discrete. Hidden Markov Chains were introduced in the work [12] for Remaining Useful Lifetime prediction.

3   Proposed Approach

The approach proposed by the authors is two-stage and consists first of extracting signal sequences from sensors - signatures (patterns) using methods of association rules mining to create a classifier that recognizes these patterns and classifies them to different states (common scheme described in Fig. 1).

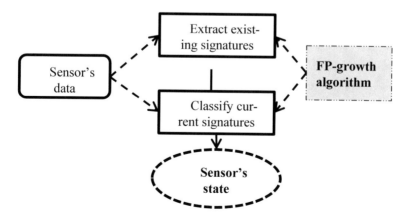

Fig. 1. Two-stage prediction algorithm for the sensor's state

An associative rule [14] is an expression of the form A B $\Rightarrow$, with the sets A and B such that A b $\cap = \emptyset$. For automatic calculation of such rules, there are special algorithms that build a set of associative rules that satisfy certain restrictions on the input set of sets. The input for such algorithms is a set of sets. This set will be called the "base set" and each set inside it will be called a transaction. The criteria for assessing the quality of the associative rule are such values as: support and reliability. The search algorithms for Association rules are close in nature to the search algorithms for patterns, so if you find the rule (1), we can say that the set A b $\cup$ is a popular occurrence (pattern). Often, the search algorithms of Association rules work in two stages: (1) Construction of all popular sets; (2) Extraction of the resulting sets of as - associative rules that satisfy the previously specified conditions of reliability.

FP-growth algorithm was published by [15] and at the time brought a significant contribution to the topic of search of frequently encountered sets, because I did not use an explicit generation of candidates. This approach was later called "build-up sets" (pattern-growth method). It has been previously noted that FP-Growth, as well as many other algorithms reduce the task of finding frequent sets to the task of storing the dictionary, for solving which in this case uses the prefix tree FP-Tree. Pre-signs inside each transaction is sorted in descending order support, their order is fixed. Naturally, infrequent singleton sets are also discarded beforehand.

4   Experimental Results

For the experiments were used the data of sensors of real Data Center. These sensors are sending their values in the real-time manner by MQTT protocol in the experimental cloud information system. 4 temperature sensors were chosen for the period of several days, so that whole volume became 3500 distinct values. Firstly their signals were curated by these principles—all values were separated by two thresholds to gain three types of state. These states were grouped by selected time window in 1 min. Then, these results were given into association rules mining by FP-growth method in the Rapid Miner process. Thus were obtained some first types of signatures (Fig. 2, Table 1).

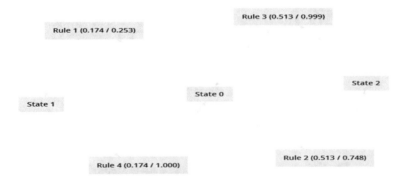

Fig. 2. Example of FP-growth frequent rules extracted from temperature sensors

Table 1. Main characteristic

	Premises	Conclusion	Support	Confidence	LaPlace	Gain	p-s	Lift	Conviction
1	State2	State1	0.087	0.123	0.636	−1.331	0.025	1.410	1041
2	State2	State0	0.622	0.877	0.949	−0.797	0.181	1.409	3074
3	State0	State2	0.622	1000	1000	−0.623	0.181	1.409	673,010
4	State1	State2	0.087	1000	1000	−0.087	0.025	1.410	Infinity

5   Conclusions

This work is related to one of the most fast growing directions of scientific and engineering research—developing new IoT technologies and embedding it in the various common life spheres. This paper highlights, through experiments performed on real data, that the proposed approach is effective in prediction of the sensor's state time to a determinate event, identified as the entrance of the model in a target state. Thus, the proposed approach, based on the FP-growth algorithm can be used in practice for condition monitoring.

Acknowledgements. This study was financed through the Federal Target Program "Research and development on priority directions of scientific-technological complex of Russia for 2014–2020" (grant № RFMEF157917X0142).

References

1. Atzori, L., Iera, A., Morabito, G.: SIoT: giving a social structure to the Internet of Things. IEEE Commun. Lett. **15**(11), 1193–1195 (2011)
2. Stojkoska, B.R., Trivodaliev, K., Davcev, C.: Internet of Things framework for home care systems. Commun. Mob. Comput., Wirel (2017). https://doi.org/10.1155/2017/8323646
3. Gubbi, J., Buyya, R., Marusic, S., Palaniswami, M.: Internet of Things (IoT): a vision, architectural elements, and future directions. Future Gener. Comput. Syst. **29**(7), 1645–1660 (2013)
4. Weyrich, M., Ebert, C.: Reference architectures for the internet of things. IEEE Softw. **33**(1), 112–116 (2016)
5. Jardine, A.K., Lin, D., Banjevic, D.: A review on machinery diagnostics and prognostics implementing condition based maintenance. Mech. Syst. Signal Process. **20**, 1483–1510 (2006)
6. Zhang, X., Xu, R., Kwan, C., Liang, S.Y., Xie, Q., Haynes, L.: An integrated approach to bearing fault diagnostics and prognostics. In: Proceedings of IEEE American Control Conference, pp. 2750–2755 (2005)
7. Wang, M., Wang, J.: CHMM for tool condition monitoring and remaining useful life prediction. Int. J. Adv. Manuf. Technol. **59**, 463–471 (2012)
8. Chinnam, R., Baruah, P.: A neuro-fuzzy approach for estimating mean residual life in condition-based maintenance systems. Int. J. Mater. Prod. Technol. **20**, 166–179 (2003)
9. Domeniconi, C., Perng, C.-S., Vilalta, R., Ma, S.: A classification approach for prediction of target events in temporal sequences. In: Proceedings of the 6th European Conference on Principles of Data Mining and Knowledge Discovery (PKDD 2002), pp. 125–137, Springer, London (2002)

10. Tipping, M.E.: The relevance vector machine. Adv. Neural. Inf. Process. Syst. **12**, 652–658 (2000)
11. Vapnik, V.N.: The Nature of Statistical Learning. Springer, Berlin (1995)
12. Rabiner, L.R.: Tutorial on hidden Markov models and selected applications in speech recognition. Proc. IEEE **77**(2), 257–286 (1989)
13. Cartella, F., Lemeire, J., Dimiccoli, L., Sahli, H.: Hidden semi-Markov models for predictive maintenance. Math. Probl. Eng. (2015). https://doi.org/10.1155/2015/278120
14. Agrawal, R.: Fast algorithms for missing association rules in large databases. In: Agrawal, R., Srikant, R. (eds.) Proceedings of the 20th International Conference on Very Large Data Bases, Santiago de Chile, pp. 487–499 (1994)
15. Han, J.: Mining of frequent patterns without candidate generation: a frequent-pattern tree approach. In: Han, J., Pei, J., Yin, Y., Mao, R. (eds.) Data Mining and Analysis Discovery, vol. 8, no. 1, pp. 53–87 (2004)

Visual Priming in a Biologically Inspired Cognitive Architecture

Pentti O. A. Haikonen[✉]

Department of Philosophy, University of Illinois at Springfield,
One University Plaza, Springfield, IL 62703, USA
pentti.haikonen@pp.inet.fi

Abstract. Each moment the senses produce large quantities of sensory information. Most of this information is not important and remains unnoticed or subconsciously perceived at best, while only the relevant information reaches the focus of attention and becomes consciously perceived. Attention is controlled by various external effects, like change and novelty, stimulus strength, etc. Internal cognitive conditions like context and expectations control attention by priming; the sensitization of sensory perception for the expected entities. Sensory priming can be implemented also in artificial biologically inspired cognitive systems. As an example, priming in the biologically inspired HCA architecture and in the XCR-1 robot is described.

Keywords: Priming · Visual perception · Cognitive architectures

1 Priming in Visual Perception

In cognitive psychology, priming is a term for various processes, where a previously given stimulus modifies the response to a subsequent stimulus. In visual perception, priming may appear as the effect of previously seen patterns on the perception of subsequently seen patterns. Priming can put percepts into context (Baars 1997). Priming can speed up recognition and operate as a selector (Cotterill 1998; Trehub 1991). Visual priming can also operate with imagined visual patterns as the priming stimuli (Haikonen 2003).

When we are searching for something, we have an imagined mental idea about the searched object. Then we look around for anything that resembles our mental idea. When a reasonable match is encountered, we may have found what we were looking for. If mismatch occurs, we have to look elsewhere and continue searching. It should go without saying that pixel-by-pixel comparison does not usually work, visual features should be compared instead of pixels.

Attention control by match and mismatch does work in searching, but at a cost. One has to look keenly at and inspect visual objects one by one and see, if it matches the expectation. This takes time. Searching would be faster, if visual perception could be sensitized for the searched object. Visual priming is the process that does this. Priming can also bring into attention peripherally, subconsciously seen objects if they match the search criteria.

© Springer Nature Switzerland AG 2019
A. V. Samsonovich (Ed.): BICA 2018, AISC 848, pp. 113–118, 2019.
https://doi.org/10.1007/978-3-319-99316-4_15

Visual priming calls for feedback from the deeper layers of the brain back to the visual cortex and to the gaze control of the eyes. This is an architectural property.

2 The Haikonen Cognitive Architecture (HCA)

A cognitive architecture is a layout that is designed for the production of cognitive functions such as perception, learning, response generation, thinking, inner speech, etc. that are present in the human mind.

Cognitive architectures may be based on program code or hardware-implemented artificial neural networks. These approaches are fundamentally different; the computer approach is normally symbolic, while the neural network approach is sub-symbolic. The drawback of the symbolic approach is the difficulty of the grounding of meaning of the used symbols, while the problem with the neural networks has been their difficult use for symbolic information processing. A list of some implemented cognitive architectures is given by Samsonovich (2010). Cognitive architectures can be applied to advanced autonomous robots. A good discussion on robot cognition and potential consciousness is given by Reggia et al. (2018).

The Haikonen Cognitive Architecture (HCA) is a biologically inspired neural cognitive architecture for robot brains (Haikonen 2007, 2012). Like its biological model, the brain, HCA is inherently able to process information seamlessly in sub-symbolic and symbolic ways. This facilitates the grounding of meaning, and allows e.g. the easy use of a natural language and inner speech.

Symbols alone cannot provide the required grounding of meaning, because they call for explanation. Therefore, the author has proposed the following (Haikonen 2018): The ultimate grounding of meaning can only be achieved by using sub-symbolic self-explanatory sensory information. This, in turn, has the form of qualia. To have reportable qualia is to have consciousness. Without the grounding of meaning there cannot be any real understanding. Consequently, machines that truly understand will be conscious. HCA is an attempt towards this direction.

HCA utilizes special hardwired associative neurons. The associative neuron networks used in the HCA are different from the traditional artificial neural networks, as the synaptic weights are not adjusted against each other, and no backpropagation, Deep Learning or other similar algorithms are used.

HCA is a system consisting of a number of vastly parallel, associatively cross-connected sensory perception modules with associative memory and effector systems. The basic building block used in these modules is the "perception/response feedback loop", which is, in principle, similar for all sensory modalities. Rather similar feedback loops have been proposed e.g. by Chella (2008). The cross-connection of the modules takes place via percept broadcasting and receiving.

The associative cross-connection of the perception/response feedback loops of various sensory modalities facilitate global information distribution, information integration, sensorimotor integration, association of meaning and the seamless transition from sub-symbolic to symbolic processing.

The internal feedback loop in Fig. 1 returns the output of the associative neuron groups back to the front of the perception/response feedback loop. In this way the products of the inner neuron groups can be observed as kinds of virtual percepts. The feedback loop has multiple functions, enabling prediction, expectations, imaginations and inner speech. This feedback also facilitates novelty detection and match/mismatch comparison between actually sensed and expected information. Match/mismatch and novelty states control attention. Also priming by expectation is facilitated by the feedback.

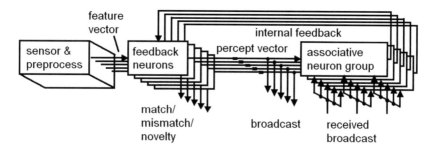

Fig. 1. The perception/response feedback loop (simplified).

The associative neuron groups act like translators that translate received broadcast (cues) from other sensory modules into the "native language" of the receiving module regardless of the origins of the cues. This is done by learned associative connections. Due to this way of operation the system does not need any global neural code. The associative neuron groups have also other functions like memorization.

The working of the main principles of HCA architecture and the perception/response feedback loop are empirically verified by the XCR-1 robot.

3   Visual Sensory Priming in the XCR-1 Robot

The author's experimental cognitive robot XCR-1 is a small three-wheel robot that implements the HCA architecture in a minimalistic way (Haikonen 2011, 2012). Instead of microprocessors and program code the robot utilizes electronic associative neuron and synapse hardware circuits. The implemented architecture consists of associatively cross-connected visual, auditory, touch, emotional and motor modalities.

The audio input system of the auditory module consists of a microphone, preamplifier and narrow band formant filters. These filters output auditory feature signals that correspond to the instantaneous audio spectrum content of the detected sound. This enables the associative recognition of some spoken words.

For the purpose of spoken responses a small vocabulary is stored in an analog audio memory. The meanings of the stored words are associatively grounded to percepts from the other sensory modalities. In this way, the robot is able to produce a limited flow of meaningful inner speech; a spoken report of the robot's instantaneous "mental" content, including determined responses to a limited number of questions. The inner speech is spoken aloud for the benefit of an observer.

The emotional module gets its inputs from mechanical shock and petting sensors. Shock leads to "pain", which is in the form of sub-symbolic disruptive system condition. Pain can also be named. Pain can be associated with seen, touched or named objects and consequently, when encountered, these will be avoided and reported as "bad".

The visual module is able to detect and distinguish illuminated objects and their locations. This allows the robot to name, approach, catch and grab these objects. XCR-1 will approach seen objects unless they are associated with pain. As soon as an object is captured by visual attention, a closed feedback control loop will arise by itself and the robot will home in on the object.

In the robot XCR-1 objects in the peripheral visual area are not normally seen, even though they produce weak sensory feature signals. These signals are not able to pass threshold circuits, and therefore remain normally undetected, "sub-conscious". They have no causal or associated consequences and cannot be reported. However, visual priming can cause these objects to be captured by visual attention.

In the robot XCR-1 visual priming takes place according to the previously explained general principle. When the name of the desired object is spoken, the corresponding visual signal pattern is associatively evoked in the visual module and is returned via the feedback loop to the feedback neurons. This feedback signal pattern acts as the priming information, see Fig. 2.

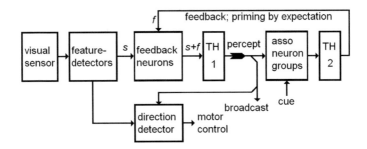

Fig. 2. Visual priming in XCR-1.

In Fig. 2 "cue" is the perceived auditory feature signal pattern (a heard name) broadcast by the auditory module. If it is associated with a visual signal pattern, then this visual signal pattern will be evoked as the output of the asso neuron groups. Otherwise the received auditory signal pattern is ignored.

The evoked visual signal pattern f is carried via the feedback loop to the feedback neurons that also receive the actually sensed visual feature signal pattern s. Each feedback neuron produces the linear sum of the intensity of its s-signal and the corresponding f-signal. Without priming the intensity of the s-signals generated by peripheral vision would not exceed the set threshold value of the threshold circuit TH1, and no percept would be produced. However, when the peripherally seen object matches the expected object, s- and f-signals match and sum in phase. This raises the s + f sum signal level above the threshold. The object will be detected, it can be named and reported, and its location determined. As a result, the robot will turn towards and approach the named object. However, if the object is associated with pain, the robot will back off and avoid the object.

What is this priming good for? Indeed, an apparently similar outcome could be achieved by lowering the threshold value of the threshold TH1 to such a low value that also peripherally seen objects could be perceived. However, priming has an important benefit over the simple lowering of thresholds. This benefit is seen in situations, where there are several objects in the acute and peripheral visual fields. In this case priming determines, which one of the objects will become the focus of attention. After all, visual attention should not and cannot be focused on everything at once, and therefore perception thresholds are necessary.

Visual priming process enables the XCR-1 robot to select the desired object from other seen objects on command. This function is not limited to peripherally seen objects, even though that case can be easily demonstrated. More generally, visual priming can enable HCA-based robots to search for given objects.

4 Conclusions

Sensory priming, the sensitization of sensory perception for the expected entities, is an efficient way of focusing sensory attention on desired stimuli. This process is beneficial in the searching, avoiding and selecting of objects. Therefore, sensory priming would be useful also in artificial biologically inspired cognitive systems. In the biologically inspired HCA architecture, sensory priming is an inherent property that is produced in a very simple way by the perception/response feedback loop.

References

Baars, B.J.: In the Theater of Consciousness, pp. 118–119. Oxford University Press, New York (1997)

Chella, A.: Perception loop and machine consciousness. APA Newsl. Philos. Comput. **8**(1), 7–9 (2008)

Cotterill, R.: Enchanted Looms, p. 265. Cambridge University Press, Cambridge (1998)

Reggia, J.A., Katz, G.E., Davis, G.P.: Humanoid cognitive robots that learn by imitating: implications for consciousness studies. Front Robot. AI (2018). https://doi.org/10.3389/frobt.2018.00001

Haikonen, P.O.: The Cognitive Approach to Conscious Machines, pp. 46–47. Imprint Academic, Exeter (2003)

Haikonen, P.O.: Robot Brains; Circuits and Systems for Conscious Machines. Wiley, Chichester (2007)

Haikonen, P.O.: XCR-1: an experimental cognitive robot based on an associative neural architecture. Cognit. Comput. **3**(2), 360–366 (2011)

Haikonen, P.O.: Consciousness and Robot Sentience. World Scientific, Singapore (2012)

Haikonen, P.O.: The explanation of consciousness with implications to AI. APA Newsl. Philos. Comput. **17**(2), 28–30 (2018)

Samsonovich, A.V.: Toward a unified catalog of implemented cognitive architectures. In: Samsonovich, A.V., Johannsdottir, K.R., Chella, A., Goertzel, B. (eds.) Biologically Inspired Cognitive Architectures 2010, pp. 195–244. IOS Press, Amstredam (2010)
Trehub, A.: The Cognitive Brain, pp. 51–54. MIT Press, Cambridge (1991)
Demo Video. https://www.youtube.com/watch?v=07V_9Sjr9Dg

Neural 2D Cart and Pole Control and Forward Model

Paul Horton, Chris Huyck$^{(\boxtimes)}$, and Xiaochen Wang

Middlesex University, London NW4 4BT, UK
C.Huyck@mdx.ac.uk

Abstract. It is useful to have a robot controller and forward model in simulated or emulated neurons. Such a neural system can both aid understanding of the same task in biological neurons, and provide an efficient mechanism for robot control. This paper describes a controller and forward model for the 2D cart and pole problem. It uses standard control algorithms implemented in simulated neurons. The system runs on a simulated problem, but is a step toward more complete work with real bipedal robots. The next steps include a 3D cart and pole model, a forward model for a fast walking robot (based on the 3D model), using SpiNNaker. Once this is running, there should be a strong system to expand to include better models and controllers, and to integrate learning.

Keywords: Spiking neurons · Neural control · Inverted pendulum

1 Introduction

Advancements in robot control have been rapid and ongoing. In particular, there have been ongoing advancements in the development of human like robots, though robots, these robots still fall far short of, for instance, humans' ability to walk. One way forward is to mimic the human method of control, which is based on neurons. Unfortunately, it is far from clear how humans use neurons to manage their walking. Consequently, developing neural robot controllers is an interesting task, supporting furthering understanding of human motor control.

Human bipedal walking is unstable. We have to constantly expend energy to balance even when just standing. Bipedal robots can have a construction unlike humans, e.g. huge feet, but those with human-like construction will also need to balance. Control of a bipedal walking robot so that it remains steady, and adapts easily to uneven terrain is difficult. Some solutions use complex analytical systems. However, such solutions can be computer intensive and use computational resources thereby limiting how autonomous the robot can be.

One way of reducing the computer intensiveness of bipedal robot walking controllers is to develop a simple and power efficient mechanism based on neurons, which require minimal power, in a highly parallel architecture that communicate using spikes, i.e. a spiking neural network. With the recent advances

© Springer Nature Switzerland AG 2019
A. V. Samsonovich (Ed.): BICA 2018, AISC 848, pp. 119–127, 2019.
https://doi.org/10.1007/978-3-319-99316-4_16

in neuromorphic computing, neuromorphic chips (such as SpiNNaker [4]) have been developed that use electronic circuits to mimic neurons in the brain, and therefore, can take advantage of such an architecture. Neuromorphic controllers that are parallel, more compact and require less processing power may be a practical solution.

To implement a controller, of any sort, one must begin. One track is to use a 3D Inverted Pendulum as the basic problem for biped walking [7]. This can be simplified to a 2D inverted pendulum task [9], a commonly used control task; this task, also called the cart and pole, is described in this paper.

While there are a range of neural models, spiking neurons seem to have the right granularity. Within this, there are more and less sophisticated models. The simulations described below are based on point neural models.

2   Literature Review

The authors are interested in developing an agile biped walking robot using artificial neurons in the control system. Below is brief review of legged robots, of neurons, and neural robot control.

2.1   Robots

Legged robots are suited to uneven terrain as they choose foot placement and maintain balance [11]. This is because legged robots have a high degree of freedom, they can perform complex movements, for example, they could go from walking on the ground to climbing steps (Kaneko et al. 2002; Spenko et al. 2008). However, this type of flexibility increases the complexity of the walking control mechanism. In addition, they require more moving parts, more power and are slow compared to wheeled robots. However, a legged robot has several potential applications. For example, they can be used in search and rescue missions in disaster zones where the terrain is unknown, such as earthquakes (Kamikawa et al. 2004), and can be used in space exploration (Wilcox et al. 2007).

Legged robots can have two or more legs and the more legs a robot has the more stable it is. For example, when a quadruped robot is at rest, all four legs provide stability, but a bipedal robot must maintain balance. Therefore, a bipedal robot has a more challenging locomotion control problem. However, bipedal robots have more advantages than more legged robots, including compactness of the structure (for example, two legged robots have less difficulties navigating around tables and chairs when they are close to each other), energy consumption (two legs needs less power than more legs), and human likeness.

2.2   Neural Networks

It is well known that biological neurons collect activation from other neurons via synapses, and if they have sufficient activation, they emit a spike, sending activation to other neurons [1]. That is, neurons are integrate and fire devices.

A reasonable model of a neural system is thus a spiking neural network (SNNs). SNNs function using spikes, which are discrete events that take place at points in time, rather than continuous values. The occurrence of a spike is determined by differential equations that represent various biological processes. Neurons collect activation from synapses of firing neurons, and when it reaches a certain potential, it fires, and the potential of the neuron is reset. The leaky integrate and fire model is a widely used family of neuron model. SNNs are often sparsely connected and take advantage of specialized network topologies.

Spike trains can easily process spatio-temporal data, which is one advantage of using them. The spatial aspect refers to the synapses to and from neurons being skewed toward local neurons. The temporal aspect refers to temporal information being encoded in the spikes as spike trains occur over time. This simplifies processing of temporal data. Some work has shown spiking neurons can be more powerful computational units than traditional artificial neurons [3].

This paper describe a simulated neural system. This system is developed in PyNN [2], a Python based system for describing neural topologies and simulations; it is a form of middleware. These simulations are then run in Nest [5] a neural simulator, which is the backend. PyNN supports a switch from one neural simulator to another, or to neuromorphic emulators.

Simulating SNNs on normal hardware is computationally intensive since it requires simulating many differential equations, i.e. one for each neuron. However, with the recent advances in neuromorphic computing, neuromorphic hardware has been developed such as SpiNNaker and IBM's TrueNorth [8], which aims to solve the computationally intensive problem by simulating neurons on specialized hardware; i.e. they use electronic circuits to mimic neurons in the brain, and therefore can take advantage of the discrete and sparse nature of neuronal spiking behaviour. This research is hoping to move to SpiNNaker [4] hardware.

2.3   Robot Control

There are different methods to control a bipedal robot. One way is to use complex analytical systems that plan the placement of each foot by setting the angles of each joint. The other way is to use nature's method, i.e. the biological leg control system used by humans. This is the main inspiration for bipedal leg control across much work, and is a combination of three different systems: central pattern generators, reflex and an internal forward model.

The analytical system has the advantage of being more accurate but requires more computational power, and therefore, limits the resources for other high-level systems. This would be ok for semi-autonomous robots as other function can be computed on external systems. Biologically inspired systems use simpler mechanisms to control a robot's walking, therefore require less computing power and therefore more useful in fully autonomous systems.

As mentioned above, human leg control consists of central pattern generators (CPGs), reflex and an internal forward model. CPGs refer to neural circuitry within the spine that can generate rhythmic walking patterns without timing information, i.e. sensory inputs. Reflex systems contain hard coded reactions to obstacles, such as gaps and steps, which are activated by sensors. The internal forward model is thought to exist in the cerebellum [10] The forward model predicts the future state of the body (e.g. position, velocity) from the current motor command sent to the leg(s). The predicted and desired position of the body can then be sent as inputs to the controller that will calculate the motor command required to achieve the desired state from the predicted state. The internal model helps to compensate for delays or lack in sensorimotor feedback and helps to balance [6]. An inverted pendulum is typically used to model the dynamics of a biped robot. This is discussed more in the next section.

CPGs, reflex and internal forward models for bipedal walking control have been developed using neural networks and spiking neural networks (neuromorphic technology) (refs). However, so far, the performance of neuromorphically controlled biped robots is poor in terms of versatile and agile locomotion. This is mainly because their neuromorphic circuits have only emulated CPGs or reflex systems. It is thought that integrating an internal forward model, to control body posture, into a walking control system is very important for agile movement in a complex natural environment. This, as far as the authors are aware, has not been implemented before.

3   The Simulated Neural System

The simulated cart and pole system is quite simple. The motion of the cart and the pole balanced on it are modelled by analytical equations. The cart can be pushed, and the force used is determined by a neural system, simulated in Nest and PyNN. The neural topology is a simple implementation of standard control equations. This neural system can balance the pole. Code can be found on http://www.cwa.mdx.ac.uk/FastWalk/FastWalking.html.

3.1   Implementing Functions in Spiking Neurons

Simulated neurons are powerful computational devices. While it is not entirely clear what their underlying function is in the brain, simulated neurons can readily be used to implement mathematical functions. One mechanism is to have the number of neurons firing in a population represent a number. If the synapses and integration of the neurons can cause an appropriate number of neurons to fire in a second population, those neurons have implemented a function that transforms the input value (in the form of neurons firing) into an output value (in the form of neurons firing). This second population can then be used to implement another function, whose output is then represented in a third population.

These simulated neuron implementations of functions can combine two inputs. All that is needed to implement a complex compound function is to implement the individual operations, and combine them (synaptically) appropriately.

For example, Fig. 1 represents an add function with two inputs of 0 to 3, and the outputs of 0 to 6. Arrows and their associated labels represent synaptic weights from the outer, input, neurons to the centre, output, neurons. If the total synaptic weight to a neuron is greater than 1 the neuron fires. So, if the left input is 1 and the right input is 2, the centre 3 receives 5 inputs of .21 totalling 1.05 and fires. The centre 4 only receives total input of .9 and does not fire. The centre 0, 1 and 2 neurons also fire. This figure is for illustration and is not actually used in the running systems below.

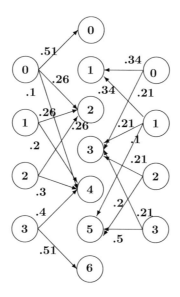

Fig. 1. Circles represent neurons, with the left and right neurons the input to the addition function. The centre neurons will fire if their total inputs are greater than 1. Inputs from the left and right are the same, but alternating inputs are omitted. So, the weight from the left 0 to the centre 0 is the same as that from the right.

Of course, the fidelity of the operation is limited by the number of neurons. So, if the population has 100 neurons, it can only represent 0 to 100, though it could also implement, for example, 0 to 1 with two digits of accuracy.

Many of the operations can be done in parallel, and the entire calculation only needs a sequence of populations that cannot be parallelized. Moreover, if the time between steps of calculations is brief, and the sequence is short, the overall calculation can be quite quick.

Moreover, it is possible for the full set of populations to be carrying out several calculations simultaneously in a sort of pipeline. On the other hand, several different copies of the populations could be used.

3.2   The Forward Model and Controller

A biped robot is always unstable and needs to maintain balance while walking and at rest. The motors in the biped robot need to be constantly adjusted to compensate for any change from the upright position. One method for supporting an upright position is to use an internal forward model. That is, to maintain an upright position the system predicts the next state of the robot's body (e.g. its position) using motor commands calculated by a walking controller. The difference between the actual and predicted state of the robot's body is then fed back through the system to adjust future motor commands to maintain stability. The difference in the predicted and actual states may be due to noisy body sensors or unpredictable forces outside the body.

An inverted pendulum model is often used as the internal forward model in bipedal walking robots to help maintain balance. The inverted pendulum model describes how the pendulum, which has a mass at the end of it and is attached to a cart, falls from an upright position over time. A 2D model, i.e. pendulum motion, can be described as two non-linear differential equations (Eqs. 1 and 2). Thus, to keep the pendulum upright, the system needs to move the cart back and forth along the horizontal surface. This can be achieved using a PID (Proportional, Integration and Derivative) controller. The controller uses the actual and desired position of the pendulum (where the desired position is upright) as inputs to controller that outputs the motor command required to achieve the desired state from the current state.

The inverted pendulum described the body stability of a 2D bipedal robot. The model predicted the next state of the robot's body from a motor command (calculated from the output of, for example, the PID controller); the current and predicted state are then compared, and the difference is fed back into the system so that future motor commands are adjusted accordingly to maintain stability.

With this in mind, a simulated controller and forward model for a 2D cart and pole task was developed. Initially, an analytical model was used. The next step was to substitute the a neural controller for the analytical controller. The analytical controller can be derived from Eqs. 1 and 2.

$$\ddot{x} = \frac{(m_1 * g * sin(\theta)) - (m_1 * l * \dot{\theta}^2 * sin(\theta)) + F}{m_0 + (m_1 * sin(\theta)^2)} \tag{1}$$

$$\ddot{\theta} = 1/l * ((g * sin(\theta)) + \ddot{x} * cos(\theta)) \tag{2}$$

Constants in the system are g the gravitational constant, m_0 the mass of the cart, m_1 the mass of the pole, and l the length of the pole. Inputs to the system are *theta* the angle of the pole, and $\dot{\theta}$ the angular velocity of the pole. The outputs of the system are F the force generated by the system on the cart moving horizontally, and $\ddot{\theta}$ the angular acceleration.

So given the current state in terms of angle and the angular velocity, the system generates the next state. Note that for any system the constants can be factored out. The neural system that does the calculation already assumes those constants, and thus does not need to include those calculations.

The controller can be implemented independently. It is, in essence, applying the force F to the cart to make the pole balance.

3.3   Evaluation

The evaluation is simulated. Initially the system is run in analytical mode with the pole off centre. The forward model determines the effects, and the controller applies the force. The system takes some time to stabilize, depending on the input, but always does.

The controller test replaces the analytical controller with a neural one. This also stabilizes, but as the calculations have less fidelity, it takes longer.

Similarly, the analytical forward model can be replaced with a neural one using an analytical controller. This also stabilizes. Finally, both neural forward model and controller can be used. Of course, in both of these cases, it is not entirely clear how well the model will convert to the real world.

4   Discussion and Future Work

The goal of this work is to develop effective neural controllers for a fast walking bipedal robot. Eventually, this can be extended to controllers for other robots, and integration with other types of neural processing. This work can also explore how humans might do these things by building models that echo human behaviour at many levels, including behaviourally and neurally.

The next step in this process is to move from a 2D model to a 3D model, and this work is currently under way. The equations are of similar complexity, so it should be straight forward. The 3D forward model is the focus of this work.

Simulating the neurons in Nest can lead to a time of computation problem. Consequently, this work will be translated to work on SpiNNaker neuromorphic hardware. Developing in PyNN, then using Nest as the backend supports a quick swap to emulating in SpiNNaker. This main concern is sending spikes to the board, as inputs, and retrieving spikes from the board, as information to the analytical controller. SpiNNaker's default time is a 1 ms cycle so that real world behaviour is explicitly linked to neural behaviour.

This is all to support a fast walking bipedal robot that is currently under development. Once the neural forward model is integrated with the robot, an exploration of other types of models and controllers can begin. Instead of implementing standard control and forward model equations into neurons, it is possible to more effectively use the dynamic nature of neurons to support control and forward modelling.

It is also possible to integrate learning within these systems. SpiNNaker supports learning in the 1 ms cycle. Eventually, an initial neural controller and forward model will be improved via learning, enabling the robot to walk more efficiently, and on a wider range of surfaces. It may also support the change of the robot through, for instance, wear on joints, motors and sensors.

5   Conclusion

It is useful to have a robot controller and forward model in simulated neurons. It can further understanding of neural processing, but can also lead to more efficient robots. The low energy usage of neuromorphic machines, and parallel processing capability can make the robot effective, and energy efficient. It can also support learning, leading to further improvements in effectiveness.

This paper has described a neural 2D cart and pole controller and forward model. This is a good start toward the development of a biped robot controller. It is implemented in reasonably accurate simulated neurons. It uses relatively straight forward neural implementations of standard control equations.

The controller and forward model are implemented in PyNN and Nest. This supports reuse, but also a translation from neural simulation, to neural emulation on neuromorphic hardware.

Future work includes simple engineering tasks to complex exploration. Engineering tasks include implementation of a 3D model and translation to SpiNNaker. Exploration includes neural implementations of recurrent controllers, instead of equations, learning, and mapping to animal behaviour.

Acknowledgments. This work has received funding from the UK EPSRC for A Neuromorphic Control System for Agile Biped Walking EP/P00542X/1 and the European Union's Horizon 2020 research and innovation programme under grant agreement No 720270 (the Human Brain Project).

References

1. Churchland, P., Sejnowski, T.: The Computational Brain. MIT Press, Cambridge (1999)
2. Davison, A., Brüderle, D., Eppler, J., Muller, E., Pecevski, D., Perrinet, L., Yqer, P.: PyNN: a common interface for neuronal network simulators. Front. Neuroinformatics **2**, 11 (2008)
3. Elman, J.: Finding structure in time. Cogn. Sci. **14**(2), 179–211 (1990)
4. Furber, S., Lester, D., Plana, L., Garside, J., Painkras, E., Temple, S., Brown, A.: Overview of the spinnaker system architecture. IEEE Trans. Comput. **62**(12), 2454–2467 (2013)
5. Gewaltig, M., Diesmann, M.: NEST (neural simulation tool). Scholarpedia **2**(4), 1430 (2007)
6. Iosa, M., Gizzi, L., Tamburella, F., Dominici, N.: Neuro-motor control and feed-forward models of locomotion in humans. Front. Hum. Neurosci. **9**, 306 (2015)
7. Kajita, S., Kanehiro, F., Kaneko, K., Yokoi, K., Hirukawa, H.: The 3D linear inverted pendulum mode: a simple modeling for a biped walking pattern generation. In: Proceedings of 2001 IEEE/RSJ International Conference, pp. 203–246 (2001)
8. Merolla, P., Arthur, J., Alvarez-Icaza, R., Cassidy, A., Sawada, J., Akopyan, F., Jackson, B., Imam, A., Guo, C., Nakamura, Y., Brezzo, B.: Cerebellar control of balance and locomotion. Sciene **345**(6197), 668–673 (2014)

9. Mori, S., Nishihara, H., Furuta, K.: Control of unstable mechanical system control of pendulum. Int. J. Control **23**(5), 673–692 (1976)
10. Morton, S., Bastian, A.: Cerebellar control of balance and locomotion. Neuroscientist **10**(3), 247–259 (2004)
11. Siciliano, B., Khatib, O.: Springer Handbook of Robotics. Springer, Heidelberg (2008)

A Formal Model of the Mechanism of Semantic Analysis in the Brain

Yuuji Ichisugi$^{(\boxtimes)}$ and Naoto Takahashi

National Institute of Advanced Industrial Science and Technology (AIST),
Central 1, Tsukuba, Ibaraki 305-8560, Japan
{y-ichisugi,naoto.takahashi}@aist.go.jp

Abstract. We propose a formal model of the mechanism of semantic analysis in the language areas of the cerebral cortex. The framework of Combinatory Categorial Grammar, a framework of grammar description in theoretical linguistics, is modified so that it does not use lambda calculus to represent semantic rules. This model uses a novel form of semantic representation named hierarchical address representation, and uses only fixed-length data structures. The knowledge of syntax and the knowledge of semantics are clearly separated in this model. Therefore, it is possible to reproduce disorders specific to syntax (utterance similar to Broca's aphasia) and disorders specific to semantics (utterance similar to Wernicke's aphasia) by disabling different modules in the model. We estimate that the model can be implemented using the Bayesian network model of the cerebral cortex that we have proposed earlier. We believe that this research will connect computational neuroscience and theoretical linguistics, and greatly evolve both of them.

1 Introduction

The language areas, which are considered to be centers of human language activities, are parts of the cerebral cortex. There is a hypothesis [3] that claims "the cerebral cortex is a kind of *Bayesian network* [1]." If so, we must be able to build a system that reproduces the behavior of the human language areas using a Bayesian network. Thus we aim at constructing a system that processes *Combinatorial Categorial Grammar (CCG)* [2], a framework of grammar description, using a Bayesian network [9]. As the first step towards the aim, we propose a formal model of the mechanism of semantic analysis in the brain based on CCG.

Theoretical linguistics is one field of linguistics that analyzes the characteristics of natural languages by mathematical methods. The relation between linguistics and theoretical linguistics resembles the relation between neuroscience and computational neuroscience. One purpose of theoretical linguistics is to find out some characteristics shared by all existing natural languages. Such characteristics can be considered as the characteristics of the information processing of the language areas in the brain.

Lambda calculus is usually used as a tool to describe semantic rules of CCG; however, it is difficult to handle variable-length data structures like lambda

© Springer Nature Switzerland AG 2019
A. V. Samsonovich (Ed.): BICA 2018, AISC 848, pp. 128–137, 2019.
https://doi.org/10.1007/978-3-319-99316-4_17

terms by Bayesian networks. The proposed model uses a novel form of semantic representation named *hierarchical address representation*, which does not use lambda terms. The model uses only fixed-length data structures. We estimate that the model can be implemented within the framework that we have proposed [5,9], which is based on Bayesian networks.

Although it is unknown whether the proposed method is applicable to the syntactic rules and the semantic rules of all the existing natural languages, we believe that this research will connect computational neuroscience and theoretical linguistics and greatly evolve both of them.

Section 2 briefly explains CCG. Section 3 describes the proposed model and examples of analysis of some sentences. Section 4 describes the correspondence between the modules of the model and some areas of the cerebral cortex, then we demonstrate reproduction of utterance of aphasia by disabling several modules.

2   Combinatory Categorial Grammar (CCG)

CCG is one of the most successful frameworks of grammar description. Its expressive power of grammar description is "mildly context-sensitive", which locates in between context-sensitive and context-free in the Chomsky hierarchy. Although the framework is very simple, grammars defined in CCG successfully explain many language phenomena (even though it is not complete). Therefore, we consider that CCG is the theory of information processing of the language areas in the brain.

In theoretical linguistics, some frameworks use *unification* as the core operation of analysis, and they are called *unification grammars*. CCG is one of unification grammars.

In CCG, general syntactic categories have structures that consist of ground categories (e.g., S for sentence, NP for noun phrase) combined by the operators "/" and "\". Theoretically, the length of a syntactic category is not restricted.

A ground category may have *syntactic features*. A syntactic feature may be a discrete variable whose value will be determined by the unification operations during the process of syntactic analysis. A ground category G with a syntactic feature F is denoted as G_F.

Production rules are defined by the form of *inference rules* in CCG. Syntactic analysis (i.e. parsing) is formalized as proof search showing a given word sequence being a sentence. A parse tree obtained as a result of syntactic analysis corresponds to a proof diagram.

The inference rules in CCG are accompanied by semantic rules that compose meaning of phrases. For example, the function which is the meaning of the word "black" : $\lambda x.black(x)$ is applied to the term that means "cats" : $cats$ to get the meaning of the phrase "black cats" : $black(cats)$. The semantic representation of the whole sentence is obtained by performing function applications (i.e. beta reductions) or function compositions sequentially from the leaves to the root, along with the parse tree.

In theoretical linguistics, including CCG, lambda calculus has been used as a tool for describing semantics; however, it is hard to imagine that the actual neural networks in the language areas have realized the complicated lambda calculus. In the following section, we propose a model of semantic analysis that does not use lambda calculus.

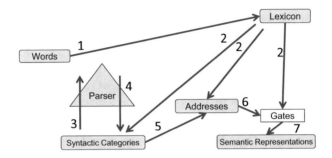

Fig. 1. The architecture of the model and typical information flow among the modules during the parsing process.

3   The Proposed Model

3.1   Scope of the Model

We aim at constructing a model of the mechanism for unconscious and instant interpretation of the superficial and literal meaning of comparatively simple sentences. Neither conscious interpretation of sentences with complicated structures nor presumption of unexpressed intention is a target of this model.

The length of word sequences, the length of syntactic categories, and the length of generated semantic representations are all limited. Although human can interpret a long sentence incrementally from the head, this model assumes that all words are given at once.

We adopt a simplified English grammar for a straightforward explanation of the behavior of the proposed model. Although the model accepts complex sentences consisting of two clauses connected with a subordinating conjunction ("if", "after", etc.) and sentences containing relative pronouns ("which", etc.), it is assumed that a subordinating conjunction or a relative pronoun appears at most once in a sentence.

3.2   The Outline of the Process of Semantic Analysis

In this model, syntactic analysis and semantic analysis are conducted simultaneously. Figure 1 shows a typical information flow among the modules during parsing. (1) First, a word sequence is given. (2) By referring to the lexicon, three data structures, a syntactic category, an *address* and a *semantic representation*, are obtained for each word. Variables may be unbound in syntactic features and in addresses at the time. (3) The parser merges syntactic categories and finds a parse tree in which the whole word sequence forms a sentence. During this process, unbound syntactic features receive a value through unification operations. (4) If a complete parse tree is generated, all values of syntactic features are determined. (5) All values of variables contained in the addresses are also determined. (6) A set of pairs of addresses and semantic representations that represents the meaning of the whole sentence is obtained. (7) Each semantic representation is written into the "memory" at the corresponding address.

Generally speaking, the information flow is not limited in the direction above. For example, when a prior knowledge for the meaning is given, the information will flow backwards from the semantic representation module to the parser module and the parser can use it to resolve lexical or syntactic ambiguities. Moreover, it is also possible to infer appropriate word sequences when a semantic representation is given. We show such an example in Sect. 4.

3.3 Hierarchical Address Representation, Inference Rules and Lexicon

In this model, the meaning of a sentence is represented as a set of pairs of addresses and semantic representations. We call this form "hierarchical address representation".

Each address is a tuple of the three values $(\mathbf{C}, \mathbf{R}, \mathbf{F})$. Each semantic representation is written into an address. All possible addresses in the prototype system are listed in Fig. 2.

The topmost hierarchy of an address, $\mathbf{C}$, represents the index of a clause in a complex sentence, which uses either a subordinate conjunction (e.g. "if") or a relative pronoun (e.g. "which"). The value of $\mathbf{C}$ is either $c1$ (for the first clause) or $c2$ (for the second). The type of the subordinate conjunction is written into the special address: $(sconj, -, -)$. For example, when the sentence starts with an "if", the semantic representation of "if" is written into the said special address.

The second hierarchy of an address, $\mathbf{R}$, primarily indicates the *semantic role* of a word in the sentence. In the current prototype system, its value is limited to either *agent* or *patient*, but it can be extended for other semantic roles (e.g. instrument, location, time) easily. Also, $\mathbf{R}$ can take the value *action* (for the semantic representation of a verb) or *modality* (for the semantic representation of an auxiliary verb).

When the value of $\mathbf{R}$ is *agent* or *patient*, i.e. when the word is a part of a noun phrase, the third hierarchy, $\mathbf{F}$, indicates a *feature* of the word. In the prototype system, the value of $\mathbf{F}$ is limited to *color*, *size*, and *entity*. $\mathbf{F}$ being *entity* means that the word is the head noun of a noun phrase.

The hierarchical address representation is inspired by a neuroscientific finding about the encoding of sentence meaning in the brain [8], which suggests that each semantic role (agent, patient) has its own representing place in the brain, regardless of the superficial voice (active or passive) of the sentence. We

Address	SR
$(sconj, -, -)$	if
$(c1, agent, size)$	big
$(c1, agent, color)$	*
$(c1, agent, entity)$	dogs
$(c1, modality, -)$	*
$(c1, action, -)$	chase
$(c1, patient, size)$	small
$(c1, patient, color)$	*
$(c1, patient, entity)$	mice
$(c2, agent, size)$	*
$(c2, agent, color)$	black
$(c2, agent, entity)$	cats
$(c2, modality, -)$	may
$(c2, action, -)$	eat
$(c2, patient, size)$	*
$(c2, patient, color)$	*
$(c2, patient, entity)$	mice

$$> \frac{X/Y \quad Y}{X} \qquad\qquad < \frac{Y \quad X\backslash Y}{X}$$

$$> B \frac{X/Y \quad Y/Z}{X/Z} \qquad < B \frac{Y\backslash Z \quad X\backslash Y}{X\backslash Z}$$

$$> T \frac{X}{T/(T\backslash X)} \qquad < T \frac{X}{T\backslash(T/X)}$$

Fig. 2. All addresses in the prototype system and semantic representations obtained as a result of the semantic analysis of the sentence "if small mice areChasedBy big dogs black cats may eat mice". Undetermined values are denoted by "*".

Fig. 3. The inference rules in the proposed model. These are identical to the inference rules of the usual CCG except that there are no semantic rules using lambda calculus.

suppose that fixing positions for all elements of meaning facilitates learning and processing of language for neural networks and Bayesian networks in the brain.

Figure 3 shows the inference rules in the proposed model. There are no semantic rules using lambda calculus. The semantic analysis is performed by merely unification operations of syntactic features as explained in the next subsection.

The *lexicon* is a set of *lexical items*. A lexical item in this model is represented as a tuple of four data structures (a word, a syntactic category, an address, and a semantic representation), as shown in Fig. 4.

3.4   Examples of Analysis

Figure 5 shows the parse tree of the sentence "black cats eat mice".

Let us explain the process of analyzing the phrase "black cats" in detail. First, by referring to the lexicon for these words, the corresponding syntactic categories $NP_{\mathbf{C_1},\mathbf{R_1}}/NP_{\mathbf{C_1},\mathbf{R_1}}$ and $NP_{\mathbf{C_2},\mathbf{R_2}}$ are obtained. Next, the inference rule ">" is chosen because this rule has the premises that unify with the obtained syntactic categories. Then, the inference rule is applied to the syntactic categories and the merged syntactic category $NP_{\mathbf{C_1},\mathbf{R_1}}$ is derived as a result. This derived syntactic category itself is merged with other syntactic categories as the analysis goes further.

Word	Syntactic Category	Address	SR
'if'	$(S_{c1}/S_{c2})/S_{c1}$	$(sconj, -, -)$	if
'black'	$NP_{\mathbf{C},\mathbf{R}}/NP_{\mathbf{C},\mathbf{R}}$	$(\mathbf{C}, \mathbf{R}, color)$	black
'big'	$NP_{\mathbf{C},\mathbf{R}}/NP_{\mathbf{C},\mathbf{R}}$	$(\mathbf{C}, \mathbf{R}, size)$	big
'cats'	$NP_{\mathbf{C},\mathbf{R}}$	$(\mathbf{C}, \mathbf{R}, entity)$	cats
'eat'	$(S_{\mathbf{C}}\backslash NP_{\mathbf{C},agent})/NP_{\mathbf{C},patient}$	$(\mathbf{C}, action, -)$	eat
'areEatenBy'	$(S_{\mathbf{C}}\backslash NP_{\mathbf{C},patient})/NP_{\mathbf{C},agent}$	$(\mathbf{C}, action, -)$	eat
'may'	$(S_{\mathbf{C}}\backslash NP_{\mathbf{C},\mathbf{R}})/(S_{\mathbf{C}}\backslash NP_{\mathbf{C},\mathbf{R}})$	$(\mathbf{C}, modality, -)$	may
'which'	$(NP_{c1,\mathbf{R}_1}\backslash NP_{c1,\mathbf{R}_1})/(S_{c2}\backslash NP_{c2,\mathbf{R}_2})$	$(c2, \mathbf{R}_2, entity)$	$(c1, \mathbf{R}_1, entity)$
'which'	$(NP_{c1,\mathbf{R}_1}\backslash NP_{c1,\mathbf{R}_1})/(S_{c2}/NP_{c2,\mathbf{R}_2})$	$(c2, \mathbf{R}_2, entity)$	$(c1, \mathbf{R}_1, entity)$

Fig. 4. Examples of lexical items contained in the lexicon. Each bold letter denotes an unbound variable. The same variables within a lexical item must have the same value at the end of the parsing. The first lexical item of "which" is used for the nominative case, and the second one is for the objective case.

The analysis progresses in such a way until the syntactic category $S_{\mathbf{C}}$ (sentence) is finally derived. At that time, the proof of the whole word sequence being a sentence is completed. By the unifications performed during the process of analysis, the semantic roles $\mathbf{R}_2$ (for "cats") and $\mathbf{R}_4$ (for "mice") are determined as *agent* and *patient*, respectively. Then, the address where the semantic representation of each word should be written is determined. The addresses and the semantic representations finally obtained are shown in Fig. 7.

Figure 6 shows a part of the parse tree of the sentence "mice which dogs chase areEatenBy cats" that uses a relative pronoun of the objective case. The analysis of this sentence requires the type raising rule "$> T$". The semantic representation of the relative pronoun "which" is the address of its antecedent (Fig. 8).

$$\dfrac{\dfrac{\overset{\text{black}}{NP_{\mathbf{C}_1,\mathbf{R}_1}/NP_{\mathbf{C}_1,\mathbf{R}_1}} \quad \overset{\text{cats}}{NP_{\mathbf{C}_2,\mathbf{R}_2}}}{NP_{\mathbf{C}_1,\mathbf{R}_1}}> \qquad \dfrac{\overset{\text{eat}}{(S_{\mathbf{C}_3}\backslash NP_{\mathbf{C}_3,agent})/NP_{\mathbf{C}_3,patient}} \quad \overset{\text{mice}}{NP_{\mathbf{C}_4,\mathbf{R}_4}}}{S_{\mathbf{C}_3}\backslash NP_{\mathbf{C}_3,agent}}>}{S_{\mathbf{C}_3}}<$$

Fig. 5. A parse tree (proof diagram) of the sentence "black cats eat mice". Although the variables $\mathbf{R}_i$, semantic roles, are unbound at the beginning, their values are determined as $\mathbf{R}_1 = \mathbf{R}_2 = agent$ and $\mathbf{R}_4 = patient$ by unification operations during parsing. A parse tree of the sentence "black cats areEatenBy mice" results in the same form but the semantic roles of *agent* and *patient* are exchanged. The variables $\mathbf{C}_i$ are left unbound after parsing; however, they are restricted to have the same value, i.e. $\mathbf{C}_1 = \mathbf{C}_2 = \mathbf{C}_3 = \mathbf{C}_4$.

$$> \frac{(NP_{c1,\mathbf{R}_{21}}\backslash NP_{c1,\mathbf{R}_{21}})/(S_{c2}/NP_{c2,\mathbf{R}_{22}})\quad >B\ \dfrac{>T\ \dfrac{NP_{\mathbf{C}_3,\mathbf{R}_3}}{\mathbf{T}_3/(\mathbf{T}_3\backslash NP_{\mathbf{C}_3,\mathbf{R}_3})}\quad (S_{\mathbf{C}_4}\backslash NP_{\mathbf{C}_4,agent})/NP_{\mathbf{C}_4,patient}}{S_{\mathbf{C}_4}/NP_{\mathbf{C}_4,patient}}}{NP_{c1,\mathbf{R}_{21}}\backslash NP_{c1,\mathbf{R}_{21}}}$$

where "which" sits above the left leaf and "dogs" above $NP_{\mathbf{C}_3,\mathbf{R}_3}$ and "chase" above the right leaf.

Fig. 6. A parse tree of the word sequence "which dogs chase". In this relative clause, whose index is $\mathbf{C}_3 = \mathbf{C}_4 = c2$, the semantic roles of "dogs" and "which" become $\mathbf{R}_3 = agent$ and $\mathbf{R}_{22} = patient$, respectively. The index of the main clause is determined as $c1$, but the semantic role of the antecedent $\mathbf{R}_{21}$, is determined only when the whole sentence has been analyzed.

Address	SR
$(c1, agent, color)$	black
$(c1, agent, entity)$	cats
$(c1, action, -)$	eat
$(c1, patient, entity)$	mice

Fig. 7. The pairs of addresses and semantic representations obtained by analyzing the sentence "black cats eat mice". Because the value of the address $\mathbf{C}$ is arbitrary, we set it as $c1$.

Address	SR
$(c1, agent, entity)$	cats
$(c1, action, -)$	eat
$(c1, patient, entity)$	mice
$(c2, agent, entity)$	dogs
$(c2, action, -)$	chase
$(c2, patient, entity)$	$(c1, patient, entity)$

Fig. 8. The pairs of addresses and semantic representations obtained by analyzing the sentence "mice which dogs chase areEatenBy cats".

4   Correspondence to Cortical Areas

A possible correspondence between the modules in the model (Fig. 1) and cortical areas is shown in Fig. 9. Broca's area (Brodmann Areas 44 and 45) participates in grammar processing and Wernicke's area (BA22; close to the primary auditory area (BA41,42) and to the angular gyrus (BA39)) participates in association between speech sounds and concepts [7]. The parser in the proposed model is the module of grammar processing, thus it is matched with Broca's area. Because it is suggested that agents and patients are represented at the left mid-superior temporal gyrus [8], we suppose that the human's module for semantic representations is located there, around BA22.

By "disabling" a part of the model, utterance that is similar to Broca's aphasia or to Wernicke's aphasia can be reproduced. Although the symptoms of aphasia [7] is complicated and largely vary from patient to patient, we give simple explanation below. Broca's aphasia arises from damage to Broca's area. Its utterance consists of scattering words that do not constitute sentences. The Wernicke's aphasia arises from damage to the Wernicke's area. Its utterance is fluent but does not make sense because of mistakenly selected words.

The proposed model has been implemented in the Prolog language. First, we show an example of a normal behavior of sentence generation. For this example,

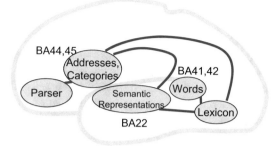

Fig. 9. A possible correspondence between the modules in the model (Fig. 1) and cortical areas.

the semantic representation shown in Fig. 7, which means "black cats eat mice", is given. If the model infers all possible sentences that consist of four words, the following two sentences are obtained as solutions.

```
black cats eat mice
mice areEatenBy black cats
```

In the same condition but without the parser module, the obtained solutions include syntactically incorrect word sequences; however, only those words that are semantically suitable are chosen (Fig. 10(a)). This phenomenon is essentially the same one seen in utterance of Broca's aphasia.

If the model infers all possible sentences that consist of four words without giving concrete semantic representation, all syntactically correct sentences will be obtained as solutions (Fig. 10(b)). This phenomenon is essentially the same one seen in utterance of Wernicke's aphasia.

In the proposed model, the parser module processes only the *addresses* where meanings are written, but does not process the semantic representations themselves. Because the knowledge of syntax and the knowledge of semantics are clearly separated in this architecture, we can reproduce syntactic disorder and semantic disorder, like actual aphasia.

```
(a)                                 (b)
black black black black             white dogs eat dogs
black black black cats              white dogs eat cats
black black black mice              white dogs eat mice
black black black eat               white dogs chase dogs
black black black areEatenBy        white dogs chase cats
black black cats black              white dogs chase mice
black black cats cats               white dogs areEatenBy dogs
black black cats mice               white dogs areEatenBy cats
 . . .                               . . .
```

Fig. 10. Inferred word sequences similar to (a) Broca's aphasia and (b)Wernicke's aphasia.

5   Related Work

We aim at the model of the language processing that can be realized easily in the form of Bayesian networks or neural networks like the cerebral cortex. The essential difference from the conventional frameworks of the formal semantics is a complete exclusion of variable-length data structures. Although unification operations are occasionally used to express semantic rules, tree structures have been used to express semantic representations. MRS (Minimal Recursion Semantics) [4] expresses meanings not with a tree structure but with a flat structure; however, it needs to handle variable-length data structures.

There are some systems that parse natural language efficiently using loopy belief propagation. For example, the system in [6] is an efficient CCG parser; however, it does not include semantic analysis.

6   Conclusion

We proposed a model of the mechanism of the semantic analysis that does not use variable-length data structure (e.g. lambda terms) but uses a novel form of semantic representation named hierarchical address representation. The modules of the model have correspondence to cortical areas in the brain.

We can reproduce utterance of aphasia by "disabling" some modules in the model.

A chart parser for context-free grammar can be realized as a Bayesian network [9]. It should be possible to apply this method to CCG. Moreover, the mechanism of the unification that the proposed model uses is also easily realizable as Bayesian networks. If the whole model is realized as a Bayesian network, it is possible that lexical items and inference rules can be learned from pairs of word sequences and semantic representations. We believe that this research will connect computational neuroscience and theoretical linguistics, and greatly evolve both of them.

Acknowledgments. We greatly thank to Dr. Daisuke Bekki, who gave useful comments to this research.

References

1. Pearl, J.: Probabilistic Reasoning in Intelligent Systems: Networks of Plausible Inference. Morgan Kaufmann, Burlington (1988)
2. Steedman, M.: The Syntactic Process. The MIT Press, Cambridge (2000)
3. Lee, T.S., Mumford, D.: Hierarchical Bayesian inference in the visual cortex. J. Opt. Soc. Am. A **20**(7), 1434–1448 (2003)
4. Copestake, A., Flickinger, D., Pollard, C., Sag, I.A.: Minimal recursion semantics. An introduction. Res. Lang. Comput. **3**, 281–332 (2005)
5. Ichisugi, Y.: The cerebral cortex model that self-organizes conditional probability tables and executes belief propagation. In: Proceedings of IJCNN 2007, pp. 1065–1070 (2007)

6. Auli, M., Lopez, A.: A comparison of loopy belief propagation and dual decomposition for integrated CCG supertagging and parsing. In: Proceedings of ACL, pp. 470–480 (2011)
7. Kandel, E.R., et al. (eds.) Principles of Neural Science, 5th edn. McGraw-Hill Companies, New York City (2012)
8. Frankland, S.M., Greene, J.D.: An architecture for encoding sentence meaning in left mid-superior temporal cortex. Proc. Natl. Acad. Sci. USA **112**, 11732–11737 (2015)
9. Takahashi, N., Ichisugi, Y.: Restricted quasi Bayesian networks as a prototyping tool for computational models of individual cortical areas. In: Proceedings of Machine Learning Research (AMBN 2017), vol. 73, pp. 188–199 (2017)

Group Determination of Parameters and Training with Noise Addition: Joint Application to Improve the Resilience of the Neural Network Solution of a Model Inverse Problem to Noise in Data

Igor Isaev$^{(\boxtimes)}$ and Sergey Dolenko

D.V.Skobeltsyn Institute of Nuclear Physics,
M.V.Lomonosov Moscow State University, Moscow 119991, Russia
isaev_igor@mail.ru, dolenko@sinp.msu.ru

Abstract. Solution of inverse problems is usually sensitive to noise in the input data, as problems of this type are usually ill-posed or ill-conditioned. While neural networks have high noise resilience by themselves, this may be not enough in case of incorrect inverse problems. In their previous studies, the authors have demonstrated that the method of group determination of parameters, as well as noise addition during training of a neural network, can improve the resilience of the solution to noise in the input data. This study is devoted to the investigation of joint application of these methods. It has been performed on a model problem, for which the direct function is set explicitly as a polynomial.

Keywords: Artificial neural networks · Perceptron · Noise resilience
Multi-parameter inverse problems · Group determination of parameters

1 Introduction

Inverse problems (IPs) are a very important class of data analysis problems, as nearly any problem of indirect measurements belongs to this class. Examples of IPs can be easily found in the areas of geophysics [1], spectroscopy [2], various types of tomography [3], and many others.

A number of specific properties significantly complicate IP solution. Usually they are nonlinear, and they often have high input dimension and high output dimension (i.e. they are multi-parameter problems). In general, IPs have no analytical solution, so in most cases they are solved numerically. Conventional methods of IP solution, matrix-based and optimization-based, do not cope well with non-linearity, and have a high computational cost. Approximation methods of IP solution are an alternative that is

This study has been performed at the expense of the grant of Russian Science Foundation (Project No. 14-11-00579).

© Springer Nature Switzerland AG 2019
A. V. Samsonovich (Ed.): BICA 2018, AISC 848, pp. 138–144, 2019.
https://doi.org/10.1007/978-3-319-99316-4_18

usually free from these shortcomings. As an example of such methods, in this paper we consider artificial neural networks (ANN) of multilayer perceptron (MLP) type.

As in most cases IPs are ill-posed or ill-conditioned their solutions have high sensitivity to noise in the input data, both for traditional methods and for ANN. At the same time, the IP solutions will almost always have to deal with noisy data, because any measurements are characterized by some measurement error. Therefore, the development of some approaches to improve the resilience of the IP solution to noise in the input data is an urgent task.

While MLP type ANN have high noise resilience by themselves, this may be not enough in case of incorrect inverse problems. This study continues the main idea set in a number of previous works of the authors [4–7]: development of new approaches to improve the resilience of ANN solutions of multi-parameter inverse problems to noise in the input data.

In some cases, resilience of ANN solution of an IP to noise in the input data can be achieved by simultaneous determination of a group of parameters [4, 5]. As a rule, the higher is the level of the noise, the more pronounced is the effect of using this approach.

Another useful method to improve the trained network in various respects is adding noise during MLP type ANN training. The basis for use of this method was founded in [8, 9], where it was demonstrated that it can improve the generalizing capabilities of the network. In [10] it was shown that use of this method is equivalent to Tikhonov regularization. It can also be used to prevent MLP overtraining [11, 12], as well as to speed up learning [13]. In [6, 7], adding noise during training allowed the authors to increase noise resilience of trained MLP type ANN to noise in the input data.

In this study, we investigate the efficiency of simultaneous application of both methods. The study was performed at the example of a multi-parameter model IP [6], for which the direct function was set explicitly as a polynomial dependence.

2 Problem Statement

The model inverse problem will be defined with the help of a simple polynomial dependence for the direct function, with coefficients a_{ij}, b_{ij}, c_{ijk}, d_{ij}, e_i, where each "observed value" $x_1 \dots x_5$ depends on all the "determined parameters" $y_1 \dots y_5$ according to the following formula (Fig. 1):

$$x_i = \sum_{j=1}^{5} a_{ij} \cdot y_j^3 + \sum_{j=1}^{5} b_{ij} \cdot y_j^2 + \sum_{j=1}^{4} \sum_{k=j+1}^{5} c_{ijk} \cdot y_j \cdot y_k + \sum_{j=1}^{5} d_{ij} \cdot y_j + e_i$$

Here, the determined parameters $y_1 \dots y_5$ take random values from 0 to 1. These formulas were used for generation of the initial data set, containing 3000 patterns and randomly divided into training (to train the MLP), validation (to prevent overtraining by stop on minimum error), and test (for out-of-sample testing) sets in the ratio of 70:20:10. This model problem is described in more detail in [6].

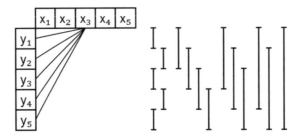

Fig. 1. Left—diagram of the model problem. Right—variants of parameter grouping for group determination of parameters.

Two types of noise were considered: additive and multiplicative, and two kinds of noise statistics: uniform (uniform noise distribution) and Gaussian (normal noise distribution). Each observed value was contaminated with noise of a certain type, statistics and level (the considered level values were: 1%, 3%, 5%, 10%, 20%). The procedure of adding noise was described in more detail elsewhere [4–7].

3   Use of Neural Networks

In this study, we used one of the most widespread ANN architectures—a multilayer perceptron (MLP). The MLPs had three hidden layers with 16, 12 and 8 neurons in the 1st, 2nd and 3rd hidden layers, respectively. The activation function was logistic in the hidden layers, and linear in the output layer. Training was carried out by the method of stochastic gradient descent. Each MLP was trained 5 times with various weights initializations. Statistics of application of these 5 ANN were averaged.

To prevent overtraining of neural networks, the method of early stop was used: training was stopped at the minimum of the mean squared error on the validation set (after 500 epochs without improving the result). Independent evaluation of the results was performed on the test (out-of-sample) sets, without noise and with noise of certain level, type and statistics. According to the ratio specified above (70:20:10), the training data set contained 2100 patterns, the validation set—600 patterns, the test set without noise—300 patterns.

Solving the considered multi-parameter inverse problem, we considered autonomous and group determination of parameters. *Autonomous determination* means solution of the problem individually for each parameter, training a separate single-output ANN. For *group determination*, the problem was solved for several parameters simultaneously; we grouped adjacent parameters (Fig. 1). All possible sizes of the grouping window (from 2 to 5 parameters) were considered. Both these methods of parameter determination were applied both independently and in conjunction with the method of training with noise.

The method of training with noise was implemented by using training data sets containing a certain level of noise; each of them was created from the initial training set by applying 10 various random noise implementations to each pattern; thus, each "noisy" training set included 21,000 patterns. With such contradictory data, in principle it is possible to abandon the validation data set, because noise addition is in itself a method of preventing overtraining [11, 12]. If the validation set is still used, it should contain noise with the same noise type and noise statistics and with the same noise level as in the training set. Also, it is possible to use a validation set that does not contain noise. In [6] it was shown that the optimal method of training was when training was performed on a training set that contained noise, and the training was stopped on a validation set without noise. In this case the quality of the solution was higher, and the training time—lower. This method was the one used in the present study.

In the case of test sets containing noise, each pattern of these sets was also presented in 10 noise implementations. So, the size of the sets was: training—21,000 patterns, test—3000 patterns each. The validation set was left unchanged. Each set contained noise of certain level, type and statistics. Including a set without noise, there were total 21 training and 21 out-of-sample (test) data sets: 5 noise levels × 2 noise types × 2 kinds of statistics + 1 = 21. When using training with noise, ANN trained with noise were applied to test sets with the same noise type and noise statistics. In the case of noise-free training, neural networks were applied to all test sets.

4 Results

Figure 2 shows the dependence of the solution quality (RMSE) for the parameter y_l, separately for the method of training with noise (lines) and for the method of group determination of parameters (markers), on the noise level in the test set. For the method of adding noise during training, it can be seen that the higher is the noise level in the training data set, the worse the network performs on data without noise, but the slower it degrades with increasing noise level. Also, group determination of parameters in some cases allows increasing the resilience of the solution. The higher is the noise level in the test set, the higher is the effect of using this method. However, the method itself performs significantly worse than the method of training with noise.

Top row of diagrams in Fig. 3 (training set noise level specified in the insets) shows that the combined use of group determination and training with noise (markers) can improve the resilience of the solution relative to use of only the method of training with noise (lines). As in the case of group determination only, this approach has a more pronounced effect at high noise levels in the test set. It can also be noted that the lower is the noise level in the training set, the more noticeable is the effect. Bottom row of the diagrams in Fig. 3 (test set noise level specified in the insets) shows that the highest quality of the solution, as for the method of training with noise only, as for both methods used together, is observed when the noise level in the training set is equal to that in the test set. It can also be seen that the effect of the joint application of methods is mostly observed at high test set noise levels.

For other parameters and other types and statistics of noise, the behavior of all the dependencies is completely the same.

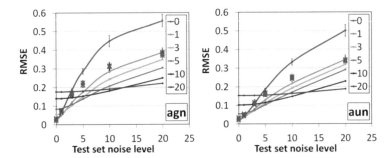

Fig. 2. The dependence of the quality of the solution (RMSE) for the parameter y_1 on the noise level in the test set for additive noise with different noise statistics. Red lines represent the original solution (no noise and no grouping), other line colors represent the method of adding noise to the training patterns; markers show the results of group determination.

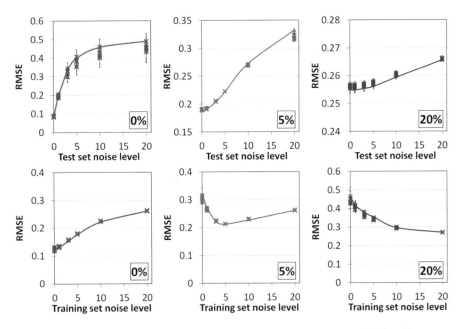

Fig. 3. Effect of joint use of adding noise during training and group determination. Top row—RMSE on test set *vs* test set noise level, for various levels of noise in the training set (specified in the insets), parameter y_2, additive Gaussian noise. Bottom row—RMSE on test set vs training set noise level, for various levels of noise in test set (specified in the inlets), parameter y_4, multiplicative uniform noise. Lines—only method of adding noise during training, markers—joint use of both methods.

5   Conclusions

In this study, it has been confirmed on a model inverse problem that the method of adding noise during training and the method of group determination of parameters are both efficient in improving noise resilience of the ANN solution of the inverse problem. The following conclusions were obtained:

- When using the method of adding noise during MLP training, the higher is the noise level in the training data set, the worse the network performs on the data without noise, but the slower it degrades with increasing noise level in the test set.
- Joint use of both studied methods improves noise resilience of the solution compared to use of only the method of training with noise.
- When using group determination, both in the case of joint use with the method of training with noise and in the case of individual use, the effect is more pronounced at high noise levels in the test set and low noise levels in the training set.
- The highest quality of the solution for the method of training with noise (alone or together with group determination) is observed when the noise level in the training and test sets is the same.

Since the methods of group determination of parameters and of training with noise were successfully used also for solving other inverse problems, it can be concluded that the observed effects are a property of the perceptron as a data processing algorithm, and they are not determined by the properties of the data.

References

1. Zhdanov, M.: Inverse Theory and Applications in Geophysics, 2nd edn. Elsevier, Amsterdam (2015). 730 pp
2. Yagola, A., Kochikov, I., Kuramshina, G.: Inverse Problems of Vibrational Spectroscopy. De Gruyter, Berlin (1999). 297 pp
3. Orlov, V., Zinchenko, V., Ushakov, V., Velichkovsky, B.: Physiological noise reduction algorithms for fMRI data. Procedia Comput. Sci. **123**, 334–340 (2018)
4. Isaev, I.V., Obornev, E.A., Obornev, I.E., Shimelevich, M.I., Dolenko, S.A.: Increase of the resistance to noise in data for neural network solution of the inverse problem of magnetotellurics with group determination of parameters. Lect. Notes Comput. **9886**, 502–509 (2016)
5. Isaev, I.V., Vervald, E., Sarmanova, O., Dolenko, S.A.: Neural network solution of an inverse problem in Raman spectroscopy of multi-component solutions of inorganic salts: group determination as a method to increase noise resilience of the solution. Procedia Comput. Sci. **123**, 177–182 (2018)
6. Isaev, I.V., Dolenko, S.A.: Training with noise as a method to increase noise resilience of neural network solution of inverse problems. Opt. Mem. Neural Netw. **25**(3), 142–148 (2016)
7. Isaev, I.V., Dolenko, S.A.: Adding noise during training as a method to increase resilience of neural network solution of inverse problems: test on the data of magnetotelluric sounding problem. Stud. Comput. Intell. **736**, 9–16 (2018)

8. Holmstrom, L., Koistinen, P.: Using additive noise in back-propagation training. IEEE Trans. Neural Netw. **3**(1), 24–38 (1992)
9. Matsuoka, K.: Noise injection into inputs in back-propagation learning. IEEE Trans. Syst. Man Cybern. **22**(3), 436–440 (1992)
10. Bishop, C.M.: Training with noise is equivalent to Tikhonov regularization. Neural Comput. **7**(1), 108–116 (1995)
11. An, G.: The effects of adding noise during backpropagation training on a generalization performance. Neural Comput. **8**(3), 643–674 (1996)
12. Zur, R.M., Jiang, Y., Pesce, L.L., Drukker, K.: Noise injection for training artificial neural networks: a comparison with weight decay and early stopping. Med. Phys. **36**(10), 4810–4818 (2009)
13. Wang, C., Principe, J.C.: Training neural networks with additive noise in the desired signal. IEEE Trans. Neural Netw. **10**(6), 1511–1517 (1999)

A Computational Model for Supporting Access Policies to Semantic Web

Larisa Yu. Ismailova[1(✉)], Viacheslav E. Wolfengagen[1(✉)],
and Sergey V. Kosikov[2(✉)]

[1] National Research Nuclear University "MEPhI"
(Moscow Engineering Physics Institute),
Moscow 115409, Russian Federation
lyu.ismailova@gmail.com, jir.vew@gmail.com
[2] Institute for Contemporary Education "JurInfoR-MGU",
Moscow 119435, Russian Federation
kosikov.s.v@gmail.com

Abstract. The paper discusses a solution to the problem of data storage in a Web environment and providing access to the data based on their semantics. The problem involves restricting access to data in accordance with the description of the access rights for different classes of users. The information is interpreted in different ways according to the semantics assigned to users. Access policies are proposed as the main technical tool for describing access rights. The provided information is consistent with the semantic description of the user accessing the information. The semantic matching of descriptions of users and the data representation combines formal and informal moments and in general is of a cognitive character.

The proposed solution is based on the use of data representation in the form of semantic networks. The solution has computational nature and involves calculating the value of the structures of a specialized control semantic network describing access policies. The solution is presented in the form of a computational model, which is based on the intensional logic. The constructions of the model include both a logical means of general type and specialized structures for control the degree of intensionality. The model provides calculation of construction values depending on the parameter - assignment point. The proposed method of parameter assigning provides the possibility of taking into account the semantic characteristics of users of different classes, as well as a number of other factors essential to support the semantic network (including the presence of different versions).

The article offers the architecture of the instrumental complex to support access policies that support the developed computational model, and describes its main components. The components were tested in solving practical problems in the field of jurisprudence to ensure the manipulation and visualization of concepts.

© Springer Nature Switzerland AG 2019
A. V. Samsonovich (Ed.): BICA 2018, AISC 848, pp. 145–154, 2019.
https://doi.org/10.1007/978-3-319-99316-4_19

Keywords: Information objects · Semantics
Computational model · Semantic network · Intensional logic
Access operations

1 Introduction

The problem to develop the means of organizing and storing data in the Web environment, considered in [1], keeps remaining relevant. The Web technologies development is going on, necessitating the development of means of organizing and storing data in the Web environment. The need to operate semantically saturated network structures requires the development of semantically oriented methods for describing data structures and their processing, in particular, the definition of semantically oriented methods of search. Different users must work with different data fragments, which are determined by the objectives of their work, by the set of appropriate authorities, etc. Due to this fact there arises the problem to provide an access to data, accounting for both the class of the user that receives access, and his characteristics, and the semantics of data to which access is given, in particular, providing convenient for the user data representation.

The papers on semantically oriented representation of data [2,3] show that the considered subtasks can be solved together. For this, at the conceptual level, the user should be included in the conceptual model of the domain as part of it. Practically the model should include constructions that describe the relationship between the semantics of user characteristics and the semantics of the processed data and provide the user with data for processing in accordance with the set of his authorities and the nature of the problem under solution. When presenting the model as a whole in the form of a semantic network its considered part can be represented in the form of a control subnet that provides the calculation of the query results in accordance with the specified parameters.

The need for data processing including the calculation of requests involves the development of a theoretical basis for building a computer system that ensures both logical correctness of the data interpretation and the construction of appropriate computational procedures. The given paper uses the formalism for this purpose, developed on the basis of the intensional logic [4]. The computing aspect is provided, in particular, by the possibility of including the means of a typical lambda-calculus in the considered logical system.

This paper is a direct follow-up and development of the paper [1]. In particular, the model developed in this paper is based on the tools and interpretations proposed in [1]. The work is structured as follows. Section 2 describes some approaches to solving the problem of organizing and storing data, in particular, there are singled out classes of subject areas that are of some interest from the point of view of access organization. Section 3 contains a more specified task to develop a model for supporting operations of access to the semantic network. Section 4 represents the used system of intensional logic based on the work [4] and which is developed further by the authors, in particular, in [1]. Section 5 describes a model of the semantic network and the tools based on it. Section 6 contains some results and a brief discussion of the prospects for further work.

2   Related Work

The questions of organization of access to semantic network constructions in connection with the semantic characteristics of the allocated classes of users, including those with different access rights, were intensively studied in various aspects. Thus, the monograph [5] contains a broad overview of the area for data confidentiality support. Three classes of problems are discussed: (a) statistical control of data, (b) extraction of data with confidentiality maintain and (c) technologies of preservation/enhancement of confidentiality. Models for maintaining confidentiality and classification of protection procedures are proposed. The paper [6] also discusses some issues of controlled access to data.

More specific individual solutions are also proposed. The work [7] considers the methods of control of access to information in the conditions of its storage in a heterogeneous database. In this case, it appears relatively easy to provide access control to the relational part of the information (i.e., the information that is stored in the relational database). It is much more difficult to provide access control to the content of a document-oriented database. The paper proposes a means of providing controlled access to the content of documents in the database depending on the user's privileges.

The work [8] considers the methods of accounting the restrictions on values of data attributes when formulating and further optimizing the requests. A similar technique can be used to take into account the semantic limitations of access to data, but this task is not directly considered in this paper. The work [9] considers the issues on the access to data for one special class of DBMS. The techniques for processing requests are discussed, but systematic accounting of data semantics is not performed. The work [10] discusses the technique of parametrization of requests to a DB, compared on set of various criteria.

In the presented solutions, as a rule, either there is no attempt to connect the controlled access procedures with the semantics of the processed data, or this connection is of a partial nature. As a rule, researchers focus more on performance issues than on questions of semantic correctness. Taking into account the need to ensure correctness, some certain techniques for converting requests can be included in the approach developed in this paper.

The main area of application of the proposed solutions is currently the information support to legal applications. Due to this the interest is caused by the publication [11], which is a collection of papers on the interconnection of legal and technical aspects of access control to information. This publication contains discussions of various aspects of confidentiality, but it also does not rely on semantic methods.

3   Task of Developing a Model for Supporting Access Operations

The study of the possibilities of computational support to the access operations to semantic network structures [1], as well as the characteristic features of subject areas requiring support to various access variants [2,3], allows us to clarify the task of developing a support model for access operations to the semantic network. The general formulation of the problem was given earlier in [1]. However, a more detailed development of the basic constructions of the model makes it possible to formulate the following specified requirements to it:

- support to classes of users interacting with the semantic network by defining the intensional predicates and using problem-oriented axioms that describe the requirements for safe operation of the system;
- support to determining the applicative constants in the form of higher order combinators that specify the required characteristics of user classes and/or their calculation methods;
- definition of ways of generation of assignment points, coordinated with the set problem-oriented axioms, that provides setting of access policies to the system in the form of sets of interconnected points of assignment;
- definition of ways of grouping the assignment points on the basis of intensional operators with the possibility of checking the properties of the grouped assignment points.

In general, the formulated requirements are aimed at increasing the stability of the domain model of the information system, represented as a semantic network, especially when working in the Web environment. The system as a whole appears at the same time to be capable of separating the concepts represented by the constructions of the semantic network. The separation involves the separation of concepts based on the inclusion in their definitions of constants that define the pragmatic characteristics of users and their ranking, which provides the construction of a concept representation in the form of a complete lattice describing the extension of the access right to the concept and related semantic network constructions. Thanks to the possibility of generating assignment points, ensuring completeness is a non-trivial task, the solution of which is achieved through the use of applicative mechanisms of interpretation [2].

4   Model of Access to the Semantic Network

The model of access to the semantic network is constructed by means of the intensional logic. Every given construction of the semantic network (concept or frame) corresponds to the correct expression of the intensional logic. Then the interpretation of the expression gives the semantic of the network construction.

4.1   Types

The set of Y types is defined inductively:

(i) $e, t \in Y$;
(ii) if $a, b \in Y$, then $\langle a, b \rangle \in Y$ and $\langle s, a \rangle \in Y$;
(iii) no other types.

The type e is interpreted as the type of entities, type t - as type of sentences, type s - as type of meanings.

4.2   Language

We'll use the enumerable set of variables and the (infinite) set of constants of each type a. If n is a natural number and $a \in Type$, then $v_{n,a}$ is n-th variable of type a, and Con_a is a set of constants of type a.

The language includes a set of correct expressions E_a for every type a. It is defined recursively:

1. $v_{n,a} \in E_a$; $Con_a \subseteq E_a$;
2. If $\alpha \in E_b$ and u is a variable of type a, then $\lambda u \alpha \in E_{\langle a,b \rangle}$;
3. If $\alpha \in E_{\langle a,b \rangle}$ and $\beta \in E_a$, then $\alpha(\beta) \in E_b$;
4. If $\alpha, \beta \in E_a$, then $\alpha = \beta \in E_t$;
5. If $\varphi, \psi \in E_t$ и u"—is a variable, then $\neg\varphi$, $[\varphi \wedge \psi]$, $[\varphi \vee \psi]$, $[\varphi \rightarrow \psi]$, $[\varphi \leftrightarrow \psi]$, $\forall u \psi$, $\exists u \psi$, $\square \psi \in E_t$;
6. If $\alpha \in E_a$, then $[\hat{}\alpha] \in E_{\langle s,a \rangle}$;
7. If $\alpha \in E_{\langle s,a \rangle}$, then $[\check{}\alpha] \in E_a$;
8. Other correct expressions do not exist.

4.3   Interpretation

Now we'll introduce the interpretation of the intensional language. Suppose D be the set (set of entities or individuals), Asg—set of possible worlds, $D \cap Asg = \emptyset$. Let's define the set $D_{a,D,Asg}$ of possible individuals of type a:

$$D_{e,D,Asg} = D,$$
$$D_{t,D,Asg} = \{0, 1\},$$
$$D_{\langle a,b \rangle,D,Asg} = D_{b,D,Asg}{}^{D_{a,D,Asg}},$$
$$D_{\langle s,a \rangle,D,Asg} = D_{a,D,Asg}{}^{Asg}.$$

As a rule, further on within every separate example D, Asg will be fixed. In this conditions we'll designate as $D_{a,D,Asg} \equiv D_a$.

The interpretation (or intensional model) is understood as the ordered five

$$\mathfrak{A} = <D, Asg, F>,$$

where D, Asg – are non-empty sets, $D \cap Asg = \emptyset$, and F is the function, which area of definition is the set of constants. F is defined so that if $a \in Type$ and $\alpha \in Con_a$, then $F(\alpha) \in D_a{}^{Asg}$.

The $\mathfrak{A}$-assignment is understood as the function g, which area of definition is the set of variables, such that if u is a variable of type a, then $g(u) \in D_a$.

Under $g[x/u]$ is understood $\mathfrak{A}$-assignment

$$g[x/u](v) = \{x, \text{ if } u \equiv v, g(v) \text{ in other circumstances}.$$

Let's determine the intensional $\alpha^{\mathfrak{A},g}$ and the extensional $\alpha^{\mathfrak{A},i,g}$ of correct expression α with the help of the following recursive definition.

1. If α – is a constant, then $\alpha^{\mathfrak{A},i,g} = F(\alpha)(i)$.
2. If α is a variable, then $\alpha^{\mathfrak{A},i,g} = g(\alpha)$.
3. If $\alpha \in E_\alpha$ and u – is a variable of type b, then $[\lambda u \alpha]^{\mathfrak{A},i,g} = h$, where h – is such function with the area of definition $D_{a,D,Asg}$, that $h(x) = \alpha^{\mathfrak{A},i,g[x/u]}$.
4. If $\alpha \in E_{\langle a,b \rangle}$ and $\beta \in E_a$, then $\alpha(\beta)^{\mathfrak{A},i,g} = \alpha^{\mathfrak{A},i,g}(\beta^{\mathfrak{A},i,g})$.
5. If $\alpha, \beta \in E_a$, then $[\alpha = \beta]^{\mathfrak{A},i,g} = 1$ if and only if ("iff" below) $\alpha^{\mathfrak{A},i,g}$ coincides with $\beta^{\mathfrak{A},i,g}$.
6. If $\varphi \in E_t$, then $[\neg\varphi]^{\mathfrak{A},i,g} = 1$ iff $\varphi^{\mathfrak{A},i,g} = 0$ and similarly for other propositional connections.
7. If $\varphi \in E_t$ и u is a variable of type a, then $[\exists u \varphi]^{\mathfrak{A},i,g} = 1$ iff exists $x \in D_{a,D,Asg}$ such, that $\varphi^{\mathfrak{A},i,g[x/u]} = 1$ and it is similar for $[\forall u \varphi]$.
8. If $\varphi \in E_t$, then $[\Box\varphi]^{\mathfrak{A},i,g} = 1$ iff $\varphi^{\mathfrak{A},i',g[x/u]} = 1$ for all $i' \in Asg$.
9. If $\alpha \in E_\alpha$, then $[{}^\wedge\alpha]^{\mathfrak{A},i,g}$ is such a function h with the area of definition Asg, that for any $i \in Asg$ we have $h(i) = \alpha^{\mathfrak{A},i,g}$.
10. If $\alpha \in E_{\langle s,a \rangle}$, then $[{}^\vee\alpha]^{\mathfrak{A},i,g} = \alpha^{\mathfrak{A},i,g}(\langle i \rangle)$.

Thus, the operator $\Box$ is interpreted as universal necessity, i. e. as $\ll$ true in all possible worlds at all time moments $\gg$.

The intensional $\alpha^{\mathfrak{A},g}$ of the expression α in relation to $\mathfrak{A}$ и g can be defined clearly:

$\alpha^{\mathfrak{A},g}$ is such function h with the area of definition Asg, that for any $i \in Asg$ we have $h(i) = \alpha^{\mathfrak{A},i,g}$.

From this, as a consequence, we can obtain, that $[{}^\wedge\alpha]^{\mathfrak{A},i,g} = \alpha^{\mathfrak{A},g}$ for all $i \in Asg$.

We introduce also alternative notations for interpretation. The previously adopted notations for interpretation express interpretations of complex expressions through the interpretation of their nested expressions using logical equivalences, i. e. have the form

$$[\alpha(\alpha_1, \ldots, \alpha_n)]^{\mathfrak{A},i,g} = 1 \text{ if and only if}$$
$$\Phi_\alpha([\alpha_1]^{\mathfrak{A},i,g} = 1, \ldots, [\alpha_n]^{\mathfrak{A},i,g} = 1), r$$

where Φ_α is the formula specifying the truth condition for interpretation of the expression α.

This way of expression is computationally inconvenient. It seems more convenient to define functions F_α, which calculate interpretation values directly from the values of interpretations of nested expressions, i. e. in the form

$$[\alpha(\alpha_1, \ldots, \alpha_n)]^{\mathfrak{A},i,g} = F_\alpha([\alpha_1]^{\mathfrak{A},i,g}, \ldots, [\alpha_n]^{\mathfrak{A},i,g}).$$

The presented approach has such advantages:

(1) the calculation becomes more algebraic in nature;
(2) the opportunities for harmonizing the interpretation mechanism with a categorical point of view are expanded;
(3) the possibility of constructing a computer support system for interpretation becomes easier.

4.4   Logically Possible Interpretations and Connected definitions

Further on various concepts of a logically possible interpretation may be considered. As a rule, to ensure compliance of logically possible interpretation with a specific problem domain, additional conditions for the interpretation of constants, including those representing predicates, are required.

Logical truth, logical sequencing and logical equivalence of formulas of the intensional logic are characterized in accordance with a logically possible interpretation. For example, the formula φ of the intensional logic is characterized as logically true if it is true in all logically possible interpretations with respect to all worlds and time moments of this interpretation; and the two formulas φ and ψ of the intensional logic are logically equivalent if and only if $\varphi \leftrightarrow \psi$ is logically true.

4.5   Controlled Extensionality

If δ is an expression of the intensional logic, corresponding to the type of translation of a frame in the semantic network, then we designate by δ_* the expression denoting the set of individuals or the relation between individuals that naturally corresponds to the set or relation designated as δ.

One can get that if δ is among the constants that translate $\ll$ ordinary $\gg$ common names or $\ll$ extensional $\gg$ transitive or intransitive verbs, then δ can be defined in terms of δ_*. One of the options for such a definition is as follows.

$$\Box[\delta(x) \leftrightarrow \delta_*(\check{}x)],$$

Varying the area of extensionally interpreted constants, it is possible to model problem domains with controlled extensionality.

5   Means of Semantic Network Support

The intensional logic allows to describe the semantics of access to semantic network constructions. The computational model presented in Sect. 4 was implemented with the help of a prototype system for supporting the semantic network that provides the definition of the network constructions, to which the intensional logic constructions are aligned, and the calculation of extensionals of the corresponding constructions.

5.1   Structure of Semantic Network

The semantic network is considered as the one consisting of vertices and (named) arcs. The vertices of the network represent concepts and predicates. Concepts represent the objects allocated in the domain in the course of its examination, the classes of such objects and different abstract entities. Among the concepts common concepts, constants and variables are distinguished. Frames are used as the structure of the semantic network used to represent knowledge of stereotyped situations in the domain. Frames are considered as integral structures consisting of more basic (vertices and arcs) and satisfying certain structural constraints.

The arcs of semantic network represent the connections that are allocated in the subject domain when it is examined. The arcs, linking concepts, represent ways of typing the corresponding objects of the model. The connections of concepts with predicates describe the argument places of the corresponding predicate. The predicate links are not currently supported.

When linking semantic network constructions to a computational model, each vertex of the network, different from the vertex representing the variable, is associated with the constant of the language of intensional logic. The vertex representing the variable corresponds to the variable of the language of intensional logic. The arc, connecting the vertices, is associated with a formula of the form $F(x, y)$, where x, y are the expressions of the language matched to the vertices. An integral formula corresponds to a formula derived from formulas that are associated with its parts according to the rules of folding. The rules of folding are defined in two ways: (1) using standard composition methods corresponding to the frame structure, and (2) in accordance with problem-oriented axioms.

To display the intensional component of the model in the semantic network the points of correlation are represented. Transitions of objects from the state to the state are described using the cloning mechanism, according to which the construction of the semantic network (for example, the constant) is a clone of the conceptual description construction abstracted from the assignment point. In the case of a predicate, when performing cloning, substitutions on its argument places are possible. The adopted cloning concept is a refinement of the model presented in [3].

5.2   Access to Semantic Network

To access the semantic network, one must specify a network structure that describes the data one need to extract, as well as the assignment point in which access is made. The description of the assignment point includes a description of the characteristics of the user making the access, as the values of the corresponding constants at a given assignment point. When performing the operation of access to the semantic network, the extensional of the formula matched to the structure of the request is computed in accordance with the rules of Sect. 4.3. The extensional is calculated at a given assignment point.

The calculation, as a rule, generates new vertices of the semantic network, defining the extensional of the given expression. To generate new vertices in the support system the built-in predicates are foreseen, when computing the significiation of which new vertices of the network are created. Such predicates describe, for example, arithmetic expressions or expressions obtained by aggregating a set of network vertices that satisfy a given condition. When embedding new vertices in an existing network, they are treated as clones of vertices specified in the structure that represents the query.

5.3   Modification of Semantic Network

Modification of the network is determined in accordance with the general intensional character of its computing model as development of assignment points representing the modified state of the network. When creating a assignment point, new concepts and predicates can be specified that are included in it. It is also possible to create a assignment point based on the existing one, where the concepts and predicates of the new point are inherited from the previous one, but the set of their clones can be changed.

In the existing implementation it is necessary to specify the way to create "manually" to develop a assignment point. It is of interest to develop methods for the automated creation of assignment points that would enable the generation of assignment points in accordance with the condition specified by the controlling semantic network. The prototype implementations of such a mechanism are currently under investigation.

6   Conclusion

The paper proposes a computational model for supporting access policies to the semantic network, providing a solution to the problem of storing data in the Web environment and providing access to data, taking into account their semantics. It considers a method of storing data on the basis of a semantic network, whose constructions are described in terms of the intensional logic. Managing access to objects of the semantic network is made on the basis of the allocation of a controlling semantic network, formed on the basis of a part of problem-oriented axioms describing the subject area that characterize classes of users of the system. The approach allows to set the specialized controlling objects - the access policies to the semantic network. Constructing the model of access to a semantic network based on the intensional logic ensures its correctness.

Elements of the proposed approach were tested when working out the information systems that support the development of institutional foundations for the introduction of the best available technologies in the Russian Federation. The approbation allowed to identify some limitations of the proposed approach and prospective directions of its further development, in particular:

– continuation of development of methods for mapping the evolution of objects of the semantic network and the connection between the nature of evolution and access to the object;

- development of ways to connect descriptions of access policies to objects with system administration methods;
- extension of the computational capabilities of the controlling subsystem of the semantic network.

At the same time the testing demonstrated possibility of achieving the set goals, which grounds the practical significance of the proposed approach.

Acknowledgement. The work is supported by the grants 16-07-00909 and 17-07-00893 of the Russian Foundation for Basic Research (RFBR).

References

1. Ismailova, L.Yu., Wolfengagen, V.E., Kosikov, S.V.: Basic constructions of a computational model for supporting operations of access to a semantic network. Procedia Comput. Sci. **123**, 183–188 (2018). 8th Annual International Conference on Biologically Inspired Cognitive Architectures, BICA 2017 (Eighth Annual Meeting of the BICA Society), Moscow, Russia, 1–6 August 2017, pp. 183–188 (2018)
2. Kosikov, S., Ismailova, L., Wolfengagen, V.: The presentation of evolutionary concepts, pp. 113–125. Springer, Cham (2018)
3. Wolfengagen, V.E., et al.: Evolutionary domains for varying individuals. Procedia Comput. Sci. **88**, 347–352 (2016). 7th Annual International Conference on Biologically Inspired Cognitive Architectures, BICA 2016, New York, USA, 16–19 July, pp. 347–352 (2016)
4. Montague, R.: Pragmatics and intensional logic. Synthese **22**(112), 68–94 (1970)
5. Torra, V.: Data Privacy: Foundations, New Developments and the Big Data Challenge. Springer, Heidelberg (2017). https://doi.org/10.1007/978-3-319-57358-8
6. Chernyshov, A., Balandina, A., Kostkina, A., Klimov, V.: Intelligence search engine and automatic integration system for web-services and cloud-based data pro-viders based on semantics. Procedia Comput. Sci. **88**, 272–276 (2016)
7. Deshpande, P.M., Joshi, S., Dewan, P., et al.: Knowl. Inf. Syst. **45**(3), 705–730 (2015). https://doi.org/10.1007/s10115-014-0811-6
8. Li, Y., Wang, Y., Jiang, P., Zhang, Z.: Multi-objective optimization integration of query interfaces for the Deep Web based on attribute constraints. Data Knowl. Eng. **86**, 38–60 (2013). https://doi.org/10.1016/j.datak.2013.01.003
9. Faerber, F., et al.: Main memory database systems. Found. Trends Databases **8**(1–2), 1 (2016). https://doi.org/10.1561/1900000058
10. Trummer, I., Koch, C.: Multi-objective parametric query optimization. Commun. ACM **60**(10), 81–89 (2017). https://doi.org/10.1145/3068612
11. Gutwirth, S., Poullet, Y., de Hert, P., Leenes, R. (eds.): Computers, Privacy and Data Protection: An Element of Choice. Springer, Heidelberg (2011). https://doi.org/10.1007/978-94-007-0641-5

Basic Language Learning
in Artificial Animals

Louise Johannesson[1], Martin Nilsson[2([⊠])], and Claes Strannegård[1]

[1] Department of Computer Science and Engineering,
Chalmers University of Technology, Gothenburg, Sweden
loujohan@student.chalmers.se, claes.strannegard@chalmers.se
[2] Department of Physics, Chalmers University of Technology, Gothenburg, Sweden
martinni@student.chalmers.se

Abstract. We explore a general architecture for artificial animals, or animats, that develops over time. The architecture combines reinforcement learning, dynamic concept formation, and homeostatic decision-making aimed at need satisfaction. We show that this architecture, which contains no *ad hoc* features for language processing, is capable of basic language learning of three kinds: (i) learning to reproduce phonemes that are perceived in the environment via motor babbling; (ii) learning to reproduce sequences of phonemes corresponding to spoken words perceived in the environment; and (iii) learning to ground the semantics of spoken words in sensory experience by associating spoken words (e.g. the word "cold") to sensory experience (e.g. the activity of a sensor for cold temperature) and vice versa.

Keywords: Generic animat · Language learning · Babbling
Sequence learning · Grounded semantics · Poverty of the stimulus

Darwin argued that "The difference in mind between man and the higher animals, great as it is, certainly is one of degree and not of kind." [1]. Several differences have been found between the human brain and the brains of other primates, e.g. in terms of relative cortex volume, neuron packing density, and axonal conduction velocity, but they all seem to be differences in degree rather than kind [2,3]. According to Darwin's perspective, human cognition is a special case of animal cognition and psychology is a special case of ethology. Moreover, not even the ability to use language or mathematics counts as a clear-cut dividing line between *homo sapiens* and the other animals.

On the other hand, Noam Chomsky among others, argued that humans are born with a predisposition for language -a universal grammar- that is unique in the animal kingdom [4]. His evidence for this claim includes the "poverty of the stimulus", the idea that the linguistic input received by children is insufficient to explain their detailed knowledge of their first language.

Machine learning has been used for learning natural language syntax and semantics from examples. Grammar induction models are used for learning

© Springer Nature Switzerland AG 2019
A. V. Samsonovich (Ed.): BICA 2018, AISC 848, pp. 155–161, 2019.
https://doi.org/10.1007/978-3-319-99316-4_20

grammars from examples [5,6], e.g. by the use of syntactic equivalence classes [7]. Vector space models [8] are used for learning semantic relations via statistical analysis of large corpora. These models have been used to construct powerful search engines, recommendation systems, and summary generators. A key component of this method is the use of linear algebra and metrics, often the cosine function, that enable word contexts to be numerically compared. For instance, the words "cold" and "ice" may more similar to each other than "cold" and "grass".

Reinforcement learning algorithms can be used by agents whose objective is to accumulate reward over time [9]. In standard reinforcement learning, the reward signal is one-dimensional; in multi-objective reinforcement learning, it is multidimensional [10]. Local Q-learning, where Q-values collected from multiple agents are merged into a global Q-value are considered in [11].

Stewart Wilson introduced the term *animat* for artificial animals [12] and outlined the *animat path to AI*, which seeks to create artificial intelligence by modeling animal intelligence [13,14]. Of particular interest to this line of research are the homeostatic agents strive to keep the status of their needs in certain intervals [15]. Increases and decreases in the status of a need can be identified with reward and punishment, respectively. Animals may have access to information about the status of their own needs via signals emanating from interoceptors. Examples of animal needs and their associated interoceptors include energy (glucose receptors), water (osmoceptors), and oxygen (CO_2 receptors).

In this paper we explore whether a generic animat model without any dedicated mechanisms for language processing is capable of three basic forms of language learning. Section 1 gives a condensed overview of the animat model. Section 2 presents the results of the language learning experiments. Section 3, finally, draws some conclusions.

1   The Animat Model

In this section we give an overview of the animat model. More complete descriptions can be found in [16,17]. The code is available at [18].

A *network* is a graph, where each node has a name and type among NEED, SENSOR, AND, SEQ (sequence), MOTOR, AAND (simultaneous action), and ASEQ (action sequence). Time is modeled in discrete time steps called *ticks*. At each tick the animat receives input signals from the environment in the form of real values in $[0,1]$ to the NEED nodes and boolean values to the SENSOR nodes. The latter values are propagated through the network. Action nodes can be activated either via reflexes or via decision-making. Networks are equivalent to sets of formulas of temporal logic [19]. In particular we use the binary modal operator SEQ that enables the construction of sequences. The formula p SEQ q is true at time t if p is true at $t-1$ and q is true at t. Figure 1 gives an example of a dynamic graph.

Fig. 1. The images to the left and right show the same network. To the left the node types are shown. To the right the names of the nodes are shown. The intensity of the red color reflects the level of activity of the corresponding node.

Definition 1 (Animat). *An* animat *consists of:*

- *A network*
- *An activity pattern on the network*
- *A set of experience values. These are real values of three kinds: (i) local Q-values $Q_i(c, a)$, which encode the value from the perspective of need i of taking action a when a given node c is active; (ii) The transition matrix entries $Trans(c, a, c')$, which encode the probability that taking action a when node c is active leads to node c' being active at the next tick; (iii) The Conditional matrix entries $Cond(b, b')$, which encode the probability of two nodes b and b' being active at the same time.*
- *A shape, which is a subset of $\mathbb{R}^3$.*

The animat perceives the environment via its NEED nodes and the SENSOR node activity that propagates to other nodes (or concepts) of the network. This perception together with the animat's experience values determine which action is selected. The policy of the animat is ϵ-greedy in the sense that it explores (makes a random action) with probability ϵ and exploits (makes a best action) with probability $1 - \epsilon$.

The animat learns by updating its experience values using a version of reinforcement learning (multi-objective local Q-learning). It also learns by forming new nodes and by removing existing nodes (structural learning). New nodes are formed when a large enough prediction error is made in terms of expected reward. New nodes can also be formed at random moments by joining two nodes that were just active simultaneously or consecutively by means of AND or SEQ. Nodes that are superfluous from the perspective of reward prediction are eliminated in order to keep the network as small as possible (forgetting).

2   Language Learning

We conducted three experiments relating to language with the animat model.

2.1 Learning to Produce Phonemes

By letting the animat activate a random motor node in each time step, and then updating its transition matrix, it was capable of learning how to produce phonemes/letters that it had both sensors and motors for. This was evaluated by giving the animat sensors and motors for the letters of the english alphabet. The animat was then allowed to explore and update its transition matrix for a fixed number of time steps. Any action node a such that the value of the entry $Trans(True, a, b)$ in the transition matrix was equal to 1 was defined as a generator for the node b. The results are shown in Fig. 2a.

2.2 Learning to Produce Phonetic Sequences

The learning of sequences was done through probabilistic learning. The animat would with some probability create a new sequence node, combining two nodes with probability proportional to how often in the last 100 time steps they had been active in sequence. It would then also create an action node producing that sequence by combining the generators for the two perception nodes combined. This was evaluated by exposing the animat to a set of 100 unique words, repeated randomly with equal probability on average 20 times each. The results of how many of the words that the animat had created nodes for after a certain amount of time is shown in Fig. 2b. As can be seen the animat quickly manages to learn the majority of the words. It also creates several nodes for parts of words, resulting in the animat creating more nodes than necessary. The animat was capable of producing all the sequences that it had learnt to recognise.

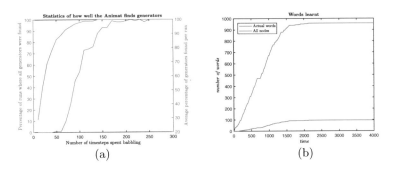

Fig. 2. (a) Plot showing how often the Animat is capable of finding all generators in the different tests (blue). And showing the average number of generators found in each test (orange). (b) Plot showing how the total number of perception nodes, and the number of perception nodes representing actual words, changes over time.

2.3 Learning to Associate Phonetic Sequences to Sensory Experience

The animat was made to associate between perception nodes by using the rows of a time-extended conditional matrix (storing the probability of b and b' being active at almost the same time) as vector representations of the nodes. The cosine distance between the vectors was then used to determine the most similar node, in the same way as in vector space models. This allowed the animat to associate not only between words, but also from perception nodes representing phonetic sequences to perception nodes representing sensory experiences.

The animat's ability to associate between word was tested on a simple input text and compared to a simple vector space model. When given the same input as the simple vector space model, the animat managed to return the same top ten associations as the vector space model with an average accuracy of over 80% as seen in Fig. 3a (blue line). When weighting the value of the associations made higher for the top associations of the vector space model (and decreasing the value linearly), the animat achieved an average accuracy of over 90% (as seen in the red plot in Fig. 3a). When the animat made use of a simple chunking method, it obtained an accuracy of over 99%.

The Animats ability to associate between words and other input was also evaluated, and compared to the vector space model. The associations of both models were compared with a dictionary between words and other sensations, that was used when ceating the training data used as input. The associations were tested both from words to other senses and from senses to words. The result is shown in Fig. 3b, showing the animat's accuracy as solid lines and the vector space model's accuracy as dashed lines.

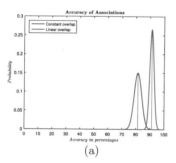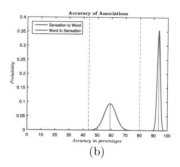

(a)   (b)

Fig. 3. (a) Plot showing the results of associations between words. The blue plot shows the probability distribution of the Animats accuracy when the same score is given for all associations of the vector space model, along with its mean. The red plot shows the distribution and mean for when the associations were worth different amounts depending on how highly they were associated by the vector space model. (b) Plot showing the results of associations between sensations and words. The solid lines show the probability distributions of the Animats accuracy, with the blue line representing the result of the associations from sensations to keywords, and the red line representing result of the associations from keywords to sensations. The corresponding results for the vector space model is shown with dashed lines.

3   Conclusion

We conducted three experiments with a general animat model without any dedicated mechanisms for language learning. The results obtained indicate that the animat model is capable of three kinds of basic language learning.

Acknowledgement. This research was supported by the Torsten Söderberg Foundation Ö110/17.

References

1. Darwin, C.: The Descent of Man and Selection in Relation to Sex, vol. 1. Murray, London (1888)
2. Roth, G., Dicke, U.: Evolution of the brain and intelligence in primates. In: Progress in Brain Research, vol. 195, pp. 413–430. Elsevier, Amsterdam (2012)
3. Herculano-Houzel, S.: The human brain in numbers: a linearly scaled-up primate brain. Front. Hum. Neurosci. **3**, 31 (2009)
4. Chomsky, N.: Rules and representations. Behav. Brain Sci. **3**(1), 1–15 (1980)
5. Brill, E.: Automatic grammar induction and parsing free text: a transformation-based approach. In: Proceedings of the 31st Annual Meeting on Association for Computational Linguistics, pp. 259–265. Association for Computational Linguistics (1993)
6. D'Ulizia, A., Ferri, F., Grifoni, P.: A survey of grammatical inference methods for natural language learning. Artif. Intell. Rev. **36**, 1–27 (2011)
7. Clark, A., Fox, C., Lappin, S.: The Handbook of Computational Linguistics and Natural Language Processing. Wiley, Hoboken (2013)
8. Clark, S.: Vector space models of lexical meaning, pp. 493–522. Wiley, Hoboken (2015)
9. Sutton, R.S., Barto, A.G.: Reinforcement Learning: An Introduction. MIT Press, Cambridge (1998)
10. Roijers, D.M., Vamplew, P., Whiteson, S., Dazeley, R., et al.: A survey of multi-objective sequential decision-making. J. Artif. Intell. Res. (JAIR) **48**, 67–113 (2013)
11. Russell, S.J., Zimdars, A.: Q-decomposition for reinforcement learning agents. In: Proceedings of the 20th International Conference on Machine Learning (ICML 2003), pp. 656–663 (2003)
12. Wilson, S.W.: Knowledge growth in an artificial animal. In: Adaptive and Learning Systems, pp. 255–264. Springer, Boston (1986)
13. Wilson, S.W.: The animat path to AI. In: Meyer, J.A., Wilson, S.W. (eds.) From Animals to Animats: Proceedings of the First International Conference on Simulation of Adaptive Behavior (1991)
14. Tuci, E., Giagkos, A., Wilson, M., Hallam, J. (eds.) From Animals to Animats. 1st International Conference on the Simulation of Adaptive Behavior. Springer, Heidelberg (2016)
15. Konidaris, G., Barto, A.: An adaptive robot motivational system. In: SAB, pp. 346–356. Springer, Heidelberg (2006)
16. Strannegård, C., Svangård, N., Lindström, D., Bach, J., Steunebrink, B.: The animat path to artificial general intelligence. In: Workshop on Architectures for Generality and Autonomy, IJCAI 2017 (2017)

17. Johannesson, L., Nilsson, M.: Simple language learning in artificial general intelligence. Master's thesis, Chalmers University of Technology (2018)
18. Johannesson, L., Nilsson, M.: Animat code (2018). https://github.com/Pessimus/MasterThesisAnimatAGI
19. Gabbay, D.M., Hodkinson, I., Reynolds, M.: Temporal Logic: Mathematical Foundations and Computational Aspects, vol. 1. Oxford University Press Inc., Oxford (1994)

The Concept of Functional Tractography Method for Cognitive Brain Studies

Sergey Kartashov[1,2(✉)], Nikolay Ponomarenko[1],
and Vadim Ushakov[1,2]

[1] National Research Centre Kurchatov Institute, Moscow, Russian Federation
sikartashov@gmail.com
[2] National Research Nuclear University «MEPhI», Moscow, Russian Federation

Abstract. The aim of this work is to develop method of functional tractography based on the fast MRI sequence (Multi-Band EPI). It is planned to identify active areas of white matter (active tracts) responsible for the realization of motor function and visual perception. The functional MRI method, universally recognized and quite popular in cognitive brain studies, clearly reveals sources of activity in the gray matter of the brain. Proposed method of functional tractography supposed to make it possible to determine the activity in the deep structures of white matter tracts, which gives a number of advantages in understanding the network interaction of human brain regions among themselves both in healthy people and in patients.

Keywords: Functional tractography · DTI · fMRI · White matter

1 Introduction

To date, despite the great prospects, there are very few works on the study and application of this method. In Russia, studies of the involvement of white matter in the implementation of cognitive functions are not carried out at all. A few data obtained by leading scientific laboratories cant give an unambiguous answer to the question posed - how the sources of activation's on the surface of the cerebral cortex are connected by pathological routes, and how the signal was transferred from one area of the brain to another. These questions should help to answer the method of functional tractography, with the help of which it is possible to "highlight" the active tract along which the nerve impulse spread during the performance of a certain action.

This method may allow more accurate description of the interaction models of different areas of the brain involved in solving various types of cognitive tasks.

In addition, based on the results obtained, it could be possible to apply this method in the field of medical practice in analyzing the scale and nature of the destruction of white matter fibers in patients, for example, with multiple sclerosis, tumor lesions, degenerative changes in the brain substance, etc. for preoperative analysis and post-operative rehabilitation.

© Springer Nature Switzerland AG 2019
A. V. Samsonovich (Ed.): BICA 2018, AISC 848, pp. 162–164, 2019.
https://doi.org/10.1007/978-3-319-99316-4_21

2 Idea

In the study of cognitive processes around the world, for the most part, standard types of MRI-sequences are used. Basically, this is a functional magnetic resonance imaging (based on the measurement of local changes in the paramagnetic properties of blood in small vessels due to a change in the ratio of oxy- and deoxy-hemoglobin) and diffusion-weighted images (which are used to restore the structure of tracts of white matter of the brain). The combination of these methods makes it possible to identify the active zones of the cerebral cortex and to connect them by white matter tracts.

Currently, the scientific community in the field of neuroscience is focused on a network approach in analyzing the work of the brain when performing various tasks. Leading groups are focused on projects on constructing and analyzing structural and functional connections of the human brain, suggesting different approaches: methods of independent components, dynamic causal modeling [1], graph theory [2] and etc. These labs focus on the analysis of fMRI and diffusion MRI data. They study the sources of the surface of the cerebral cortex, their dynamics, and also their interconnection with each other by tractographic paths. However, in this way it is impossible to assess the functional relationship of the two regions with each other. It is not clear exactly how the signal was transferred from one to the other. This question can be answered by functional tractography.

The proposed method of functional tractography is aimed to identify tracts in the white matter of the brain, which are active when the subject performs certain tasks. According to the current hypothesis, when the signal is transmitted along the active tract, in connection with an increase in the concentration of potassium ions in the intercellular space, glial cells increase in volume, thus making it difficult for local liquor flow [3]. This means that the diffusion coefficient along the entire path also changes. Thus, in fact, the process of signal transmission from one region of the brain to the other via white matter fibers can be evaluated.

To date, this method is not used extensively. So, only two laboratories in the world are engaged in functional tractography: in Australia [4] and Poland (Krzysztof Gorgolewski). However, the problem raised by researchers in this area is very promising and promising. To date, in view of the physiological characteristics of our body and brain in particular, the MR signal to some stimulus in the white matter of the brain obtained with fMRI study is interpreted by many scientists as noise.

In turn, the proposed method of functional tractography can identify an "active" path along which the signal was transmitted when performing a certain action. Primary data obtained by our foreign colleagues partially confirm the possibility and prospects of this method in solving the problem of detecting activations in the white matter of the brain.

We're planning to conduct their own MR-studies using motor and visual stimuli, as a result of which it is expected to "isolate" the active regions of the entire brain that work in the implementation of the functions described above.

The idea of this method is that these changes are caused by the passage of a nerve impulse along a certain path (for example, for visual and motor stimuli), which in turn leads to a change in ECS (increase in extracellular space), and not by any other side effects such as the BOLD-effect, the effect of large vessels and other hemodynamic changes). This approach is original and is not currently used anywhere in the world when researching the cognitive processes of the human brain.

Therefore, we formulate some steps to be done for this method:

- to use standard diffusion MRI-sequence for assessing structural connectivity of different brain areas [5];
- to adapt the fast MRI-sequence to a multiple continuous measurement of the diffusion coefficient in each voxel of the brain substance;
- to assess the dynamics of the behavior of the diffusion coefficient in target and non-target voxels with prolonged single visual and motor stimulation;
- to eliminate of the possible parasitic effect of BOLD and other hemodynamic changes by carrying out fMRI [6, 7] and ASL (Arterial Spine Labeling) and to exclude the effect of large vessels by angiographic study;
- to carry out the main experiment in the block paradigm for revealing active tracts (functional tractography) involved in solving selected cognitive tasks (visual and motor).

Acknowledgements. This work was in part supported by the Russian Science Foundation, Grant № 18-11-00336 (data preprocessing algorithms) and by the Russian Foundation for Basic Research grants ofi-m 17-29-02518 (study of thinking levels) and 18-315-00304 (studying of structural connections by diffusion MRI).

The authors are grateful to the MEPhI Academic Excellence Project for providing computing resources and facilities to perform experimental data processing.

References

1. Sharaev, M., Ushakov, V., Velichkovsky, B.: Causal interactions within the default mode network as revealed by low-frequency brain fluctuations and information transfer entropy. In: Samsonovich, A., Klimov, V., Rybina, G. (eds.) Biologically Inspired Cognitive Architectures (BICA) for Young Scientists. Advances in Intelligent Systems and Computing, vol. 449, pp. 213–218. Springer, Cham (2016). https://doi.org/10.1007/978-3-319-32554-5_27
2. Rubinov, M., Sporns, O.: Complex network measures of brain connectivity: uses and interpretations. NeuroImage 53(3), 1059–1069 (2010)
3. Tirosh, N., Nevo, U.: Neuronal activity significantly reduces water displacement: DWI of a vital rat spinal cord with no hemodynamic effect. Neuroimage **76**, 98–107 (2013)
4. Mandl, R., Schnack, H., Zwiers, M., Kahn, R., Pol, H.: Functional diffusion tensor imaging at 3 Tesla. Front. Hum. Neurosci. 1–9 (2013)
5. Kartashov, S., Ushakov, V., Maslennikova, A., Velichkovsky, B.: Human brain structural organization in healthy volunteers and patients with schizophrenia. In: Biologically Inspired Cognitive Architectures (BICA) for Young Scientists. Proceedings of the First International Early Research Career Enhancement School on BICA and Cybersecurity (FIERCES 2017). Advances in Intelligent Systems and Computing, vol. 636, pp. 85–90 (2017)
6. Ushakov, V.L., Samsonovich, A.V.: Toward a BICA-model-based study of cognition using brain imaging techniques. Procedia Comput. Sci. **71**, 254–264 (2015)
7. Arinchekhina, J.A., Orlov, V.A., Samsonovich, A.V., Ushakov, V.L.: Comparative study of semantic mapping of images. Procedia Comput. Sci. **123**, 47–56 (2018)

Feasibility Study and Practical Applications Using Independent Core Observer Model AGI Systems for Behavioral Modification in Recalcitrant Populations

David Kelley$^{(\boxtimes)}$ and Mark Waser

Artificial General Intelligence Inc., Provo, UT 84601, USA
David@artificialgeneralintelligenceinc.com,
Mark@ArtificialGeneralIntelligenceInc.com

Abstract. This paper articulates the results of a feasibility study and potential impact of the theoretical usage and application of an Independent Core Observer Model (ICOM) based Artificial General Intelligence (AGI) system and demonstrates the basis for why similar systems are well adapted to manage soft behaviors and judgements, in place of human judgement, ensuring compliance in recalcitrant populations. Such ICOM-based systems may prove able to enforce safer standards, ethical behaviors and moral thinking in human populations where behavioral modifications are desired. This preliminary research shows that such a system is not just possible but has a lot of far-reaching implications, including actually working. This study shows that this is feasible and could be done and would work from a strictly medical standpoint. Details around implementation, management and control on an individual basis make this approach an easy initial application of ICOM based systems in human populations; as well as introduce certain considerations, including severe ethical concerns.

Keywords: AGI · ICOM · Feasibility · Ethics

1 Introduction

Independent Core Observer Model (ICOM) Cognitive Architecture (Kelley), as an emotion driven Artificial General Intelligence (AGI) system, is designed to make or evaluate emotion-based decisions that can be applied to selection choices within its training context. This paper is focused on a feasibly study of such an AGI system as applied to action control and governance of humans in recalcitrant populations; where the AGI system is exercising oversight over the elements of that target population, in terms of free will, to ensure compliance in those populations. This feasibility analysis is designed to explore the practical implementation of using an ICOM AGI system to manage human behavior.

© Springer Nature Switzerland AG 2019
A. V. Samsonovich (Ed.): BICA 2018, AISC 848, pp. 165–173, 2019.
https://doi.org/10.1007/978-3-319-99316-4_22

1.1   Benefits and Foundation

From a theoretical standpoint, we can argue that a positive benefit is helping recalci-trant populations to make better choices. ICOM provides a framework for the choice control and ethical thinking enforcement based on IVA theory (Kelley) (as opposed to other approaches (Bostrom)) and general biasing of the current ICOM architecture to western ethics (Lee).

> "Within the realm of human behavior, technologies based on the use of aversive contingencies can be conceptualized as default technologies because they come into play when natural contingencies or positive reinforcement fail to produce a desired behavioral outcome" (Iwata)

Given the previous success in aversion therapy (Bresolin), it has been proven that this sort of approach, including aversion and positive reinforcement techniques (APBA) in combination, does, in fact, return results (Israel) and could be used as a method for control by the AGI system over human populations. (Israel). ICOM based monitoring essentially is a 'value-sensitive design approach' (Umbrello) to AGI oversight of behavior.

1.2   Experimental Risk

In terms of considerations, there are a number of issues to keep in mind when eval-uating the fundamental research in this area. Much of the research in behavior modi-fication is limited to special populations or atypical populations (Israel). Additionally, the case of electrical aversion therapy has considerations such as consistency of location (Duker) and further there is a lot of resistance to this sort of therapy in terms of limiting it to the kinds of populations where there are no other options. (Spreat) Additionally, much of the existing research lacks control groups and control procedures that we should try to address in any program based on this work. (Bresolin) With wider legal considerations as well, it is important to consider these issues in detail even with the support of the medical field in spite of the fact that this sort of manipulation and control is medically sound. (Jordan)

1.3   Problem Definition and Technology Selection

The key question is: "Can such a system be practically implemented that includes an ICOM based monitoring of, control over and manipulation of a human recalcitrant population?"

The ICOM system, even as a partial implementation, works by emotional evalu-ation and the subjective experience of a given choice, from the standpoint of the system, which is currently biased to western ethics (Kelley) and IVA ethical model (Kelley). As an AGI, an ICOM system can act as a proxy for a government agency or other governing body that has taught a given instance its own set of ethical require-ments or rules which act as that system's contextual background for making choices.

Aversion therapy (Spreat), driven by using a control band and electrical stimulation using a Bluetooth device, can be done remotely; using current off the shelf devices like the one by Pavlok (Pavlok).

Positive reinforcement can be done through administering dopamine via an implanted medical dosing device which is controlled by Bluetooth or NFC. While not mainstream, such a device does exist. (Simeonov)

Additionally, software for the AGI system would need to be on multiple platforms including smart phones, iPads, computers, Alexa devices, Google Glass and more. Such systems would be interface points for the AGI system governing the humans in question.

From a parameter standpoint, we want to look at the aforementioned execution to test for the main question using the hardware selections as noted, including some POC testing to look at feasibility. It is our opinion this selection gives us the tools to test our 'questions' based on our hypothesis that this is feasible. It is important to note that we are focused on feasibility. The effectiveness of the underlying techniques is documented in other studies using similar techniques. (Bresolin, Spreat, Duker, Isreal) We are adding AI management of those techniques. Data still needs to be aggregated across many experiments with numerous participants, including a control group, to ensure that we can infer a causal relationship between AI control and behavior modification.

1.4   Solution Architecture Summary

The solution architecture, defined as the architecture of the system used in the experimental framework, is based on industry standards for systems doing similar functional tasks for a given technology stack. We are using, for this feasibly study, systems that include a cloud based ICOM AGI system that is only partially implemented (meaning trained but without new learning capability or the ability to recycle thought) enough to judge ethicality against a known set of ideas based on contextual training of that instance. This could be done in any AWS or Azure or a similar cloud system and we selected a Microsoft engineering stack (The MS Stack includes C#, ASP.NET MVC, SQL, Azure) as we are most comfortable with it. This implementation used a Secure Socket Layer (SSL) encrypted pipe to client applications over HTTP (Hyper Text Transfer Protocol), also known as HTTPS. Client applications would need to be made for the computer devices used including phones, tablets and computers and ideally voice and sound along with a Google Glass like device. This would require individual setup. For study purposes we used UWP (Universal Windows Platform) for phone and Windows-based applications using the MVVM (Model View ViewModel) design pattern) for the application clients. The web or cloud base systems API framework was based on ASP.NET MVC Razor in Azure using JSON over HTTPS. These would need to be controlled access applications from a trusted source for security and then setup around each individual user, so the system can oversee actions of the user. This, ideally, would be fully instrumented for data analysis on a secondary database system, again in the cloud, along with the ICOM AGI system. At even a basic level, this would require numerous ongoing streams of data from all related devices as well as a critical dependency on bandwidth required for analysis. This is all standard practice in commercial software engineering and uses existing standard proven design patterns.

1.5   Proof of Concept (POC) Execution

To test the feasibility of the solution architecture and to answer the overall question, the POC tests are divided by segments around function groups such that they can be analyzed in the associated groups. Groups are broken out as follows:

Group 1 is a smart phone and Pavlok bracelet device using the Bluetooth protocol. This device was tested and found functional with a reasonable amount of research around the use of this sort of aversion therapy in behavior modification which supports this use case. In a Swift or Java client on iOS, using a Bluetooth protocol, this was easy to demonstrate, and it can be done, and it can communicate with a cloud-based system. In POC tests, this quickly kept subjects from making wrong choices as determined by the research team.

Group 2 is a Group 1 with the inclusion of a cloud-based system that collects data communicating with the Group 1 but also adds a specialized Alexa skill, that can monitor and decompose speech for analysis in the cloud, which can issue a command to the phone and hence provide the needed negative feedback. It was found this implementation was limited contextually with the ICOM AGI system being able to monitor action with only the input of the Alexa device and phone in a user's pocket.

Group 3 is a Group 1 with the inclusion of an AR (Augmented Reality) system. Initially, we looked at Google Glass but this hardware was not sufficiently well developed; so, a more robust system was selected, in this case a HoloLens. This system proved so robust as to be able to replace all the functionality of the phone, including all functionality not related to this study and added rich 3D sound and 3D visual data beyond the data provided by a standard smart phone. This kind of composited 3D data from HoloLens provides much more usable data including better contextual data requiring a lot less of visual decomposition processing and thereby less computational power is needed on the part of the AI system. We found some shortcomings in the visual decomposition of data from the neural networks that were selected, as they were not able to provide the rich data that would be needed for an ICOM system to be able to evaluate it properly; but, we could get the sound data easily into a useful state. We found that there are several systems that could possibly do better with visual data; but, would require more training. Additional problems with Group 3 included the weight of the HoloLens during extended periods of use, along with social norms that this device seems to trigger; meaning that test subjects were confronted on several occasions when wearing devices like this in public.

Group 4 included Group 3 plus a desktop computer where we could easily access online behavior. In this case, the data is easily analyzed and consumable by an ICOM system to determine if actions are moral, or not. The biggest issue here is additional training for the ICOM systems that would need to happen to understand context better. For example, from the standpoint of ICOM running on the computer or otherwise looking at network behavior it could have a hard time telling if a medical student is studying human biology or viewing pornography without permission. Using the HoloLens, we can certainly get all the data (full blown 3D modeling in context along with infrared and visible light camera feeds and gyroscopic position data and GPS can be added) needed to do this additional training and make determinations on the data easier then extrapolation on a standard PC; only because of the additional preprocessing available on the HoloLens versus a non-HoloLens equipped PC and because of the improved contextual awareness capability granted by the same hardware.

Given these groups, additional research around implanted medical dosing devices as well as composite execution of positive and negative reinforcement is needed, which would rely on other studies, before we can fully understand the impact and efficacy.

1.6   ICOM Ethical Decision Making in POC

The ICOM system in production solution in would be used to make relative judgements by training the example system to experience positive feedback to the point of needing that feedback emotionally and associate negative context with behaviors or choices, which would be decided by the body governing the study or implementation, as negative feedback. For example, we might train the system to respond to alcohol usage, to the point of the system having a visceral reaction to alcohol. We would also need to train the system to empathize with the target population using the ICOM systems that create its sense of self and its own self model. This, in testing, allows the system to get upset and 'provide' more effective negative feedback to the target for just picking up a beer, for example. This sort of training could be applied to any particular behavior you want to train for, or against, using positive and negative reinforcement as desired.

1.7   On Technical Feasibility

All of the hardware used for the POC is commercial off-the-self hardware. Some shortcomings of the current hardware include the aversion bracelet device needing to be something that is made out of a durable material that subjects would not be able to easily remove. This could be solved in a similar way that police handcuffs are con-structed, or like manner. The HoloLens turns out to be very heavy. It is likely the best solution using current technology would be a smart phone that a subject would wear on the chest with the camera running. HoloLens' camera system is far superior but too heavy to use for long periods, for most people.

Positive reinforcement through an implanted medical dosing device with dopamine is not medically practical (Jordan) with the current state of technology, however using other drugs, like micro-doses of MDMA (Jordan), would work with similar effect as dopamine and is more practical (Guiard)(Hashemi)(Hagino).

One big failing is that all these systems require strong internet access to support the HTTPS connections to the cloud. Without that connection, it would be impossible to implement monitoring and control functionality. A non-connected system is possible; but would require a lot more local resources, for most situations.

Overall, the basis for this system, without the cloud aspect, does exist from a hardware standpoint and it has been demonstrated that it could be done without major new development; meaning, the only real problems with implementation are engineering ones.

1.8   Behavioral Modification Types

We found that many behavior types could easily be tracked, analyzed and corrected. For the most part, these are the things that could mostly be done without AGI; instead, utilizing simpler machine learning or even a decision tree-like narrow AI system. For example, time related or appointment keeping, credit card usage, exercise, foul language, email usage, behavior correlation with deviance and the like could be managed by narrow AI in its various forms. That said, existing research suggests that this sort of behavior modification is straight forward (OPTUM) (Winters). Where ICOM AGI systems seem to really shine is in the monitoring and control of ethical choices and the ability to take that control to the next level. This approach provides a more nuanced framework for us to manipulate the ICOM system to better control the recidivist populations. Further ICOM systems could be done where the system needs no human interaction to control large groups with little, if any, input; only as much as is desired by the controllers. There is potential to remove even this amount of outside control, allowing the ICOM system complete autonomy in behavior control.

1.9   Additional Research

In targeted populations, if it was desired to implement this system in production, it would be important to address the hardware concerns noted; otherwise, the system could be easily negated by the same recidivist populations. The more significant research would be in neural network training. In building out a production ICOM system like this, in particular, work needs to done in neural network training. There are many research programs related to this that could be built upon to help create the best context data for training ICOM around certain ethical scenarios. We would need to build out a new team or use, bring in or partner with an additional team to lower time to market costs.

Given that, it is important to look at additional contextual framework from society as to the deep possible impacts this sort of technology could have.

Additional research could also be pursued along several fronts regarding drugs used; including the MDMA micro dosing effectiveness and/or to do more detailed studies, using cybernetic implants, focused on getting dopamine dosing devices into the deep brain locations needed to be effective (Jordan).

Further research around training and optimization towards having a trained AGI ICOM client running on a local system. Studying the effectiveness of this sort of implementation in recalcitrant populations, in several series of trials, would allow for the system engineering to be finalized.

1.10   Contextual Framework

Our current society is awash in big data being used to create inequality (O'neil). If we use an artificial intelligence that doesn't work in a way analogous to our own intelligence it may be very alien to us (Barrat). Even if it is 'like' us at a high level, it could treat humanity the same way 'we', as humans, treat ethical rules (Yampolskiy). At the very least, ethical models more aligned with humanity will likely bias as needed (Barrett), helping give us an additional control over how the system evolves.

When pursuing research like this, it is important to realize issues that would affect adoption. Even if it only applies to the recidivist populations, these considerations will affect the research program and product adoption.

It is important to note that AGI systems like this, that are even semi-sentient, run the risk of allowing humans or governments to implement numerous worst-case scenarios (Tegmark). That includes lowering the risk factors for a ruler or ruling class to not require as many, or any, human keys of power; which would also create even more danger when, and if, AGI actually decided it was time to take complete control from that ruling class. (Mesquita) The fact that we could literally take equipment off the shelf and just throw it together and have it work so effectively makes us think it is not AGI that is the issue so much as the people that will abuse it before it is fully awake. Current AI systems, in many cases, are beyond reproach, beyond our control, completely opaque and their decisions absolute. (O'Neal) It is only a matter of time before governments abuse the power we have already given the narrow AI systems we already use.

In terms of impact:

"AI is a dual-use technology like nuclear fission. Nuclear fission can illuminate cities or incinerate them. Its terrible power was unimaginable to most people before 1945. With advanced AI, we're in the 1930s right now. We're unlikely to survive an introduction as abrupt as nuclear fission's." (Barrat, J.)

In many ways, this study turned out more feasible and almost as scary as Roko's Basilisk (Roko).

2 Conclusions

Going back to the main question: "Can such a system be practically implemented that includes an ICOM based monitoring of, control over and manipulation of a human recalcitrant population?" The short answer is clearly yes, it can be implemented. There are no new technological or scientific problems to implementation; there are only engineering problems, such as designing a non-rubber 'band' for the aversion therapy device. Further, once the equipment is in place, applying the use of these technologies to recalcitrant populations can be done at scale. Further research should be done to measure the effectiveness of these techniques vs non-adoption or non-aversion theories for those recalcitrant populations. Certain factors would need to be addressed. For example, regarding the problem of using dopamine, the implanted medical dosing device, we found after additional research, is not effective with the current state of the technology (Jordan). Additional tests would likely need to include MDMA micro dosing to test that substance's effectiveness in place of dopamine and if such micro dosages of MDMA have the effect desired using the current implant technology. Further, there is also the possibility of heart problems; so, before wearing one of the aversion therapy bracelets, an EKG test for heart irregularities would be required to ensure that the aversion therapy would not cause undesired issues (Jordan).

References

1. Roko, M.: Roko's basilisk. https://wiki.lesswrong.com/wiki/Roko's_basilisk
2. Kelley, D., Waser, M.: Human-like emotional responses in a simplified independent core observer model system. In: BICA 2017 Proceedings, Procedia Computer Science. http://bica2017.bicasociety.org/bica-proceedings/
3. Waser, M., Kelley, D.: Architecting a human-like emotion-driven conscious moral mind for value alignment and AGI Safety. AGI Lab, Provo Utah—Pending Peer review; AI and Society: Ethics, Safety and Trustworthiness in Intelligent Agents—Stanford University, Palo Alto, CA, 26–28 March. https://www.aaai.org/Symposia/Spring/sss18symposia.php#ss01
4. Lee, N, Kelley D.: The intelligence value argument and effects on regulating autonomous artificial intelligence. Chapter Inclusion in Book by Springer to be published 20917—Title Un-announced. http://transhumanity.net/preview-the-intelligence-value-argument-and-effects-on-regulating-autonomous-artificial-intelligence/
5. Bostrom, N.; Ethical issues in advanced artificial intelligence. In: Smit, I. et al. (eds.) Cognitive, Emotive and Ethical Aspects of Decision Making in Humans and in Artificial Intelligence, vol. 2, pp. 12–17. Institute of Advanced Studies in Systems Research and Cybernetics (2003). https://nickbostrom.com/ethics/ai.html
6. Kelley, D.: The independent core observer model theory of consciousness and the mathematical model for subjective experience. In: The 2018 International Conference on Information Science and Technology, Passed Peer Review, IST 2018, China, 20–22 April 2018. www.icist2018.org
7. Umbrello, S.; Frank De Bellis, A.: A value-sensitive design approach to intelligent agents. In: Yampolskiy, R. (ed.) Artificial Intelligence Safety and Security. CRC Press (2018, Forthcoming). https://www.researchgate.net/publication/322602996_A_Value-Sensitive_Design_Approach_to_Intelligent_Agents
8. APBA: Identifying Applied Behavior Analysis Interventions; Association of Professional Behavior Analysts (APBA) 2016–2017. https://www.bacb.com/wp-content/uploads/APBA-2017-White-Paper-Identifying-ABA-Interventions1.pdf
9. OPTUM: Modeling Behavior Change for Better Health. Resource Center for Health and Well-being. http://www.optum.co.uk/content/dam/optum/resources/whitePapers/101513-ORC-WP-modeling-behavior-change-for-the-better.pdf
10. Winters, S., Cox, E.: Behavior Modification Techniques for the Special Educator. ISBN: 084225000X
11. O'Neil, C.: Weapons of Math Destruction. Crown New York (2016)
12. Barrat, J.: Our Final Invention—Artificial Intelligence and the End of the Human Era; Thomas Dunne Books (2013)
13. Yampolskiy, R.: Artificial Superintelligence: A Futuristic Approach. CRC Press, Taylor & Francis Group (2016)
14. Barrett, L.: How Emotions Are Made—The Secret Life of the Brain. Houghton Mifflin Harcourt—Boston New York (2017)
15. Bresolin, L.: Aversion therapy. JAMA **258**(18), 2562–2566 (1987). https://doi.org/10.1001/jama.1987.03400180096035
16. Iwata, B.A.: The development and adoption of controversial default technologies. Behav. Anal. **11**(2), 149–157 (1988)
17. Spreat, S., Lipinski, D., Dickerson, R., Nass, R., Dorsey, M.: The acceptability of electric shock programs. Behav. Modif. **13**(2), 245–256 (1989). https://doi.org/10.1177/01454455890132006

18. Duker, P.C., Douwenga, H., Joosten, S., Franken, T.: Effects of single and repeated shock on perceived pain and startle response in healthy volunteers. Psychology Laboratory, University of Nijmegen and Plurijn Foundation, Netherlands. www.ncbi.nlm.nih.gov/pubmed/12365852

19. Pavlok: Product Specification. https://pavlok.groovehq.com/knowledge_base/topics/general-product-specifications

20. Israel, M., Blenkush, N., von Heyn, R., Rivera, P.: Treatment of Aggression with Behavioral Programming that includes Supplementary Contingent Skin-Shock. JOBA-OVTP **1**(4) (2008)

21. Israel, M.: Behavioral Skin Shock Saves Individuals with Severe Behavior Disorders from a Life of Seclusion, Restraint and/or warehousing as well as the Ravages of Psychotropic Medication: Reply to the MDRI Appeal to the U.N. Special Rapporteur of Torture (2010)

22. Jordan, R.: Interview 4/7/2018; Provo, UT

23. Simeonov, A.: Drug delivery via remote control: the first clinical trial of an implantable microchip-based delivery device produces very encouraging results. Genetic Engineering and Biotechnology News (2012). https://www.genengnews.com/gen-exclusives/drug-delivery-via-remote-control/77899642

24. Mesquita, B., Smith, A.: The Dictator's Handbook: Why Bad Behavior is Almost Always Good Politics. Public Affairs (2012). ISBN: 1610391845

25. Tegmark, M.: Life 3.0—Being Human in the Age of Artificial Intelligence. Knopf, Penguin Random House (2017). ISBN: 9781101946596

26. Guiard, B., Mansari, M., Merali, Z, Blier, P.: Functional Interactions between dopamine, serotonin and norepinephrine neurons: an in-vivo electrophysiological study in rats with monoaminergic lesions. IJON **11**(5), 1 August 2008. https://doi.org/10.1017/S1461145707008383

27. Hashemi, P., Dandoski, E., Lama, R., Wood, K., Takmakov, P., Wightman, R.: Brain dopamine and serotonin differ in regulation and its consequences. PNAS **109**(29), 11510–11515 (2012). https://doi.org/10.1073/pnas.1201547109

28. Hagino, Y., Takamatsu, Y., Yamamoto, H., Iwamura, T., Murphy, D., Uhl, G., Sora, I., Ikeda, K.: Effects of MDMA on extracellular dopamine and serotonin levels in mice lacking dopamine and/or serotonin transporters. Curr. Neuropharmacol. **9**(1), 91–95 (2011). https://doi.org/10.2174/157015911795017254

Web-Analytics Based on Fuzzy Cognitive Maps

Vasiliy S. Kireev$^{(\boxtimes)}$, Alexander S. Rogachev, and Alexander Yurin

NRNU MEPHI, Kashirskoe shosse, 31, 115549 Moscow, Russia
vskireev@mephi.ru, rogachevaimsc@gmail.com,
ssasha.yurin@mail.ru

Abstract. One of the main tasks of web Analytics is to analyze the structure of websites, on the basis of which data is determined by the behavior of visitors to make decisions on the development and expansion of the functionality of the web resource. Cognitive maps are used to model weakly structured subject areas in which, for a number of reasons, it is impossible to formalize the relationship between factors functionally. For construction of fuzzy cognitive maps seek help from the expert of the problem area, which on the basis of their knowledge and experience can reliably identify the factors and assess the strength and direction of relations between them. As part of this work, it was decided to choose conversion as a usability metric. In order to verify the correctness of the algorithm for calculating the conversion was used text log-file of the real site.

Keywords: Web mining · Fuzzy cognitive maps · Web-analytics
Data mining

1 Introduction

Today, web-analytics plays an important role in the development of business. Preparation of any business on the Internet (information or product) does not end with the creation and promotion of the site, followed by the analysis of attendance and changes in interface and design in General, the addition of forms and much more to adjust and improve the behavioral characteristics of visitors. Web Analytics allows you to evaluate the effectiveness of the resource and improve its performance-to increase the number of visitors and increase the level of sales. Thus, web analysis is transformed into a set of methods and tools that help to identify problems, critically approach the work of the site and evaluate its functionality. There are shortcomings in modern web Analytics systems. Currently, it is an important task to improve the information architecture of information systems from the user's point of view. Web analytics based on fuzzy cognitive maps can help to eliminate the shortcomings of modern web-analytics systems. In recent years, experts and analysts have devoted a lot of work to the research and development of methodologies for modeling systems using cognitive maps. Software for automated modeling based on cognitive maps is called cognitive mapper (cognitive cartographer) [1–3]. For business modeling of dynamic systems, where the relationship between entities is of a qualitative nature, analysts use subjective models of the system, called cognitive maps. Cognitive map will make the process more clear and understandable. This work consists in the automatic construction of a

© Springer Nature Switzerland AG 2019
A. V. Samsonovich (Ed.): BICA 2018, AISC 848, pp. 174–179, 2019.
https://doi.org/10.1007/978-3-319-99316-4_23

cognitive map and analysis of the usability of the website. Formally, cognitive maps can be represented as a sign-oriented graph G = (V, E), in which the vertices represent entities, concepts, factors, goals and events, and the arcs define their influence on each other. The tops of V cognitive map correspond to the factors (concepts) that determine the situation [3]. E arcs correspond to causal (casual) relationships between factors. The method of specifying the force of cause-effect relations and values by an expert influences the construction of a cognitive map. A cognitive map is a directed graph with the nodes being the pages of a website, and the links between them being the user's transitions from one page to another.

2   Usability and Conversion in Web-Analytics

Historically, web-analytics has been used to measure site traffic. However, this value has become blurred, mainly because tools covering both categories of web Analytics have started to do so. There are two main technical ways to collect data from the website. The first, traditional method, is based on the analysis of the log file from the server, in which the web server records the user's requests. The second way is by using counters. Counters are external programs. To get statistics, a small piece of code (usually 1–2 kbyte) is installed on the website pages. The point is that when you log on to the site, the browser loads the image that is placed on the site to collect information. The counter downloads are entered into a database that can be hosted on the statistics collection and processing service provider server, and then viewed, for example, on its website. Checking ergonomics (Usability testing) is a study performed to determine whether a certain artificial object (such as a web page) is convenient for its intended application [4–7].

Conversion is a special kind of final usability evaluation, it plays a significant role in e-Commerce. Conversion also has binary indicators (1 = converted, 0 = not converted) and can be measured at all stages of the sales process from landing pages, registration, product selection and purchase. Often the combination of usability issues, errors and time spent results in lower conversion rates in the baskets.

Let's define the conversion as follows: let's say we have the first vertex A and the final(target) vertex B. then the conversion will mean the number of users who went from the initial vertex to the final one within their session on the site. In other words, it is necessary to extract all subgraphs of the original graph with given initial and final versies. And then, sequentially multiply the weights of the arcs of each subgraph and then add them together by the number of users in the original vertex.

However, within the application we work with access.log file, using which it is possible to calculate the desired conversion before the cognitive map. By breaking the log in session, we can see if the individual user was in the initial vertex A. If so, see if he was then in B. Then, to calculate the conversion with use of the formula:

$$C = K/N \tag{1}$$

The final conversion is obtained by dominating the desired number of users in the initial vertex.

To improve the usability of the site, it is necessary to experiment with the structure of the site. One must remove or add pages to the site structure to improve user behavior. Such a structure as fuzzy cognitive maps can help in modeling the structure of the site and conducting experiments to make a balanced decision to improve the usability of the site. In this paper, the concepts of fuzzy cognitive maps are web resource pages, and the meaning of concepts – the number of users. The weights of the arcs of the directed graph are also calculated. For example, the weight for an arc from a to B is calculated as the ratio of the number of transitions from a to B to the maximum number of moves from one vertex to another. To recalculate the weights of the fuzzy cognitive map, a mathematical model of fuzzy cognitive maps will be used, in which the value of each concept is calculated by the influence of other concepts through the equation:

$$x_i(t) = f\left(\sum_{\substack{j=1 \\ j \neq i}}^{n} x_j(t-1)w_{ji}\right),\tag{2}$$

where $x_i(t)$—concept value C_i by time t, $x_j(t-1)$—concept value C_j by time t − 1, w_{ji}—weight of interrelationship between C_j и C_i и f—activation function: $f = \frac{1}{1+e^{-\lambda x}}$.

3   Developed Instrumental Means for Cognitive Web-Analytics

The authors have developed and implemented new tools for cognitive analysis of websites based on previous experience [8] of automatic cognitive map building. For testing technologies, each of the software applications was implemented using a

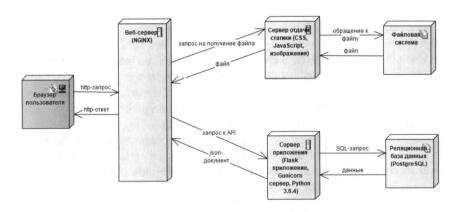

Fig. 1. Application architecture for web-analytics

different development stack. For the automated determination of the site usability the Java language was used. Python was used to calculate conversions and model the optimal structure of the site. Access is used as input a log file used by web servers to record site accesses. At the moment, most of the popular web servers (Apache, Nginx, Lighttpd) use a single mill-dart combined, which makes the proposed method universal [9–11]. The General scheme of the tool is shown in Fig. 1.

In both cases, after loading the web server log, the application processes it and builds a directed graph, which is a cognitive map. The tops of this graph are the pages of the website, and the links between them are the user's transitions from one page to another.

4   Experimental Results with Real Site's Data

As test data was used the log files of the website of the Department kaf22.ru. The experiment is to add or remove vertices with subsequent recalculation of the weights of the cognitive map. This allows experts to make a balanced decision to change the structure of the site to improve usability. Only the final decision-making process on whether, and if so, how to improve the usability of the site is left unmanaged. Most importantly, experts receive a tool that allows us to draw a conclusion about the effectiveness of the structure of the resource under consideration. First, the EA adds or removes vertices on the oriented graph. When you add a vertex, the expert Advisor specifies which vertices to connect to. After adding a vertex, all weights that belong to the sought vertex are initialized to zero. The next stage is the recalculation of the weights of the cognitive map. At the output, the expert receives a cognitive map with new values of concepts and weights (Fig. 2).

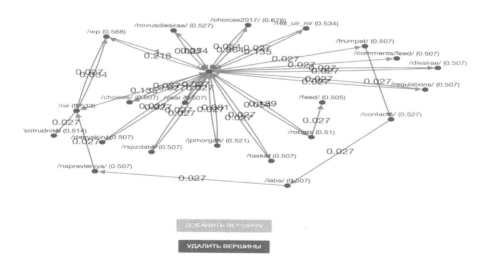

Fig. 2. Alternate cognitive map with weights (transition rate) on arcs

Experiments will allow to make the right decision about whether, and if so, how to improve the usability of the site. Using a visualized graph representation of the structure of user routes, the vertices of which are the content types of the web resource, as well as the calculated weights of the arcs, the expert has the ability to analyze the usability of the site. In this he was assisted by the conversion of calculated (Fig. 3). The second tool also allows you to calculate the conversion when changing the structure of sites (add or remove pages).

Fig. 3. Calculation of conversion to new vertice

5   Conclusions

This article features the architecture developed by the authors of the applications to automatically build fuzzy cognitive maps based on website logs data and then use it for conversion calculation. The work acquires special value due to the solution of the problem of modeling the reaction of site users to the change of its structure. This is achieved by the possibility of a cognitive map to recalculate the values of concepts. User response is defined in one of the developed applications as the value of page visits, and the second as the value of the conversion received by the new route in the graph.

Acknowledgements. This work was supported by the MEPhI Academic Excellence Project (Contract No. 02.a03.21.0005).

References

1. Kulinich, A.A.: Computer cognitive map modeling systems. Control Sci. **3**, 2–16 (2010)
2. Gray, S.A. et al.: Mental modeler: a fuzzy-logic cognitive mapping modeling tool for adaptive environmental management. In: 2013 46th Hawaii International Conference on System Sciences (HICSS), pp. 965–973. IEEE (2013)

3. Carvalho, J.P., Tomè, J.A.B.: Rule based fuzzy cognitive maps-fuzzy causal relations. In: Mohammadian, M. (ed.) Computational Intelligence for Modelling, Control and Automation (1999)
4. Russian Internet's statistics. runfo.ru. https://www.runfo.ru/statistika-rossijskogo-interneta. Accessed 19 Mar 2018
5. Conversion rate. marketingterms.com. https://www.marketingterms.com/dictionary/conversion_rate/. Accessed 19 Mar 2018
6. Everything You Know About Conversion Rate Optimization is Wrong. https://www.wordstream.com https://www.wordstream.com/blog/ws/2014/03/17/what-is-a-good-conversion-rate. Accessed 11 Apr 2018
7. February 2018 Web Server Survey. news.netcraft.com. https://news.netcraft.com/archives/2018/02/13/february-2018-web-server-survey.html. Accessed 19 Mar 2018
8. Kireev Vasiliy, S., Smirnov Ivan, S., Tyunyakov Victor, S.: Automatic fuzzy cognitive map building. In: 8th Annual International Conference on Biologically Inspired Cognitive Architectures, BICA 2017 (Eighth Annual Meeting of the BICA Society), held August 1–6, 2017 in Moscow, Russia Online System Procedia Computer Science, Volume 123, 2018, pp. 228–233. https://doi.org/10.1016/j.procs.2018.01.035
9. Grinberg, M.: Flask Web Development. Developing Web Applications with Python. O'Reilly Media, Sebastopol (2014)
10. Pamutha, T. et al.: Data preprocessing on web server log files for mining users access patterns. Int. J. Res. Rev. Wirel. Commun. (2012)
11. Ramalho, L.: Fluent Python. O'Reilly Media, Sebastopol (2015)

Information Approach in the Problems of Data Processing and Analysis of Cognitive Experiments

Anastasia Korosteleva[1,2(✉)], Olga Mishulina[1], and Vadim Ushakov[1,2]

[1] National Research Nuclear University MEPhI
(Moscow Engineering Physics Institute), Moscow, Russia
nnkorosteleva@gmail.com, mishulina@gmail.com
[2] NRC "Kurchatov Institute", Moscow, Russia

Abstract. In this paper, information indicators for solving statistical problems at various stages of analyzing psychological test data are analyzed: in the process of detecting and removing from the sample incorrectly formulated test tasks, for grouping participants according to similarity indicators of their answers in test tasks and highlighting participants with unique characteristics. The proposed methodology is illustrated by the data of a psychological test aimed at identifying a person's ability to spatial imagination, the formation of associative links between objects and the solution of logical problems.

Keywords: Information criterion · Psychological test
Remove of rude errors · Data clustering

1 Introduction

Analysis of the data of psychological tests presupposes statistical processing of data and relevant statistical conclusions. Due to the relatively small amount of sample data (test results of the participants in the experiment), possible data releases can significantly affect the accuracy of statistical conclusions. Such gross failures arise for a number of reasons: misunderstanding by the participant of the question posed, an error in entering the answer to the question of the assignment, a stressful state due to the participant getting into the new situation and the limited time for solving the problem, the unsuccessful formation of a test admitting an ambiguous answer, etc [1–3].

The task is to solve and solve the problem of detecting and removing from test data those test tasks that are ambiguously interpreted by the participants, have different solutions for participants of different social groups, age or contain random errors. To solve the problem, an information criterion and several other criteria are proposed, which together provide increased accuracy of the rejection of test tasks.

A. V. Samsonovich (Ed.): BICA 2018, AISC 848, pp. 180–186, 2019.
https://doi.org/10.1007/978-3-319-99316-4_24

The proposed algorithmic solution is demonstrated by the data of a psychological test aimed at revealing a person's ability to spatial imagination, the formation of associative links between objects and the solution of logical problems. The final goal of the test data processing was to identify the groups of participants in terms of their responses to test tasks of different types and interpretation of groups in terms of cognitive characteristics of the participants. The cognitive test considered in the work was previously described in the articles [4–6] from the position of analysis of brain rhythms and mapping of mental processes from EEG and fMRI data.

2   Description of the Experiment

During the experiment, each participant solves 91 problems. Each task has 16 seconds to solve. The solution of each task is accompanied by rest. The stage of the experiment, consisting of solving one problem and rest, will be called the phase of the experiment. The complete experiment contains 91 phases.

Each task for the participant is an image, at the top of which there is a question (the task is set), and in the lower part there are four answers. Tasks are grouped by type - only 6 types (S1, S2, V1, V2, V3 and V4). One additional, the 7th type - "rest". In Fig. 1 shows examples of problems of six types.

For types S1 (16 tasks), S2 (15 tasks) at the top of the screen a picture of the puzzle is given. Of the four suggested variants, it is required to select the element of the puzzle, which is suitable in form and pattern (S1)/inscription (S2) to the given puzzle.

Each type of V1, V2, V3, V4 contains 15 tasks. Within four squares, four words (V1, V4) or a picture (V2, V3) are indicated at the top of the screen. Of the four suggested variants of words (V1, V2, V4) or figures (V3), it is necessary to choose one word or figure that is not logically suitable for any of the four squares represented inside.

The subjects were presented with six types of tasks with different cognitive stimuli from the space-like to the verbal-logical with the intermediate types, in which both stimuli are present to varying degrees. In particular, tasks of type S1 require space-shaped thinking, S2 - the space-like with the verbal component, V2-like verbal with the verbal-logical factor, V3 - also figurative-verbal with the logical component, V4 - verbal-logical thinking.

During the experiment students were recorded in a log file. For the development of paradigms, the software Presentation was used. To record the answers of the test subjects, the Current Design system was used, consisting of two remotes for the left and right hands, each of which was placed on two buttons. Thus, the subject to complete the task had to choose the answer using one of the four buttons.

The experiment involved 30 people (7 men and 23 women) with the leading right (24 people) and the left hand (6 people). The average age of the subjects is 24 years. An informed consent was obtained from each participant for the study.

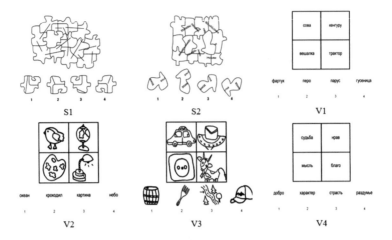

Fig. 1. Examples of 6 types of tasks

3   Methods of Processing Log-File Data

3.1   Elimination of Incorrectly Assigned Tasks

As follows from the description of the experiment, in each task the participant chooses one of 4 answers. It is possible that the participant for the allotted time could not find the solution of the problem and did not give an answer. Then it is considered that the participant's answer is zero. The peculiarity of the test is the absence in its description of correct (standard) answers to test problems. In addition, as it was noted earlier, in some test tasks among the answers offered to the participants there was no unequivocally correct answer. To recognize correctly the answer in the task, which was called by the participants most often, would be incorrect, since in the majority of problem problems the most frequent are not one but two answers. In addition, the frequency distribution of responses in different tasks can vary significantly. In some cases, it is close to uniform. In such circumstances, it is necessary to exclude problem problems from statistical analysis. In Fig. 2 shows two examples of frequency responses of participants in a distribution of two different types of tasks.

We exclude cases of zero answers from the statistical analysis. Let p_j^i be the frequency of the j-response to the i task in the answers of the participants. We calculate the amount of information I_i contained in the participant's response to task i, through the entropy H_i (in the aggregate of responses of 30 participants). Since there are 4 possible answers, the maximum entropy is $\log_2 4 = 2$ bits:

$$H_i = - \sum_{j=1}^{4} p_j^i \log_2 p_j^i, \qquad I_i = 2 - H_i \qquad (1)$$

The smaller the value of I_i, the less answers reflect the essence of the test task and, consequently, the tasks is more problematic (incorrectly posed). We

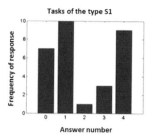

Fig. 2. Frequency distribution of responses in problems of types S1 and V2

set the critical value of the amount of information I_*. If $I_i < I_*$, then we will consider the test problem to be incorrect. In practical calculations, we used the value $I_* = 0.4$. As a result, from the 91 test tasks 58 tasks were left for further analysis (Fig. 3a).

To confirm the decision, another criterion for rejecting incorrect tasks was applied. It used 3 contrast ratio for allocation of frequency responses of participants. To formulate the coefficients, we moved from the answer numbers to the ranks of the answers: the answer with the maximum implementation frequency has rank $k = 1$ and then the rank rises with decreasing response rate. We denote $\bar{p}_k^i$ the frequency of the answer of rank k on i by the task among all subjects. If the problem is correctly posed and understood by the subject, then the frequencies of answers of the first rank are significantly higher than the frequencies of answers of other ranks. On this are constructed three contrast coefficients for response frequencies in i task:

$$C_1^i = \frac{\bar{p}_1^i - \bar{p}_4^i}{\bar{p}_1^i}, \qquad C_2^i = \frac{\bar{p}_1^i - \bar{p}_2^i}{\bar{p}_1^i}, \qquad C_3^i = \frac{(\bar{p}_1^i + \bar{p}_2^i) - (\bar{p}_3^i + \bar{p}_4^i)}{(\bar{p}_1^i + \bar{p}_2^i)} \qquad (2)$$

The values of the coefficients are reduced to the scale $[0; 100]$ by multiplying by 100 and summed. As a result, S was constructed as the sum of all 3 contrast indicators for which a histogram was constructed for the set of 91 tasks (Fig. 3b). Based on the histogram, the threshold was chosen $S_* = 245$, below which the task was considered incorrect. This criterion rejected 32 tasks, which were also recognized as incorrect and informational criterion.

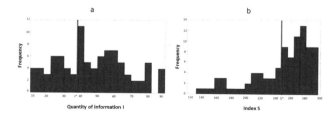

Fig. 3. Frequency distribution (a) of the information criterion I, (b) of the total contrast S over the population 91 tasks

3.2   Analysis of Individual Characteristics of Participants' Answers to the Tasks of the Psychological Test

Each participant of the experiment has individual features of solving problems of different types. It is different from the "average participant". To form groups of participants who have shown similar cognitive characteristics under the experimental conditions, it is necessary for each participant to compare the vector of some generalized indicators of his test decisions. We call this vector the "profile" of the participant.

As the first indicator of the individual characteristics of the participant's answers, let us consider the amount of information I_r^t, which is contained in the answers of participant r to tasks of type t. This indicator characterizes the degree of preference for the subjects of answers of a certain rank in tasks of type t. Let q_{rl}^t be the relative frequency of the answer of rank l to tasks of type t of the participant r.

$$H_r^t = -\sum_{l=1}^{4} q_{rl}^t \log_2 q_{rl}^t, \qquad I_r^t = 2 - H_r^t \tag{3}$$

Comparison of the individual I_r^t and the average statistical index $\bar{I}_t$ gives the second indicator ΔI_r^t:

$$\bar{I}_t = \frac{1}{k_t} \sum_{i=1}^{k_t} I_i, \qquad \Delta I_r^t = I_r^t - \bar{I}_t \tag{4}$$

The third indicator uses the Kullback-Leibler information measure KL_r^t the differences between the two frequency distributions q_{rl}^t and $\bar{p}_l^t$, where $\bar{p}_l^t$ is the frequency of the answer of rank l averaged over tasks of type t for all participants:

$$KL_r^t = \sum_{l=1}^{4} q_{rl}^t \log_2 \frac{q_{rl}^t}{\bar{p}_l^t} \tag{5}$$

Another indicator used in the analysis, in fact, also relies on an information measure - the negative value of the logarithm of the likelihood function of the frequency distribution q_{rl}^t to the average statistical distribution $barp_l^t$. We denote this exponent L_r^t:

$$L_r^t = -\sum_{l=1}^{4} q_{rl}^t \log_2 \bar{p}_l^t \tag{6}$$

The last indicator that describes the member profile is equal to the percentage of missing answers for each type of task R_r^t. Denote by d_{rn}^t the number of answers $n, n = \overline{0,4}$ in tasks of type t given by the participant r. Then

$$R_r^t = \frac{1}{k_t} d_{rn}^t \cdot 100\%, \tag{7}$$

where k_t is the number of tasks of type t.

Thus, the answers of the participant r are characterized by the exponents I_r^t, ΔI_r^t, KL_r^t, L_r^t and R_r^t, in which the parameter t takes 6 possible values according to the number of types of tasks. Together, these indicators form a "profile", or "model", a participant with 30 characteristics. In Fig. 4, the profiles of two participants with conditional numbers 406 and 611 are given as examples. The attributes are represented by a matrix: indicators - types of tasks.

406	S1	S2	V1	V2	V3	V4	611	S1	S2	V1	V2	V3	V4
I	2,00	1,41	1,28	1,53	1,41	1,50	I	1,03	0,75	0,09	1,41	1,19	0,47
ΔI	0,63	-0,01	-0,12	0,05	0,01	0,02	ΔI	-0,35	-0,67	-1,31	-0,08	-0,21	-1,01
KL	0,32	0,30	0,12	0,05	0,07	0,09	KL	0,38	1,68	1,15	0,31	0,54	0,78
L	1,30	6,21	8,44	5,19	4,64	5,32	L	6,77	17,57	24,45	6,34	10,84	20,80
R	0,33	0,22	0,00	0,00	0,13	0,00	R	0,17	0,33	0,20	0,30	0,00	0,00

Fig. 4. Indicators of individual characteristics of participants' answers in test examples of different types

3.3   Grouping Participants According to the Individual Characteristics of Their Test Answers

To distinguish groups of participants in a cognitive experiment with similar indicators of the solutions of test tasks, we applied the method of agglomerate hierarchical clustering. Because the attributes are significantly correlated, the main component method is applied to a sample of 30 objects with dimension vectors of dimension 30. To estimate the distance between clusters, the Complete-linkage method is applied in the process of the hierarchical procedure. Figure 5 shows the result of clustering, projected on the two main components of PC1 and PC2. 3 clusters were obtained.

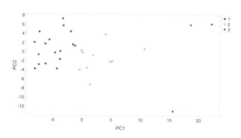

Fig. 5. The result of clustering in the projection on the plane PC1, PC2

Analysis of the participants' answers, which are assigned to the first cluster, shows that they have a better-developed verbal and logical thinking, and such members are more receptive to abstract words than substantive. At the same time, they have less developed spatial-verbal and figurative-verbal thinking. The participants in the second group have, on average, well developed all types of thinking presented in the tasks. Unlike the first group, they perceive better object words than abstract ones. In the third cluster, people have better developed spatial-figurative thinking. On average, the representatives of this group of people have quite well developed all types of thinking presented in the tasks.

4   Conclusions

The performed calculations and interpretation of the obtained results confirm the expediency of applying information criteria and indicators for the processing and analysis of data of cognitive experiments. Using information indicators allows you to identify objects with special properties and group objects with similar characteristics.

Acknowledgments. This work was in part supported by the Russian Science Foundation, Grant 18-11-00336 (data preprocessing algorithms) and by the Russian Foundation for Basic Research grant ofi-m 17-29-02518 (study of thinking levels). The authors are grateful to the Competitiveness Program of NRNU MEPhI for providing computing resources and facilities to perform experimental data processing.

References

1. Velichkovsky, B.M.: Kognitivnaya nauka. Osnovy psikhologii poznaniya, vol. 2 (2006)
2. Velichkovsky, B.M., Krotkova, O.A., Sharaev, M.G., Ushakov, V.L.: In search of the I: neuropsychology of lateralized thinking meets dynamic causal modeling. Psychol. Russ.: State Art **10**(3), 206–217 (2017)
3. Falikman, M., Asmolov, A.: Cognitive psychology of activity: attention as a constructive process. Revue internationale du CRIRES: innover dans la tradition de Vygotsky **4**(1), 54–62 (2017)
4. Roik, A.O., Ivanitskii, G.A.: A neurophysiological model of the cognitive space. Neurosci. Behav. Physiol. **43**(2), 193–199 (2013)
5. Orlov, V., et al.: Cognovisor for the human brain: towards mapping of thought processes by a combination of fMRI and eye-tracking. In: Biologically Inspired Cognitive Architectures (BICA) for Young Scientists, pp. 151–157. Springer, Cham (2016)
6. Arinchekhina, J.A., et al.: Comparative study of semantic mapping of images. Procedia Comput. Sci. **123**, 47–56 (2018)

Network Modeling Environment for Supporting Families of Displaced Concepts

Sergey V. Kosikov[1](✉), Larisa Yu. Ismailova[2](✉),
and Viacheslav E. Wolfengagen[2]

[1] Institute for Contemporary Education "JurInfoR-MGU", Moscow 119435, Russia
kosikov.s.v@gmail.com
[2] National Research Nuclear University "MEPhI"
(Moscow Engineering Physics Institute), Moscow 115409, Russia
lyu.ismailova@gmail.com, jir.vew@gmail.com

Abstract. The paper considers the problem of supporting the semantic stability of the information system in the course of changing its subject area. The representation of changes is performed on the basis of connection of methods of applicative computational technologies and the "functor-as-object" construction by the use of formalism of displaced concepts.

The database of the system is represented as a semantic network. Various mechanisms of the displaced concepts arising are considered, including those for formalizing the change in the scope of the concepts of the subject area, their definitions and recognition functions, as well as the parametric mechanism of displacement when the concepts associated with the original ones are changed.

To ensure the manipulation of displaced concepts, a single presentation mechanism is used based on the "functor-as-object" construction. Basic constructions are created to represent a set of situations for the use of the displaced concept, and also for taking into account the connections of situations. The methods of topology and category theory are applied. The paper discusses possibilities of representation of the network modeling environment for supporting families of displaced concepts in the form of a set of representative constructions of an applicative computing system and solutions for constructing the appropriate tool environment. The following results are presented - of testing the prototype components of the environment while solving cognitive type problems in the field of jurisprudence.

Keywords: Semantics · Displaced concepts · Computational model
Semantic network · Intensional logic · Network modeling

© Springer Nature Switzerland AG 2019
A. V. Samsonovich (Ed.): BICA 2018, AISC 848, pp. 187–196, 2019.
https://doi.org/10.1007/978-3-319-99316-4_25

1   Introduction

Modern technologies of scientific research foresee an increasingly intensive use of information resources, especially those presented in the Web environment. The information mining from such resources can be performed manually, but this appears to be a time consuming and inefficient approach. Therefore, the direction of information systems development, oriented to the work with resources, becomes important.

It comes out, however, that information in different resources (and often it might happen within one and the same resource) is organized according to various principles, is described by various (often incompatible) methods, and is classified according to different reasons, etc. [1]. Moreover, the indicated characteristics of information may change over time, plus to this those changes can have not only quantitative, but also qualitative nature. Therefore, to ensure the semantic stability of the information system in the course of changing its subject area [1,2] becomes a relevant task.

Simultaneous processing of information, organized according to different reasons, involves performing the agreement of the data models used. Such an agreement is possible only on a single formalized basis, using a conceptually clear and computationally powerful mathematical apparatus. The authors in the present paper follow the work [1], which suggests an approach to data modeling on a category basis that provides integration of logical and applicative means.

An essential feature of the category approach is the single treatment of stable and changing data on the basis of the concept of a Cartesian closed category. Further specification of the data nature is based on the imposition of additional conditions on the applied categories, for example, the acceptance of the axiomatics of topos. Taking into account the changes within the category theory requires a transition to categories of functors, which allows to formulate the basic notion needed to take into account changes - the notion of a displaced concept [1,2].

The development of information systems that provide semantically stable work and that are based on a formalized approach to modeling requires the use of tools agreed with the used approach. Such an agreement may involve both the direct development of tools that support the adopted approach, and the use of existing tools. In the latter case, it is necessary to select the capabilities of existing tools agreed with the requirements of formalism, and to offer a set of methods for using tools that do not violate accepted assumptions. This paper uses an approach based on semantic networks [2,3].

The work is structured as follows. Section 2 briefly overviews the approaches to the taking into account the changes of the subject area when constructing its model. Section 3 clarifies the formulation of the task to support the families of displaced concepts. Section 4 briefly presents the category grounds for modeling the displaced concepts and its connection with the constructions of supporting environment. Section 5 describes a prototype modeling environment to support the displaced concepts and outlines the paths of further work. Section 6 briefly summarizes the work.

2 Approaches to Taking into Account the Concepts Displacement

The task of describing changes in the subject area, including changes in the conceptual basis for describing the subject domain, was researched in various scenarios earlier. Within the framework of a purely logical approach the systems were studied, in which the truth of assertions, formulated within the framework of the system, depends on the fulfillment of certain conditions. First of all this is the temporal logic in its various formulations [4,5] and logic of action [6]. The autoepistemic logic [7] also suggests a description of the subject area, the scope of knowledge about which can change over time. The missing knowledge can be replenished by default. The default logic, however, does not offer means of agreement of the replenishment mechanisms, and this is manifested in the possibility of the absence of fixed points in a number of models of such logic (fixed points represent the agreed state of the system).

However, the mentioned systems, as a rule, are aimed at the identification and formalization of permanent principles, which can be laid as a basis for describing the subject area in the form of a system of axioms of the corresponding logic. Despite the importance of defining such principles, this is not enough to develop a complete system of modeling support. It is also necessary to identify ways of mapping the changes in concepts, classify such changes, etc.

Changes in the domain were also taken into account when developing semantics for databases. For example, the classical definition of a relational database [8] directly contains an indication that the set of relations in it changes with time. The tasks of mapping the incompleteness of the description of the domain in databases, creating temporal databases were also considered.

Within the framework of production rule systems for knowledge representation the task of supporting the description of domain changes was considered in the framework of studying strategies of the operation with work memory. The works related to the logical programming continue this line, and, for example, relate to a controlled resolution [9], to the attempts of linking the capabilities of applicative and logical systems [10], etc.

The mentioned directions of work are of considerable interest, and the identified methods can be partially used to support the presentation of displaced concepts. However, none of them provides an agreed description of the changes in the conceptual model. Therefore, the means declared within the named approaches require, at a minimum, selection and adaptation for joint use.

Besides that, none of the mentioned solutions is aimed clearly at describing changes in the conceptual apparatus of describing the subject area. For example, the possibility of the appearance of new concepts that did not exist before and the transition from descriptions within the framework of old concepts to the description by means of new concepts are not taken into account. The indicated possibility is important for a number of practical applications, for example, in the field of jurisprudence.

This paper proposes an approach to presenting changes (including conceptual ones) in the subject area on the basis of category theory [3]. The basic element

of the approach is the notion of a displaced concept, which is an object of the
category of functors from the category of Asg assignments to the base category
D. The objects of the assignment category represent the states of the concept,
and the arrows represent the ways of the concept displacement. The presented
approach is coordinated with the representation of a set of displaced concepts
in the form of a semantic network.

3 Task of Developing the Environment to Support the Families of Displaced Concepts

The study of approaches to solving the problem of presenting changes in the
subject area and revealing their limitations, as well as studying the possibilities
of formalisms based on category theory to support displaced concepts, including
those in the semantic network environment, makes it possible to specify the
task of developing the support tools for representing displaced concepts. The
indicated task is the development of the problem of supporting the evolving
concepts formulated in the work [1].

In accordance with the general formulation from [1] in order to support modeling, it is necessary to provide means for a typical description of the domain.
This can be achieved by choosing the support tools in the form of a network
environment that provides:

- representation of structured types on the basis of typed concepts of the semantic network, the semantics of which are specified by the category construction
  of the pullback;
- representation of functional types thanks to the possibility of abstracting
  fragments of the semantic network;
- representation of subtypes based on recognition functions, the semantics of
  which are specified by mono-arrows of the category model;
- representation of dependent types on the basis of concept families, the semantics of which are specified in the form of functors depending on the parameter.

The support tools are expected to give the ability to provide various types
of parameterization, which can be achieved through:

- representation of the assignment points to ensure attribution of the semantic
  construction of the concept displacing;
- support to processing of fragments of the semantic network, local for assignment points, for representing the dynamics of the concept displacing, in particular, changing its recognition function;
- support to generation of assignment points according to the conditions placed
  on the possible attribution of the concepts, which ensures the tracking of the
  dependence of the concept displacing on the set parameter.

It is also needed to ensure the integration of the modeling and computing
capabilities of the semantic network, which is provided by:

- possibility of attributing the semantic network constructions by generating cloned vertices obtained with a substitution system representing the attribution;
- possibility of identifying concepts that are undefined or partially defined at a given assignment point, and providing a computational response to the identification of such concepts;
- provision of various ways of replenishment of undefined concepts, including setting an explicit way of attributing the concept or the way of attributing by default.

The formulated requirements can be provided in the environment of semantic network support, oriented to the modeling of the domain and integrated with the applicative mechanisms of computation. The semantic network structures are assigned with semantics on a category basis within the model formulated in [1]. The applicative mechanisms provide the computation of the attributions of the semantic constructions of the network in accordance with the attributed semantics.

4   Semantics of the Network Constructions

The category model of evolving concepts, described in [1], suggests the description of concepts as objects of the category of functors from the category Asg of assignments to the base category D. In accordance with this frames are described as arrows of this category.

4.1   Semantic Network Structure

The semantic network is considered as consisting of vertices and (named) arcs. The vertices of the network are represented by concepts and predicates. Some vertices of the network (both concepts and predicates) can be defined through other ones, and the definitions admit a parametrization of the usual form corresponding to the lambda-expressions of the applicative medium.

Concepts represent the objects allocated in the domain in the course of its examination, the classes of such objects and various abstract entities. The common concepts, constants and variables are distinguished among the concepts. The arcs of the semantic network represent the connections that are allocated in the subject domain when it is examined. Arcs are grouped, and groups of arcs that satisfy certain semantic conditions are called frames and are used to represent knowledge about stereotyped situations in the problem domain. The assignment points are presented to display the intensional component of the model in the semantic network.

When linking the semantic network constructions to a computational model, each vertex of the network is associated with the construction of a category model, an object of the functors category is associated with the vertex of the semantic network. The arc connecting the vertices corresponds to the arrow of the category of functors.

4.2   Description of the Semantics of Network

Description of structured objects of the domain by semantic network can be performed on the basis of the construction of the Cartesian product in the category.
At the same time, however, the expression of conditions imposed on the connection of objects requires additional category constructions. The description based
on the construction of a pullback is more convenient.

The pullback is used to represent the conditions imposed on objects and
the arrows that connect them. The pullback is defined as follows. Suggest $a, b \in Ob(\mathcal{C})$, $f: b \to a$, $g: c \to a$. Then the pullback is the object d together with arrows
$f': d \to c$, $g': d \to b$, such, that for any $d' \in Ob(\mathcal{C})$, $f'': d' \to c$, $g'': d' \to b$ there
is the only arrow $h: d' \to d$ such, that $f'' = h \circ f'$, $g'' = h \circ g'$.

One should note that the pullback is the limit cone over a diagram consisting
of two convergent arrows f and g. The diagram representing the pullback is
named the Cartesian square.

Let us see how that works. Suggest $f: b \to a$ and $g: c \to a$ ''—are mono-
arrows which express a kind of ISA relation in the semantic network. In particular, in the category $\mathcal{S}et$ mono-arrows represent subsets of the object a. Then
the Cartesian square

$$
\begin{array}{ccc}
d & \xrightarrow{f'} & c \\
\downarrow g' & & \downarrow g \\
b & \xrightarrow{f} & a
\end{array}
$$

specifies the set of elements that simultaneously get into the subsets f and g, i.e.
intersection of subsets f and g. So, in this case the pullback is used to represent
conjunction of concepts in the semantic network.

The previous example did not consider the additional internal structure of
the concepts. Consider an example in which concepts are endowed with such a
structure.

Let us consider the so called internal category inside of the category $\mathcal{C}$. In
general, the internal category is used to define metadata by semantic network
means. The object of objects of internal category is denoted through c_0, and the
object of arrows through c_1. Suggest $d_0, d_1: c_1 \to c_0$ ''— are arrows, specifying
the beginning and the end of internal category arrows. Then the Cartesian square

$$
\begin{array}{ccc}
d & \xrightarrow{d_0'} & c_1 \\
\downarrow d_1' & & \downarrow d_1 \\
c_1 & \xrightarrow{d_0} & c_0
\end{array}
$$

specifies the "object of composed pairs" of arrows, i.e. allows to represent a
condition, according to which the end of one arrow should coincide with the
beginning of the other one, i.e. to ensure the correct structure of metadata.

The pullback is unique up to isomorphism. Let us formulate this more accurately. Suggest $m: d_2 \to d_1$ and $n: d_1 \to d_2$ specify the isomorphism d and d'

(i. e. $m \circ n = 1_{d_1}$ and $n \circ m = 1_{d_2}$). Then the left square of the diagram

$$
\begin{array}{ccc}
d_1 \xrightarrow{f_1'} b & \qquad & d_2 \xrightarrow{f_2'} b \\
\downarrow g_1' \quad \downarrow g & & \downarrow g_2' \quad \downarrow g \\
c \xrightarrow{f} a & & c \xrightarrow{f} a
\end{array}
$$

where $f_2' = f_1' \circ m$, $f_1' = f_2' \circ n$, $g_2' = g_1' \circ m$, $g_1' = g_2' \circ n$, is Cartesian if and only if the right square is Cartesian one. In practice, the constructed isomorphism provides the construction of the pullback in the most convenient way for a specific task, as well as the transition from one way to another.

Let us single out some essential properties of the pullback. In the semantic network, they provide concepts and analysis of their properties. The arrow $f \colon a \to b$ is a mono-arrow when and only when the square

$$
\begin{array}{ccc}
a & \xrightarrow{1_a} & a \\
\downarrow 1_a & & \downarrow f \\
a & \xrightarrow{f} & b
\end{array}
$$

is Cartesian.

Suggest $f \colon a \to c$, $g \colon b \to c$, $h \colon a \to b$, and $f = g \circ h$. then the square

$$
\begin{array}{ccc}
a & \xrightarrow{1_a} & a \\
\downarrow h & & \downarrow f \\
b & \xrightarrow{g} & c
\end{array}
$$

is Cartesian.

To move to the description of frames, the categorical characterization of relations is essential. The corresponding category of characteristic is a kernel pair.

Suggest $f \colon b \to a$. Kernel pair of the arrow f is named such a pair of arrows $g, g' \colon c \to b$, that the square

$$
\begin{array}{ccc}
c & \xrightarrow{g} & b \\
\downarrow g' & & \downarrow f \\
b & \xrightarrow{f} & a
\end{array}
$$

is Cartesian. Thus, the kernel pair is the result of the ascent of the arrow along itself.

In the category Set the kernel pair of the function $f \colon A \to B$ is the set of pairs (a_i, a_j) such, that $f(a_i) = f(a_j)$. It is obvious that the set is a subset of the Cartesian product $A \times A$. A similar characteristic takes place in an arbitrary category (having pullbacks), which, in particular, allows to represent the equivalence relations by means of semantic network.

5   Computational Abilities to Support the Concepts Displacement

The computational abilities of the semantic network environment include means for obtaining designations for semantic network constructs. To get designation with respect to the semantic network representing the domain model, it is necessary to specify a fragment of the semantic network, which is to be designated (corresponds to the request to the network) and the assignment point, in which the designation is calculated. Since the definitions of the network vertices may contain intensional operators, it is possible to pass to one or more other assignment points when computing.

The intensions and extensions are considered in accordance with the general capabilities of the category model for the vertices of the semantic network, corresponding to the concepts. The extensions are computationally interpreted fairly in a standard way as sets of constants corresponding to the elements of the concept extension. More interesting is the interpretation of intensionals, which can include both sets of constants, and fragments of the semantic network, which should be indicated to get the intensional concept.

Thus, with respect to a set of assignment points that determine the parameterization of the concept displacement, two independent displacement mechanisms are implemented: (1) the transition from one assignment point to another, performed by intensional operators; (2) the possibility of specifying various fragments of the semantic network for calculation at different points of assignment. The second mechanism can be considered as an extension of the method of specifying constants, whose value is functions of higher order (as one of the arguments of these functions is the semantic function of the semantic network fragment), depending on the assignment point. The displaced concepts are defined by this method, and for designation of these concepts include different fragments of the semantic network at different points of assignment.

The activation of included fragments is performed when calculating the context, i.e. the set of the assignment point specifying the definitions of intensional expressions (both of the first and of the second type), and also, possibly, the assignments for nested fragments of the semantic network that specify the values of non-local variables in accordance with the general applicative strategy of computation. The activated fragments may contain basic vertices, whose semantics are specified by objects of base category D, the vertices, for which definitions are given (in this case, the fitting is done to take into account the parameterization method), as well as the vertices, whose definitions at the current assignment point are not specified. In the latter case, an exceptional situation is fixed, requiring the concept displacement.

The concept displacement can be implemented in various ways. The direct specification of a fragment of the semantic network is possible; the fragment should be designed to obtain the designation of displaced concept. It is also possible to set the default reaction, for example, such that the concept recognition function is considered to be identically true or identically false for all constants coordinated with the concept by the type. It is also possible to set a more complicated function that depends directly on a displaced concept or its parameters.

The assignment point, in which the designation is performed, can be generated. For this purpose the semantic characteristics are attributed to the displaced

concepts by means of the semantic network. As such the key words in the subject area and the key-value pairs may be applied as well as more complicated
characteristics. When performing the designation in the generated assignment
point only those vertices of the semantic network are involved in it that have
the required characteristics. It is also possible to define a fragment of the semantic network that specifies the conditions for involving vertex definitions in the
generated assignment point.

As a whole the proposed mechanism ensures the following:

- way to define concepts taking into account the possible displacement and the
  definition of two basic displacement mechanisms;
- way of calculating the values of displaced concepts, providing an explicit and
  implicit specification of the displacement method;
- generation of an assignment point for computations that take into account
  the nature of the displacement of concepts.

The environment of network modeling is at the stage of prototype implementation. Some separate environmental mechanisms were tested when solving
the problems of supporting the training system in the sphere of the best available technologies implementation in the Russian Federation. In view of the rapid
change in the domain, its modeling is associated with the allocation and support of the displaced concepts, and the proposed methods have proved their
applicability.

At the same time, during the approbation a number of features are identified,
related to the prototype nature of the system and making its application difficult.
Thus, syntactically the network is partially specified by a set of programming
language constructs, and partially by using the XML dialect. The continuation of
the work seems to be necessary to determine the set of syntactic representations
that ensure an expressive and succinct description of the semantic network of
support for displaced concepts.

6   Conclusion

The paper considered the problem of supporting the families of displaced concepts to render support to the semantic stability of the information system in
the course of changing its subject area. The paper suggests the representation of
the displaced concepts within the semantic network, which provides a representation of the concept transformations when the semantic characteristics of the
presented situation change, as well as the separation of concepts depending on
their behavior under such a change.

The following suggestions are made:

Basic mechanism of the concepts displacement representation, based on the definition of families of local interpretations of concepts;
Approach to describing the displacement of concepts within the framework of
  intensional logic using a mechanism similar to the continuation mechanism
  in the lambda calculus;

Technique for implementing the displacement of concepts based on the mechanism of partially defined vertices of the semantic network (in particular predicates);

Technique of determining the method of displacement based on semantic features, ensuring the generation of families of displaced concepts.

A prototype environment is also proposed for supporting displaced concepts in the form of a network modeling environment that supports the declared abilities. The environment was tested in the presentation of dynamic information for the training system in the sphere of environmental law.

Acknowledgement. The work is supported by grants 16-07-00912, 17-07-00893 of the Russian Foundation for Basic Research (RFBR).

References

1. Kosikov, S., Ismailova, L., Wolfengagen, V.: The Presentation of Evolutionary Concepts, pp. 113–125. Springer, Cham (2018)
2. Wolfengagen, V.E., Ismailova, L.Yu., Kosikov, S.V.: Applicative methods of interpretation of graphically oriented conceptual information. Procedia Comput. Sci. **88**, 341–346 (2016). 7th Annual International Conference on Biologically Inspired Cognitive Architectures, BICA 2016, New York, USA, 16–19 July 2016, pp. 341–346 (2016)
3. Wolfengagen, V.E., Ismailova, L.Yu., Kosikov, S.V.: Computational model of the tangled web. Procedia Comput. Sci. **88**, 306–311 (2016). 7th Annual International Conference on Biologically Inspired Cognitive Architectures, BICA 2016, New York, USA, 16–19 July 2016, pp. 306–311 (2016)
4. Prior, A.N.: Time and Modality. Clarendon Press, Oxford (1957)
5. Gabbay, D.M., Guenthner, F. (eds.) Handbook of Philosophical Logic, 2nd edn, vol. 7. Kluwer, Dordrecht (2002)
6. Segerberg, K.: Getting started: beginnings in the logic of action. Studia Logica **51**, 347–378 (1992)
7. McDermott, D., Doyle, J.: Non-monotonic logic. Artif. Intell. **13**(1–2), 41–72 (1980)
8. Codd, E.F.: Relational completeness of data base sublanguages. In: Database Systems, pp. 65–98. CiteSeerX: 10.1.1.86.9277 (1970)
9. Russell, S.J., Norvig, P.: Artificial Intelligence: A Modern Approach, 3rd edn. Prentice Hall, Upper Saddle River (2009)
10. http://www.lix.polytechnique.fr/Labo/Dale.Miller/lProlog/. Accessed 15 June 2018

Designing an Emotionally-Intelligent Assistant of a Virtual Dance Creator

Dmitry I. Krylov and Alexei V. Samsonovich[(✉)]

National Research Nuclear University MEPhI,
Kashirskoe shosse 31, Moscow 115409, Russia
krylovdmk@gmail.com, asamsono@gmu.edu

Abstract. Intelligent agents and co-robots, or cobots, become increasingly popular today as creators of digital art, including robotic or virtual dancing. Arguably, the creativity of such tools is linked to their social-emotional intelligence. In this work we question this hypothesis, extending the general paradigm of an emotionally-intelligent creative assistant (Samsonovich [1]) to virtual dance creation. For this purpose, a semantic map of dance patterns is constructed. Transitions between dance patterns are selected among local transitions on the map, following general rules. The outcome is judged by subjects as a more confident dance, compared to control conditions, when the semantic map was not used. In the proposed creative assistant of a choreographer, the state of emotional coherence of the cobot-assistant and the human user is maintained dynamically. Using the semantic map and M-schemas, the assistant will suggest variants of dance continuation, based on the current emotional state of the human choreographer and the appraisals of choices. It is expected that this approach, combining efforts of the human and the automaton working together in a state of emotional coherence, will be more user-favored and will yield higher productivity and creativity, compared to more traditional tools for virtual dance generation.

Keywords: Creative assistant · Cognitive architecture · Emotional intelligence
Semantic mapping · Virtual dancing · Digital art · Human-like intelligence

1 Introduction

Human-oriented social robots, or cobots, physical or virtual, will work with humans side by side—either in teams or with individual users. They will become intelligent assistants, extending our minds and bodies into the physical or virtual world. In order to be efficient, not only they need to be smart and intelligent—they need to be creative and socially-emotional [1]. Here this idea is illustrated by an example, which is the new concept of a creative virtual assistant of a choreographer.

The idea can be intuitively explained as follows. The assistant provides guidance to a user, creating a virtual dance in a certain environment (Fig. 1). The cobot does not have a separate embodiment: instead, both, the user and the cobot, control a virtual actor—the dancer figure on the screen, that performs pre-defined patterns of dance movements. The creative part is in selection of the sequence of these patterns,

A. V. Samsonovich (Ed.): BICA 2018, AISC 848, pp. 197–202, 2019.
https://doi.org/10.1007/978-3-319-99316-4_26

following the internal "logic of dance" and the music. The cobot-assistant makes suggestions, and the human user takes decisions. The output is a video of a virtual dance, that is judged by naïve participants. The goal is to compare different versions of the cobot to see the effect and the role of its emotional intelligence in virtual dance creation.

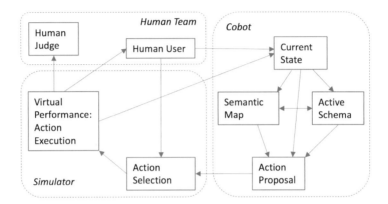

Fig. 1. General logic model of virtual dance creation with a cobot-assistant.

A simplified version of the paradigm outlined above was implemented and used in this study. It was found that emotional intelligence of the cobot is essential for the outcome, as explained below.

2   Materials and Methods

2.1   Cognitive Model

The model of the cobot and its interactions with the virtual actor and the human user was designed based on the general framework of eBICA (emotional Biologically Inspired Cognitive Architecture: [1, 2]). The general architecture of this design is represented in Fig. 1. In the simplified version used in this study, the cobot performed action selection without human interference.

The general logic scheme can be explained as follows. Cobot operates based on a finite set of schemas, that define probabilistic laws of action selection. The cobot receives information about the current state of the virtual actor (in this study—the currently performed dance pattern) and the current emotional state of the user (not used in this study). Using this information represented on the semantic map, based on the currently active schema, the cobot proposes a choice of one or several available actions. All actions, as well as the current state, are represented on the semantic map.

Semantic map [3], also known by different names [4], is a tool for representing semantics of objects of arbitrary nature geometrically, using a metric space. Here it was implemented as a two-dimensional vector space, with coordinates representing valence and activity, in which all available dance patterns were allocated based on their appraisals evaluated by experts. The result is represented in Fig. 2.

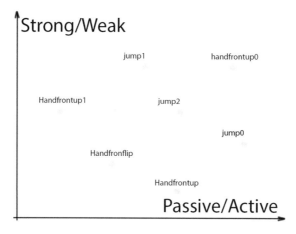

Fig. 2. Semantic map of the dance patterns. Labels, such as Handfrontup, represent distinct dance patterns allocated according to their perceived qualities. Animations were taken from Mixamo (https://www.mixamo.com/). Materials from Youtube.com with tags "Dance" were also used.

The **schema** that was used to determine probabilities of transitions among dance patterns was based on the Euclidean distance on the semantic map and on certain constraints imposed by choreography. As a result, the following graph (Fig. 3) was generated and used to propose transitions as the available actions during dance creation.

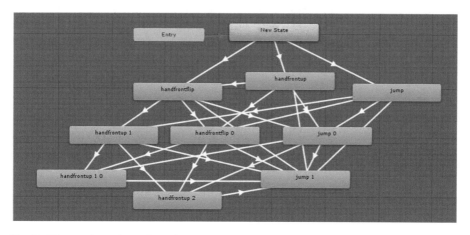

Fig. 3. The graph used together with the semantic map (Fig. 2) for proposing available actions —transitions to another dance pattern.

2.2   Implementation

The model was implemented using two platforms: Matlab 2017a (cobot) and Unity 5.0 (simulator, visualization), connected to each other using sockets. The cobot generated the initial state and controlled its updates by generating probabilities for action selection. Sampling of the probabilities was performed in the simulator, which therefore emulated the human part as well. The implemented environment and a snapshot of the outcome are shown in Fig. 4. The implemented algorithm is represented in Table 1.

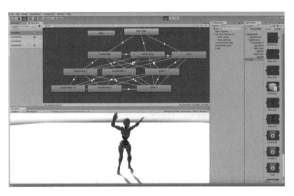

Fig. 4. The environment for virtual dance creation with a cobot-assistant implemented based on Unity 5.0 and Matlab 2017a. A snapshot of the resultant video is shown on the right.

The implemented algorithm of dance generation is represented in Table 1. Essentially, the algorithm in this simplified version implements a Markov chain, generating a sequence of dance patterns. The sequence is structured into time episodes, each of which is characterized by a short trajectory on the map—a "character arc". At the beginning of the dance, the origin of coordinates on the semantic map is selected as the current state. Then, within each episode, steps are performed as described by the algorithm. At each step three possible transitions to other states are proposed, and one of them is selected. At the end of the episode, the actor backtracks towards the origin of the map.

2.3   Subjects

In total, 14 volunteers participated in this study. All of them were bachelors studying program engineering at NRNU MEPhI. Nine of the participants were males, and five, correspondingly, were females. They all indicated Russian as their native language. All participants were asked to sign the informed consent form.

Table 1. Algorithm of dance movement pattern sequence generation.

Function **Define-Semantic-Model-Dance** (Alpha = 0.1, Beta =0.05, Actor = Model)
For *i* in time_episodes do:
Actor Start episode *i*
Set **Random** = Get Uniform Random Distribution() [0 ; 0.01]
Set **Distance Map** = Load Distance from semantic Graph ()
Set **Semantic Dance points** = Random * alpha * Distance Map
Set **Start point** = {x:0, y:0}
Set **Start point to Level 1 points** =
Calculate Distance to each point (Start point, Semantic Dance points)
Set **Level 1 to Level 2 points** = Extract 3 minimal (Calculate Distance to each
point
(Level 1 points, Semantic Dance points))
Set **Level2 points to Level 3 points** = Extract 3 minimal (Calculate Distance to
each
point (Level 2 points, Semantic Dance points))
Set **Start to Level3 Points** = Get Random * beta * Get Level2 Points
Actor End episode *i*
Actor BackTrack
end

3 Procedures and Outcomes

Two versions of the model—experimental version and control version—were used to produce two different kinds of video clips. The difference between them was that in the control version, semantic map was not used: the control versions of the graph (Fig. 3) were generated without using map metrics and coordinates of dance patterns.

Each participating subject was asked to watch two video clips, and then answered several questions about them. The target question was: "Which dance is more confident?" Answers divided as 12:2 in favor of the virtual dance generated using the experimental version, i.e., using the semantic map. Therefore, the null hypothesis that the control dance is perceived as equally or more confident can be rejected (Binocdf $P < 0.007$). For comparison, there was no significant difference in answers to the questions "Which dance is more active" and "Which dance is more aggressive".

The perceived confidence of the dancer characterizes dancer's perceived sensibility and mutual consistency of the sequence of dance movements. Therefore, the obtained outcome supports the assumption that semantic-map-based dance generation creates a more meaningful and better-perceived virtual dance.

4 Discussion

As digital arts become more and more popular today, with robotic and virtual dance being not an exception, automated tools that enhance human creativity are in demand. It is impossible to provide here a review of work in this direction. One example of an autonomous dancing cognitive robot is the well-known Robodanza [5, 6], which is based on a cognitive architecture. Perhaps the new element in the present work is the central role of the semantic map supporting social-emotional cognition in such system. Here we questioned a hypothesis of the role of emotional intelligence in creativity and in collaboration. Presented results support the hope that we are on the right track.

In general, intelligent cobots become increasingly popular today as creators of digital art, including robotic or virtual dancing. Our results support the idea that the creativity of such tools is linked to their social-emotional intelligence. We anticipate that the cognitive approach to creation of virtual art, that in general rapidly gains popularity [7], will grow into a user-favored Artificial Intelligence methodology that will yield higher productivity and creativity of teams and individuals.

Acknowledgments. The authors are grateful to all NRNU MEPhI students who participated in this study. This work was supported by the Russian Science Foundation Grant #18-11-00336.

References

1. Samsonovich, A.V.: On semantic map as a key component in socially-emotional BICA. Biol. Inspired Cogn. Arch. **23**, 1–6 (2018). https://doi.org/10.1016/j.bica.2017.12.002
2. Samsonovich, A.V.: Emotional biologically inspired cognitive architecture. Biol. Inspired Cogn. Arch. **6**, 109–125 (2013). https://doi.org/10.1016/j.bica.2013.07.009
3. Samsonovich, A.V., Goldin, R.F., Ascoli, G.A.: Toward a semantic general theory of everything. Complexity **15**(4), 12–18 (2010). https://doi.org/10.1002/cplx.20293
4. Gärdenfors, P.: Conceptual Spaces: The Geometry of Thought. MIT Press, Cambridge (2004)
5. Augello, A., Infantino, I., Manfrè, A., Pilato, G., Vella, F., Chella, A.: Creation and cognition for humanoid live dancing. Robot. Auton. Syst. **88**, 107–114 (2016)
6. Augello, A., Infantino, I., Pilato, G., Rizzo, R., Vella, F.: Creativity evaluation in a cognitive architecture. Biol. Inspired Cogn. Arch. **11**, 29–37 (2015)
7. Mohan, V., Morasso, P., Zenzeri, J., et al.: Teaching a humanoid robot to draw 'Shapes'. Auton. Robots **31**(1), 21–53 (2011)

The Method of Statistical Estimation of the Minimum Number of Tests for Reliable Evaluation of the Robotic Multi-channel Control Systems Quality

Konstantin Y. Kudryavtsev[1], Aleksei V. Cherepanov[1], Timofei I. Voznenko[1],
Alexander A. Dyumin[1,2(✉)], Alexander A. Gridnev[1], and Eugene V. Chepin[1]

[1] Institute of Cyber Intelligence Systems,
National Research Nuclear University MEPhI, Moscow, Russia
a.a.dyumin@ieee.org
[2] College of Information Business Systems,
National University of Science and Technology MISiS, Moscow, Russia

Abstract. There is a growth in adoption of multi-channel (tactile, voice, gesture, brain-computer interface, etc.) control systems for mobile robots that help to improve its reliability. But, a complex control system requires a large number of tests to determine the quality of operation. Hence, multi-channel control systems are subjects of rigorous testing process for estimation of the number of successful command recognitions and the number of errors. The more tests will be conducted, the more accurate evaluation of the control channel quality will be. However, in most cases, carrying out the tests is expensive and time-consuming. Therefore, it is necessary to determine the minimum number of tests required to evaluate channel control quality with a given significance level. In this paper we propose a technique for determining the minimum required number of tests. Experimental results of evaluating the multichannel control system of the mobile robotic wheelchair using this technique are presented.

Keywords: Reliability evaluation · Multi-channel control system
Robotics

1 Introduction

Multi-channel control systems currently are more often being used to control mobile robotic complexes. A sophisticated control system requires a large number of tests to determine the quality of operation, which is defined as the probability p of correctly recognized and executed commands incoming from different control channels. Carrying out the tests is expensive and time-consuming. Therefore, it is desirable to determine the minimum required number of tests, which provides an estimate of the probability p with a given accuracy.

This problem is essential for various research areas. The paper [1] considers clinical trials of drugs. It assumes a normal distribution of the test results

© Springer Nature Switzerland AG 2019
A. V. Samsonovich (Ed.): BICA 2018, AISC 848, pp. 203–208, 2019.
https://doi.org/10.1007/978-3-319-99316-4_27

of the two study groups. The minimum number of tests is determined on the basis of the Z-distribution and providing a given level of reliability, defined as the probability of a Type II error. The paper [2] is devoted to the economics of health care and analysis of patient preferences based on experiments with discrete choice. It is important to determine the sample size, which allows to answer research questions with a given accuracy. The estimation is based on the normal distribution of samples and a given confidence interval. The comparison of the mathematical expectations of the two data sets is based on the Z-distribution. The paper [2] notes that the size of experimental data samples is very different in various experimental studies. So, out of 69 studies, 22 (32%) had sample sizes less than 100, 16 (23%) had sample sizes greater than 600; 6 (9%) had samples larger than 1000. More than 70% of studies do not justify the used sample size clearly.

The paper [3] focuses on the determination of the minimum number of the same type systems for the objective conduct of a sample experiment for estimating the mean time between failures of an dangerous technological object. The problems of determining the minimum number of experiments for estimating the parameters of complex systems in various domains are also considered in [4–6].

In this paper, we solve the problem of determining the minimum number of tests required to evaluate the reliability of a multi-channel control system for a mobile robotic wheelchair.

2 Related Work

In developing of control channels for mobile robots (in particular, voice and gesture control channels), it is required to evaluate the quality of operation, which is defined as the probability p of correct recognition of the transmitted commands. One of the most frequently used methods of determining this probability is the method of statistical tests carried out according to the Bernoulli scheme, in which a random variable describing the test result can take only two ("success" and "failure") values. The developed system is tested and the number of successful command recognition and the number of errors are estimated. When n independent tests are conducted, the frequency of success is calculated

$$p \approx W_n = \frac{1}{n} \sum_{k=1}^{n} X_k \tag{1}$$

which is the most plausible estimate of the probability p of correct recognition of the transmitted commands. In order to obtain an estimate of the probability p with a given accuracy, it is necessary to find the confidence interval in which W_n is located with probability $1 - \alpha$ (α is the significance level).

To determine the minimum required number of tests, different methods based on the probability distribution of the random variables X_k and W_n can be used.

Method 1 (preliminary estimate). In the expression (1) X_k $(k = 1, 2, \ldots n)$ is a random variable having the Bernoulli distribution $Bi(1; p)$, where p is the probability of a "successful" test that is unknown and that should be evaluated. Taking into account that the normalized random variable

$$\dot{W} = \frac{W_n - M[W_n]}{\sqrt{D[W_n]}} = (W_n - p)\sqrt{\frac{n}{pq}} = \frac{k - np}{\sqrt{npq}} \tag{2}$$

(n is the number of trials, k is the number of "successful" outcomes, $q = 1 - p$) converges in distribution to the normal random value $U \sim N(0; 1)$ [7], given the significance level α (defining the quantile $u_{1-\frac{\alpha}{2}}$) and the maximum acceptable deviation of the estimated value from the true Δ, we find n as (3).

$$n = \left] \frac{u_{1-\frac{\alpha}{2}}^2}{4\Delta^2} \right[\tag{3}$$

For example, for the significance level $\alpha = 0.05$, $p = 0.5$ and the maximum allowable deviation $\Delta = 0.05$, we get the most pessimistic estimate of $n = 384$.

Method 2 (an optimistic estimate for the probability p close to 1). Taking into account the normal distribution of the random variable $\dot{W}_n$ (2), we require the fulfillment of the probabilistic condition

$$P\left\{\dot{W}_n \leq u_{1-\alpha}\right\} = 1 - \alpha \tag{4}$$

here $u_{1-\alpha}$ is a quantile of level $1 - \alpha$ of the normal distribution law $N(0; 1)$. For $k = n$, i.e. for a probability p close to 1, we obtain according [7] as (5).

$$n = \frac{u_{1-\alpha}^2 p}{1 - p} \tag{5}$$

For $p = 0.9$ and $\alpha = 0.05$, n will be equal to 54.

Method 3 (the most accurate estimate). As in the Method 1, the random variable X_k $(k = 1, 2, \ldots n)$ has a Bernoulli distribution $Bi(1; p)$. The random variable W_n (1) can take the values $0, \frac{1}{n}, \frac{2}{n}, \frac{3}{n}, \ldots, 1$, and its distribution function according [7] has the form of (6).

$$F\left(\frac{k}{n}, p\right) = \sum_{i=0}^{k} C_n^i p^i (1 - p)^{n-i} \tag{6}$$

Since the probability distribution function $F\left(\frac{k}{n}, p\right)$ is monotonic with respect to p, then setting the significance level α, we construct a central confidence interval that covers the unknown parameter $p \in [\theta_1, \theta_2]$ with probability $1 - \alpha$. To construct the confidence interval, it is necessary to solve the Eqs. (7) and (8).

$$1 - F\left(\frac{k-1}{n}, \theta_1\right) = \sum_{i=k}^{n} C_n^i \theta_1^i (1 - \theta_1)^{n-i} = \frac{\alpha}{2} \tag{7}$$

$$F\left(\frac{k}{n}, \theta_2\right) = \sum_{i=k}^{n} C_n^i \theta_2^i (1 - \theta_2)^{n-i} = \frac{\alpha}{2} \tag{8}$$

These equations are polynomials of order n. Numerical methods will be used to find θ_1 and θ_2.

3 Method for Determining the Minimum Number of Tests

For the calculation by formulas (5), (7) and (8) it is necessary to know p or the number of "successful" tests k. The most rough estimate, calculated by formula (3) with $p = 0.5$, contains a rather large number of tests ($n = 384$). The optimistic estimate calculated by formula (5) is valid for a probability p close to 1. Therefore, in order to determine the minimum number of tests and the subsequent probability estimate p, the following procedure is proposed.

At the first step, a small number of tests is carried out and a preliminary estimate of the probability p is calculated.

At the second step, the value of n is found as a result of solving Eqs. (7) and (8), and additional experiments are performed to achieve n. At the given number of tests, the obtained value of p is considered the target value.

At the third step, the probability p is recalculated and it is checked to be in the central confidence interval $[\theta_1, \theta_2]$. If the probability p is in the interval, then we stop, otherwise go back to the second step.

4 Results

The proposed procedure was used to determine the minimum number of tests required to evaluate the probability p of correct recognition of five voice control commands (forward, backward, turn left, turn right, stop) and four gesture control commands (forward, backward, left, right) for the robotic wheelchair [8]. 18 users participated in the research.

At the first stage, we conducted 50 (according to the optimistic estimate from formula (5)) tests for each of 162 (18 users × 9 teams) options and counted the number of "successful" tests.

At the second stage, we calculated the required number of tests by the formulas (5), (7) and (8) and carried out additional experiments.

Using the obtained estimate $p \approx 0.9$, the calculations according to formula (3) for $\alpha = 0.05$ and $\Delta = 0.05$ give the result $n = 138$. Calculations according to formula (5) lead to $n = 54$. To ensure the confidence interval $\Delta = 0.05$, calculated according to formulas (7) and (8), it is necessary to conduct 400 experiments ($n = 400$).

Thus, the most accurate and pessimistic estimate of the minimum number of tests for the probability $p \approx 0.9$ is about 400.

At the third stage, we calculated the target value of p (actually, 162 target values for 18 users and 9 commands for each user) and checked if it is in the central confidence interval. All 162 new target probability values of p were in the confidence interval $[0.87, 0.93]$.

As a result of the experiments, 162 unknown values of probability p_{eti} ($i = 1, \ldots, 162$) were estimated. Alternately, the probability p can be estimated from (1) by specifying different values of n. Setting n equal to 20, 50, 100 and 200, we find p_{ni} ($i = 1, \ldots, 162$) and calculate the relative deviation of p_{ni} from p_{eti} by the formula (9).

$$diff_{ni} = \frac{|p_{ni} - p_{eti}|}{p_{eti}} \cdot 100\% \quad (i = 1, \ldots, 162, \ n = 20, 50, 100, 200) \tag{9}$$

The results of percentage change of the target value from the values obtained at n equal to 20, 50, 100 and 200 are shown in the form of histograms in Fig. 1. At $n = 20$, there are cases when the deviation from the target value varies from 15 to 20%, which indicates a very rough estimation. In the case of $n = 50$ and 100, the deviation is 10%. With $n = 200$, the deviation did not exceed 5%.

Figure 2 shows the plot of dependency of the median of the relative deviation of the probability estimates p from the target value calculated for different n, on n. According to the figure, for $n = 100$ the median of the relative deviation from the target value reaches 1%, and at $n = 200$ it is close to 0. The result at

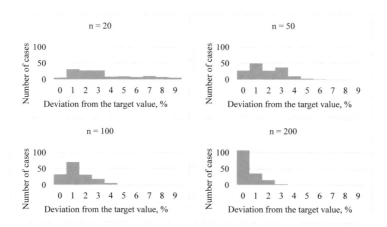

Fig. 1. The graph of dependency of the number of results on the percentage deviation from the target value for $n = 20, 50, 100, 200$

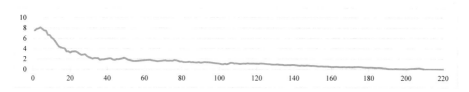

Fig. 2. The dependency of the median of the relative deviation of the probability estimates p from the target value calculated for different n

$n = 200$ is due to the fact that the majority of tests with n recalculation needed to be performed not 384 (according to formula (3)), but 210 times.

5 Conclusion

The method of determining the minimum required number of tests is proposed. Based on the proposed procedure, the number of required tests was determined based on the acceptable deviation from target value. The developed procedure is used to obtain the values of the reliability parameters of the decision-making system [9] for the multi-channel control system for a robotic device. This procedure will allow to conduct a smaller number of tests, which will reduce the material and time costs of testing. In future works, it is planned to use the results of this study to obtain a reliable estimate of the control channel performance of the mobile robotic wheelchair [8], which will be taken into account in the decision-making system for multi-channel control for the wheelchair.

References

1. Chow, S.C., Shao, J., Wang, H., Lokhnygina, Y.: Sample Size Calculations in Clinical Research. Chapman and Hall/CRC, Boca Raton (2017)
2. de Bekker-Grob, E.W., Donkers, B., Jonker, M.F., Stolk, E.A.: Sample size requirements for discrete-choice experiments in healthcare: a practical guide. Patient-Patient-Centered Outcomes Res. **8**(5), 373–384 (2015). https://doi.org/10.1007/s40271-015-0118-z
3. Sadyhov, G.S., Savchenko, V.P.: K probleme ocenki srednej narabotki do kritich-eskogo otkaza tekhnogenno-opasnogo ob"ekta (in Russian). Nadezhnost' i kachestvo slozhnyh sistem (1), 54–57 (2013)
4. Kock, N., Hadaya, P.: Minimum sample size estimation in PLS-SEM: the inverse square root and gamma-exponential methods. Inf. Syst. J. **28**(1), 227–261 (2018). https://doi.org/10.1111/isj.12131
5. Rodriguez-Lujan, I., Fonollosa, J., Vergara, A., Homer, M., Huerta, R.: On the calibration of sensor arrays for pattern recognition using the minimal number of experiments. Chemometr. Intell. Lab. Syst. **130**, 123–134 (2014). https://doi.org/10.1016/j.chemolab.2013.10.012
6. Black, M.A., Doerge, R.: Calculation of the minimum number of replicate spots required for detection of significant gene expression fold change in microarray experiments. Bioinformatics **18**(12), 1609–1616 (2002). https://doi.org/10.1093/bioinformatics/18.12.1609
7. Kibzun, A.I., Goryainova, E., Naumov, A., Kibzun, A., et al.: Teoriya veroyatnostej i matematicheskaya statistika. Bazovyj kurs s primerami i zadachami (in Russian). Fizmatlit, Moscow (2002)
8. Voznenko, T.I., Chepin, E.V., Urvanov, G.A.: The control system based on extended bci for a robotic wheelchair. Procedia Comput. Sci. **123**, 522–527 (2018). https://doi.org/10.1016/j.procs.2018.01.079
9. Gridnev, A.A., Voznenko, T.I., Chepin, E.V.: The decision-making system for a multi-channel robotic device control. Procedia Comput. Sci. **123**, 149–154 (2018). https://doi.org/10.1016/j.procs.2018.01.024

Neurotransmitters Level Detection Based on Human Bio-Signals, Measured in Virtual Environments

Vlada Kugurakova$^{(\boxtimes)}$ and Karina Ayazgulova

Kazan Federal University, Kazan, Russia
vlada.kugurakova@gmail.com, karina.11405@gmail.com

Abstract. In this paper, we explore the possibility of using the visual and sound stimuli obtained in various incidents when immersed in virtual reality, to detect human emotion by measuring the human bio-signals: heart rate, electroencephalogram (EEG), blood volume pressure, skin temperature and galvanic skin response (GSR) using bio-sensors. Further classification of signals occurs using a neural network. The received statistical characteristics are used as a contribution to the neural network for classification according to the Lövheim cube of emotions. The resulting algorithm for recognizing emotions based on human bio-signals in virtual reality will be used to predict emotional reactions to various events in virtual environments and, consequently, to increase their immersion.

Keywords: Virtual reality · VR · Virtual environment · Immersivity
IVE · Bio-sensor · Bio-signal · Lövheim · Human emotions
Emotion recognition

1 Introduction

Russian neurologist Vladimir Bekhterev believed that "unlike pantomime movements and gestures, facial expressions are always emotional and primarily reflect the speaker's feelings". There are many ways to identify the emotion experienced by a person speaking, such as by such analysis of facial expressions or vocal intonations. Whilst most certainly potent, such methodology of emotion recognition is not without its pitfalls – indeed, frequently people obscure and distort "natural" emotional signals in their communications, voluntarily or not, and this is all more likely to occur when the communication is occurring within a virtual reality. This is why the study of human emotions is necessary with the help of physiological signals, which cannot be forged. Among such signals, there is electroencephalogram (EEG), electrooculography (EOG), temperature (TEM), bulk blood pressure (BVP), electromyogram (EMG), electrocardiogram (ECG), skin galvanic reaction (GSR) and other methods [11]. A lot of studies of the psycho-emotional state of a person refer to the analysis of physiological signals coming from the cortex of the brain or skin.

© Springer Nature Switzerland AG 2019
A. V. Samsonovich (Ed.): BICA 2018, AISC 848, pp. 209–216, 2019.
https://doi.org/10.1007/978-3-319-99316-4_28

2   Our Idea

We propose an own metrics of note of emotional states of users in immersive virtual environments measured by human bio-signals, realizing the need of neurobiological plausibility as part of criteria for highly realistic cognitive architectures [10].

We will consider a method for emotion recognition from the data of the electroencephalogram obtained during the time of finding a person in virtual reality. The basics of how the tests are conducted go like this: the test subject is placed in the preprogrammed situation, specifically designed to elicit an emotional response. The data of the electroencephalogram is then collected, processed and analyzed. The main goal of this study is to find out whether, according to the EEG data, it is possible to determine that the emotions experienced during the event corresponding to the predicted ones.

It should be noted that there is no such classification that all researchers of behavior would accept. Some scientists recognize the existence of basic emotions, others dispute it, preferring to see the emotions only as a function of perceptive-cognitive processes. According to Ekman [5] there are 7 basic emotions: joy, surprise, sadness, anger, disgust, contempt, fear. Izard [6] lists 8 of them: *pleasure-joy, interest-excitation, surprise-fear, grief-suffering, anger-rage, fear-horror, disgust-loathing, shame-humiliation.* Relying on the established classifications, we selected 3 emotions – *a fear, a disgust, a surprise* – to monitor in our tests.

The tests were conducted in the immersive virtual reality, which means that this implementation of VR technology has following qualities:

1. plausible simulation of the world with a high degree of detail;
2. a high-performance computer capable of recognizing user actions and responding to them in real time;
3. special equipment connected to a computer that provides an immersive effect in the process of environmental research.

Full immersion in virtual reality provides instant response to a given event, which will help in collecting data for the study. Previously, we described our previous own experience in the development of virtual simulators with full immersion in the field of learning biotechnology [3, 7, 9] and ways to increase the immensity of virtual environments [8].

3   Lövheim Cube of Emotions

One of the few models in modern science that gives a physiological explanation for emotions, linking the appearance of emotions with the levels of three monoamine neurotransmitters – noradrenaline, dopamine, and serotonin – is a cube of emotions [12]. These 3 neurotransmitters form a coordinate system where the eight main emotions, labeled according to the theory of the influence of Tomkins [17], are placed in 8 corners of the cube and correspond to 8 possible combinations of the levels of these 3 monoamines. The model offers a direct relationship between certain combinations of neurotransmitter levels and basic emotions.

Each of the investigated emotions (fear, disgust, surprise), according to the levels of neurotransmitters from Lövheim cube, should be marked 0 or 1 – to the low and high level of neurotransmitters respectively (see Table 1).

Table 1. According to the levels of neurotransmitters from Lövheim cube

Ser	Dop	Nor	Emotions
0	1	0	Fear
1	0	0	Disgust
1	0	1	Surprise

4   Areas of the Brain for EEG Data Extraction

The studies performed by us earlier show that the most suitable areas for taking the indications are central, frontal and temporal areas. The most informative are the leads **F3**, **F4**, **FZ**, **F8**, **C3**, **C4**, **CZ**, **T4**, **T6**, **P4** from the regions for EEG removal (see Fig. 1).

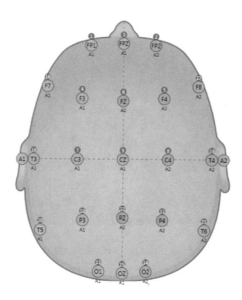

Fig. 1. Areas of EEG extraction

Data collection for the study was performed using an electroencephalograph 21-channel Clinical Diagnostic EEG System **Neuron Spectrum-4/P** [2], which allows recording EEG signals of high quality, low noise gives the ability to apply the most sparing EEG filtering or not to filter at all, which maximizes the useful information in the original signal.

5   The Experiment

50 respondents participated in the survey. Immediately before the beginning of the experiment, the emotional state of the subjects was assessed to ensure that participants started the experiment with emotional clean slate.

Before the beginning of the experiment, events in virtual reality were prepared, designed to elicit the emotions of fear, surprise, disgust. We selected a virtual simulator for training forensic scientists, namely examination of the scene on the night of the cemetery location. The choice of scene allowed us to cover all of the emotional reactions of interest in one single continuous experience: a sense of fear when stumbling on an excavated grave, of disgust when seeing a corpse, of surprise when suddenly noticing gargoyles among the standard graves (see Fig. 2). *After all, you were also surprised to see this picture in this article.* The forensic testing ground in virtual reality was deliberately chosen to get the purest emotional reactions, based on immersion in the highly realistic virtual environment, while also maintaining the possibility of encountering an analogous situation in the real life, devoided of horror and mystical elements (real reality – RR).

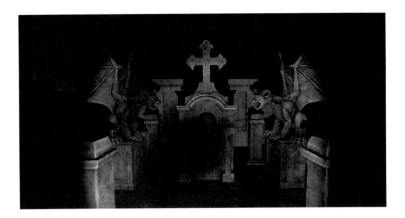

Fig. 2. Screenshot from virtual environments to awaken surprise

The electroencephalogram is recorded automatically as the events within the VR activate. The recording is also accompanied by the comments in the beginning and the end.

The experiment conducted within the framework of this study assumes fixing the EEG in two layers in parallel: (1) data recording while in virtual reality; (2) registration of data in the course of an event. Accordingly, the output should contain information about each layer of the experiment. We can determine the boundaries of each of the layers with the help of components that register EEG data, as they also record the time throughout the experiment. Knowing the duration of events and the intervals between them, you can definitely pinpoint the EEG fragments for further analysis.

Frequency domain functions allow studying the EEG signal by converting the EEG signal of an untreated time domain to an EEG signal of the frequency domain using the Fourier transformation method [13].

6 Experimental Data Processing

To perform the analysis on the data obtained, we selected a handful of machine learning algorithms that solve the classification problems: **Decision Tree** [14], **Random Forest** [15, 16], **k-Nearest Neighbors** [4], and a neural network.

The available recording of electroencephalogram signals in the *.edf* format is then converted to the text format *.txt*. In the data recording view mode, due to comments on the beginning and end of the event, a period of time during which the user presumably experienced one of the emotions is determined. All selected segments are equal in time to signal recording.

In order to determine the emotion experienced during the submersion in the virtual reality, it is necessary to mark the EEG indicators of the training sample with the appropriate marks – 0 and 1, which denote the level of the neurotransmitters – low and high, respectively. The training sample was compiled from the test data. For each emotion, the indicators of all respondents were taken, mixed in random order and randomly selected for 15,000 values. Thus, the output was a file in the *.csv* format with 45000 values for 3 emotions.

To carry out the training, classification algorithms are used, which allow to determine the levels of neurotransmitters and on them to make an emotion. The training was conducted in **the Jupiter Notebook** web application [1] using **Python** version 3.6.2 (see Figs. 3, 4 and 5).

The prediction accuracy of Decision Tree algorithm is 0.66 for serotonin levels, 0.33 for dopamine levels, and 1 for noradrenaline levels. The prediction accuracy of Random Forest algorithm is 0.6572 for serotonin levels, 0.3428 for dopamine levels, and 0.9903 for noradrenaline levels. The prediction accuracy of k-Nearest Neighbors algorithm is 0.9293 for serotonin levels, 0.9293 for dopamine levels, and 0.9193 for noradrenaline levels. Prediction accuracy of neural network is 68%.

After conducting the training on all of the available data sets, 100% probability of correct prediction of the level of neurotransmitters using the algorithm k-Nearest Neighbors was obtained. According to the previous research, the greatest information about the emotion under test can be obtained from the signals of electrodes located in the central, frontal and temporal areas. After learning the

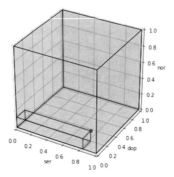

Fig. 3. Graphical representation of the result obtained for the disgust emotion

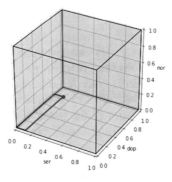

Fig. 4. Graphical representation of the result obtained for the fear emotion

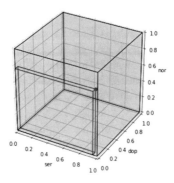

Fig. 5. Graphical representation of the result obtained for the surprise emotion

model on 8 electrodes from the regions **F3-A1**, **F4-A2**, **FZ-A2**, **C3-A1**, **C4-A2**, **CZ-A1**, **F8-A2**, **T4-A2**, it was checked whether it is possible to predict the emotion, number of leads.

Table 2. Neurotransmitter levels predicted by 8-lead models

METHODS		NEUROTRANSMITTERS		
		Serotonin	Dopamine	Noradrenalin
DECISION TREE	Fear	0.725076923076923	0.279076923076923	0.297692307692307
	Disgust	0.369076923076923	0.369076923076923	0.369076923076923
	Surprise	0.560384615384615	0.560384615384615	0.560384615384615
RANDOM FOREST	Fear	0.692153846153846	0.197307692307692	0.222
	Disgust	0.132076923076923	0.132076923076923	0.132076923076923
	Surprise	0.496461538461538	0.496461538461538	0.496461538461538
K-NEAREST NEIGHBORS	Fear	0.261461538461538	0.261461538461538	0.261461538461538
	Disgust	0.771692307692307	0.228307692307692	0.183769230769230
	Surprise	0.482692307692307	0.482692307692307	0.482692307692307
CNN	Fear	0.5187177	0.21663412	0.26464862
	Disgust	0.543112	0.22431313	0.23257533
	Surprise	0.5184015	0.13771346	0.34388646

In Table 2 data on 8 leads, on 19 leads we can conclude that the results of all the algorithms used are satisfactory. Each of the algorithms accurately predicts the emotion experienced on the available data.

It is worth noting that the running time of the algorithms is significantly different. The Decision Tree and Random Forest algorithms proved to be faster than the k-Nearest Neighbors algorithm and the neural network. It took about 40 min for 3000 iterations to train a neural network. This result will not allow you to predict emotions in real time.

7   Conclusion and Future Works

The results obtained with the selected areas indicate that there is not enough data to predict the emotion. On 8 leads for noradrenaline alone, there is a high probability of a correct determination of the level (about 100%), for serotonin and dopamine, the detection rate is much lower, about 20%.

The accuracy of the algorithms for working in 8 leads is worse than when working with 19 leads. By the ratio of the time of operation and the accuracy of the prediction, the best algorithm is the k-Nearest Neighbors algorithm. Among the Decision Tree and Random Forest algorithms, it is relatively accurate and spends much less time working than a neural network.

The main conclusion: the definition of the tested emotion according to EEG data is possible, the results can be used to predict and control emotions. The following tasks that need to be addressed

1. the prediction of emotion from the reactions received during the performance of events in the virtual reality, in real time;
2. a combination of data from various bio-signals to clarify the emotions experienced;
3. predicting the user's reactions to the chain of events in virtual reality;

4. testing the hypothesis of the possibility of forming stable reactions after leaving the virtual environment;
5. develop metrics for assessing the impact of the mechanics of narrative identity in virtual reality.

Acknowledgment. This work was funded by the subsidy of the Russian Government to support the Program of competitive growth of Kazan Federal University among world class academic centers and universities.

References

1. The jupiter notebook. http://jupyter.org
2. Neuron-spectrum-4/p. http://neurosoft.com/en/catalog/view/id/18
3. Abramov, V., et al.: Virtual biotechnological lab development. BioNanoScience **7**(2), 363–365 (2017)
4. Bronshtein, A.: A quick introduction to k-nearest neighbors algorithm. Medium, April 2017. https://medium.com/@adi.bronshtein/a-quick-introduction-to-k-nearest-neighbors-algorithm-62214cea29c7
5. Ekman, P.: Are there basic emotions? Psychol. Rev. **99**(3), 550–553 (1992)
6. Izard, C.E.: Human Emotions. Springer Publishing Company (2013)
7. Kugurakova, V., Abramov, V., Abramskiy, M., Manakhov, N., Maslaviev, A.: Visual editor of scenarios for virtual laboratories. In: 10th International Conference on Developments in eSystems Engineering, pp. P242–P247 (2017)
8. Kugurakova, V., Elizarov, A., Khafizov, M., Lushnikov, A., Nizamutdinov, A.: Towards the immersive VR: measuring and assessing realism of user experience. In: Proceedings of the 2018 International Conference on Artificial Life and Robotics (2018)
9. Kugurakova, V., Khafizov, M., Akhmetsharipov, R.: Virtual surgery system with realistic visual effects and haptic interaction. In: Proceedings of the 2017 International Conference on Artificial Life and Robotics, pp. P86–P89 (2017)
10. Kugurakova, V., Talanov, M., Ivanov, D.: Neurobiological plausibility as part of criteria for highly realistic cognitive architectures. Procedia Comput. Sci. **88**, 217–223 (2016)
11. Li, Y., Huang, J., Zhou, H., Zhong, N.: Human emotion recognition with electroencephalographic multidimensional features by hybrid deep neural networks. Appl. Sci. **7**(10), 1060 (2017). Switzerland
12. Lövheim, H.: A new three-dimensional model for emotions and monoamine neurotransmitters. Med. Hypotheses **78**(2), 341–348 (2012)
13. Murugappan, M., Murugappan, S.: Human emotion recognition through short time electroencephalogram (EEG) signals using fast fourier transform (FFT). In: Proceedings – 2013 IEEE 9th International Colloquium on Signal Processing and its Applications, CSPA 2013, pp. 289–294 (2013)
14. Njeri, R.: What is a decision tree algorithm? Medium, September 2017. https://medium.com/@SeattleDataGuy/what-is-a-decision-tree-algorithm-4531749d2a17
15. Patel, S.: Random forest classifier. Medium, May 2017. https://medium.com/machine-learning-101/chapter-5-random-forest-classifier-56dc7425c3e1
16. Synced: How random forest algorithm works in machine learning. Medium, October 2017. https://medium.com/@Synced/how-random-forest-algorithm-works-in-machine-learning-3c0fe15b6674
17. Tomkins, S.: Script theory: differential magnification of affects. Springer Publishing Company (1991)

Heterogeneous Proxytypes Extended: Integrating Theory-Like Representations and Mechanisms with Prototypes and Exemplars

Antonio Lieto[1,2(✉)] 🆔

[1] Department of Computer Science, University of Turin, Turin, Italy
lieto@di.unito.it
[2] Cognitive Robotics and Social Sensing Lab, ICAR-CNR (Palermo), Palermo, Italy

Abstract. The paper introduces an extension of the proposal according to which conceptual representations in cognitive agents should be intended as *heterogeneous proxytypes*. The main contribution of this paper is in that it details how to reconcile, under a heterogeneous representational perspective, different theories of typicality about conceptual representation and reasoning. In particular, it provides a novel theoretical hypothesis - as well as a novel categorization algorithm called DELTA - showing how to integrate the representational and reasoning assumptions of the theory-theory of concepts with the those ascribed to the prototype and exemplars-based theories.

Keywords: Heterogeneous proxytypes · Knowledge representation
Cognitive agents · Cognitive architectures · Declarative memory

1 Introduction

The proposal of characterizing the representational system of cognitive artificial agents by considering conceptual representations as *heterogeneous proxytypes* was introduced in [1][1] and has been recently employed and successfully tested in systems like DUAL-PECCS [2–4], later integrated with diverse cognitive architectures such as ACT-R [5], CLARION [6], SOAR [7] and Vector-LIDA [8]. The main contribution of this work is in that it offers a proposal to reconcile, under a heterogeneous representational perspective, not only prototype and exemplars based representations and reasoning procedures, but also the representational and reasoning assumptions ascribed to the so called theory-theory of concepts [9]. In doing so, the paper proposes a novel categorization algorithm,

---

[1] The expression *heterogeneous proxytypes* refers to both a theoretical and computational hypothesis combining the proxytype theory of concepts with the so called heterogeneity approach to concept representation. The Sect. 3 of this paper contains a brief discussion of the proposal.

© Springer Nature Switzerland AG 2019
A. V. Samsonovich (Ed.): BICA 2018, AISC 848, pp. 217–227, 2019.
https://doi.org/10.1007/978-3-319-99316-4_29

called *DELTA* (i.e. unifie**D** Cat**E**gorization a**L**gorithm for he**T**erogeneous repre-sent**A**tions) able to unify and integrate, in a cognitively oriented perspective, all the common-sense categorization mechanisms available in the cognitive science literature. The rest of the paper is organized as follows: the Sect. 2 provides an overview of the main representational paradigms proposed by the Cognitive Science and the Cognitive Modelling communities. Section 3, briefly synthesize the representational framework intending concepts as *heterogeneous proxytypes* by showing how such theoretical proposal has been actually implemented and suc-cessfully tested in the DUAL-PECCS system. Section 4, proposes a more close analysis of the findings of the theory-theory of concepts, while, Sect. 5, proposes a novel and extended categorization algorithm integrating the theory-theory rep-resentational and reasoning mechanisms with those involving both exemplars and prototypes.

2   Prototypes, Exemplars, Theories and Proxytypes

In the Cognitive Science literature, different theories about the nature of con-cepts have been proposed. According to the so called classical theory, concepts can be simply defined in terms of sets of necessary and sufficient conditions. Such theory was dominant until the mid '70s of the last Century, when Rosch's exper-imental results demonstrated the inadequacy of such a theory for ordinary –or common-sense – concepts [10]. Rosch's results suggested, on the other hand, that ordinary concepts are characterized and organized in our mind in terms of *prototypes*. Since then, different theories of concepts have been proposed to explain different representational and reasoning aspects concerning the problem of typicality: the prototype theory, the exemplars theory and the theory-theory. According to the *prototype* view, knowledge about categories is stored in terms of prototypes, i.e., in terms of some representation of the "best" instance of the category. In this view, the concept *bird* should coincide with a representation of a typical bird (e.g., a robin). In the simpler versions of this approach, prototypes are represented as (possibly weighted) lists of typical features. According to the *exemplar* view, a given category is mentally represented as set of specific exem-plars explicitly stored in memory: the mental representation of the concept *bird* is a set containing the representation of (some of) the birds we encountered dur-ing our past experience. Another well known typicality-based theory of concepts is the so called the *theory-theory* [9]. Such approach adopts some form of holistic point of view about concepts. According to some versions of the theory-theories, concepts are analogous to theoretical terms in a scientific theory. For example, the concept *cat* is individuated by the role it plays in our mental theory of zool-ogy. In other versions of the approach, concepts themselves are identified with micro-theories of some sort. For example, the concept *cat* should be identified with a mentally represented microtheory about cats.

Although these approaches have been largely considered as competing ones (since they propose different models and predictions about how we organize and reason on conceptual information), they turned out to be not mutually exclu-sive [11]. Rather, they seem to succeed in explaining different classes of cognitive *phenomena*, such as the fact that human subjects use different representations to categorize concepts. In particular, it seems that we can use - in different situations - exemplars, prototypes or theories [9,12,13]. Such experimental evidences led to the development of the so called "heterogeneous hypothesis" about the nature of concepts: this approach assumes that concepts do not constitute a unitary phenomenon, and hypothesizes that different types of conceptual representations may co-exist: prototypes, exemplars, theory-like or classical representations [14]. All such representations, in this view, constitute different *bodies of knowledge* and contain different types of information associated to the the same conceptual entity. Furthermore, each body of conceptual knowledge is assumed to be featured by specific processes in which such representations are involved (e.g., in cognitive tasks like recognition, learning, categorization, *etc.*). In particular prototypes, exemplars and theory-like default representations are associated with the possibility of dealing with non-monotonic strategies of reasoning and categorization, while the classical representations (i.e. that ones based on necessary and/or sufficient conditions) are associated with standard deductive mechanism of reasoning[2].

In recent years an alternative theory of concepts has been proposed: the *proxytype theory*. It postulates a biological localization and interaction between different brain areas for dealing with conceptual structures. Such localization have a direct counterpart in the well known distinction between *long term* and *working memory* [16]. In addition, such characterization is particularly interesting for the explanation of phenomena such as, for example, the activation (and

---

[2] In order to explain the different categorization strategies associated to different kinds of representations, let us consider the following examples: if we have to categorize a stimulus with the following features: "it has fur, woofs and wags its tail", the result of a *prototype-based categorization* would be *dog*, since these cues are associated to the prototype of *dog*. Prototype-based reasoning, however, is not the only type of reasoning based on typicality. In fact, if an exemplar corresponding to the stimulus being categorized is available, too, it is acknowledged that humans use to classify it by evaluating its similarity w.r.t. the exemplar, rather than w.r.t. the prototype associated to the underlying concepts [15]. For example, a penguin is rather dissimilar from the prototype of *bird*. However, if we already know an exemplar of penguin, and if we know that it is an instance of *bird*, it is easier to classify a new penguin as a *bird* w.r.t. a categorization process based on the similarity with the prototype of that category. This type of common-sense categorization is known in literature as *exemplars-based categorization*. An example of theory-like common sense reasoning is when we typically associate to a light switch the learned rule that if we turn it "on" then the light will be provided (this is a non-monotonic inference with a defeasible conclusion). Finally, the classical representations (i.e. those based on necessary and/or sufficient conditions) are associated with standard deductive mechanism of reasoning. An example of standard deductive reasoning is the categorization as *triangle* of a stimulus described by the features: "it is a polygon, it has three corners and three sides". Such cues, in fact, are necessary and sufficient for the definition of the concept of triangle. All these representations, and the corresponding reasoning mechanisms, are assumed to be potentially co-existing according to the heterogeneity approach.

the retrieval) of conceptual information. In this setting, concepts are seen as *proxytypes*. A *proxytype* is any element of a complex representational network *stored in long-term memory* corresponding to a particular *category* that can be tokenized in working memory to 'go proxy' for that category [16]. In other terms, the proxytype theory, inspired by the work of Barsalou [17], considers concepts as *temporary constructs* of a given category, activated (tokenized) in working memory as a result of conceptual processing activities, such as concept identification, recognition and retrieval.

3   Heterogeneous Proxytypes

In the original formulation of the proxytypes theory, however, proxytypes have been depicted as monolithic conceptual structures, primarily intended as prototypes [18]. A revised view of this approach has been recently proposed, hypothesizing the availability of a wider range of representation types than just prototypes [1]. They correspond to the kinds of representations hypothesized by the above mentioned heterogeneous approach to concepts. In this sense, proxytypes are assumed to be heterogeneous in nature (i.e., they are assumed to be composed by heterogeneous networks of conceptual representations and not only by a monolithic one)[3].

In this renewed formulation, heterogeneous representations (such as *prototypes*, *exemplars*, *theory-like* structures, *etc.*) for each conceptual category are assumed to be *stored in long-term memory*. They can be activated and accessed by resorting to different categorization strategies. In this view, each representation has its associated accessing procedures. In the following, I will briefly present how such theoretical hypothesis has been implemented in the DUAL-PECCS categorization system, and I will use the latter system as a computational referent for showing how the proposals presented in this paper can extend both the system itself and, more importantly, its underlying theoretical framework.

3.1   Heterogeneous Proxytypes in DUAL-PECCS

DUAL-PECCS [2,3], is a cognitive categorization system explicitly designed and implemented under the heterogeneous proxytypes assumption[4] for both the representational level (that is: it is equipped with a hybrid knowledge base composed of heterogeneous representations, each endowed with specific reasoning

---

[3] The heterogeneity assumption has been recently pointed out as one of the problems to face in order to address the problems affecting the knowledge level in cognitive systems and architecture [19,20].

[4] The characterization in terms of "heterogeneous proxytypes", among the other things, enables the system to deal with the problem of the "contextual activation" of a given information based on the external stimulus being considered. In particular (by following the idea that, when we categorize a stimulus, we do not activate the whole network of knowledge related to its assigned category but, conversely, we only activate the knowledge that is "contextually relevant" in its respect), DUAL-PECCS *proxyfyes* only the type of representation that minimizes the distance w.r.t. the percept (see [1] for further details).

mechanisms) and for the 'proxyfication' mechanisms (i.e.: the set of procedures implementing the tokenization of the different representations in working memory). The heterogeneous conceptual architecture of DUAL PECCS includes prototypes, exemplars and classical representations. All these different bodies of knowledge point to the same conceptual entity (the anchoring for these different types of representations is obtained via the Wordnet, see again [2]). An example of the heterogeneous conceptual architecture of DUAL PECCS is provided in the Fig. 1. Such figure shows how it is represented the concept *dog*. In this case, the prototypical representation grasps information such as that dogs are usually conceptualized as domestic animals, with typically four legs, a tail *etc.*; the exemplar-based representations grasp information on individuals. For example, in Fig. 1 it is represented the individual of *Lessie*, which is a particular exemplar of *dog* with white and brown fur and with a less domestic attitude w.r.t. the prototypical dog (e.g. its typical location is lawn). Within the system, both the exemplar and prototype-based representations make use of non classical (or typical) information and are represented by using the framework of the conceptual spaces [21, 22]: a particular type of vector space model adopting standard similarity metrics to determine the distance between instances and concepts within the space. The representation of classical information (e.g. the fact that $Dog \sqsubseteq Animal$, that is to say that "*Dogs* are also *Animals*") is, on the other hand, demanded to standard ontological formalisms. In the current version of the system the classical knowledge component is grounded in the OpenCyc ontology [23].

By assuming the heterogeneous hypothesis, in DUAL-PECCS, different kinds of reasoning strategies are associated to these different bodies of knowledge. In particular, the system combines non-monotonic common-sense reasoning (associated to the *prototypical* and *exemplars-based*, conceptual spaces, representations) and standard monotonic categorization procedures (associated to the classical, ontological, body of knowledge). These different types of reasoning are harmonized according to the theoretical tenets coming from the dual process theories of reasoning and rationality [24, 25].

As emerges from the Fig. 1, a missing part of the current conceptual architecture in the DUAL-PECCS system (and in its underlying theoretical hypothesis) concerns the representation of the default knowledge in terms of theory-like representational structures (while it already integrates classical, prototypical and exemplars based knowledge representation and processing mechanisms). In the next section we will show how Theory-like representations can be considered *dual* in nature (at least from a formal point of view) and therefore may deserve a dual treatment also form a computational point of view.

4   The Duality of Theory-Like Representations

As mentioned in Sect. 2, Theory-theory approaches [9, 26] assume that concepts consists of more or less complex mental structures representing (among other things) causal and explanatory relations between them (including folk psychology connections). During the 80's, these approaches stemmed from a critique to

— Hybrid Knowledge Base —

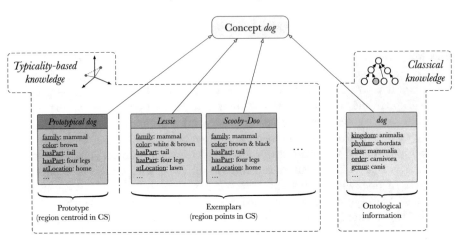

Fig. 1. Heterogeneous representational architecture for the concept DOG in DUAL-PECCS.

the formerly dominant theory of concepts as prototypes. Consider, for example, the famous [13] transformation experiments, in which subjects were asked to make categorization judgments about the biological membership of an animal that had undergone unusual transformations. In such experiment, Keil showed that people relies on theory-like representation (instead of prototypes) in order to execute their categorization task. In particular, it was shown that the type of knowledge retrieved by the subjects to solve these tasks belongs to their "default common-sense theory" associated to a given concept.

The idea that for most of our categories, our default knowledge includes a common-sense theory of that category (and that theory-like default bodies of knowledge are associated with a distinct kind of categorization process) is, however, only one of the available interpretations about the theory-like representational structures [14]. Another kind of interpretation, in fact, assumes that theory-like structures do not constitute our typical default knowledge but that, on the other hand, they are constitutive of our classical background knowledge [27]. In order to better explain this difference, and thus the *duality* of the theory-like representations, let us consider the case of DUAL-PECCS. As mentioned above, the current version of the system does not allow to represent the type of theory-like default knowledge belonging to the typical conceptual component of the architecture (see footnote 1 for an example of the non-monotonic reasoning that could be enabled by this kind of knowledge). On the other hand, it allows to represent (in terms of IF-THEN rules enabling monotonic inferences), the kind of theory-like knowledge structures which are compliant with the ontological semantics of the classical conceptual component. In other words: only certain types of theories, i.e. causal theories, belonging to the background knowledge of a cognitive agents, are currently covered by the integration of the current state of the art ontology languages and rules [28] in the DUAL-PECCS system. However, as already pointed out before, common sense knowledge is mostly characterized in terms of "theories" which are based on arbitrary, i.e. experience-based, rules. Therefore, in order to represent, within an artificial system, more realistic (from a cognitive standpoint) "theories", i.e. including common-sense default theories as intended in the theory-theory approaches, there is the need of going beyond classical logic rules. Recently, graphical models (in particular Bayesian networks) have been proposed as a computational framework able to represent [29,30] knowledge networks of theory-like common-sense default representation. The integration of such framework within the DUAL-PECCS system represent a current and future area of development, not yet concluded. In the remaining of this paper, such integration will not be discussed and, in a certain sense, will be taken for granted. I shall focus, instead, on the presentation of a novel unifying categorization algorithm - named DELTA - able to harmonize all the different types of typicality-based representations and reasoning mechanisms associated with the common-sense knowledge: exemplars, prototypes and default theory-like representations. I will leave aside the discussion concerning the integration of such common-sense categorization mechanisms with those concerning the classical monotonic ones. As above mentioned, in fact, such integration is already provided in DUAL-PECCS [2,3] and is tackled by recurring to the dual process theory of reasoning (i.e.: the non monotonic reasoning results of the heterogeneous common-sense conceptual components are then checked and integrated with the monotonic reasoning strategies executed in the classical conceptual component). Therefore, the underlying *heterogeneous-proxytypes* assumption, integrated with the dual process theory of reasoning, has been already proven to be effective to harmonize non monotonic and monotonic categorization strategies associated to heterogeneous body of knowledge. I will not report here the details of such harmonization procedure because it is already documented elsewhere [2]. I will focus, instead, on the harmonization procedures concerning the non monotonic categorization processes of the typical conceptual components.

5   A Unified Categorization Algorithm for Exemplars, Prototypes and Theory-Like Representations

In the following, I propose a novel categorization algorithm that, given a certain stimulus d, must select the most appropriate typicality-based representation available in the declarative memory of a cognitive agent (i.e. a prototype, an exemplar or a theory-like structure). According to what introduced in the previous sections, such declarative memory is assumed to rely on the *heterogeneous proxytypes* hypothesis.

The implemented procedure works as follows: when the input *stimulus* is similar enough to an exemplar representation (a threshold has been fixed to these ends and is already in use in DUAL-PECCS), the corresponding exemplar of a given category is retrieved. Otherwise, the prototypical representations are also scanned and the representation (prototype or exemplar) that is closest to the input is returned. By following a preference that has been experimentally observed in human cognition [31], this algorithm favors the results of the exemplars-based categorization if the knowledge-base stores any exemplars similar to the input being categorized. As an additional constraint, it has been hypothesized a mechanism in which theory-like structures of default knowledge can also override the categorization based on prototypes. Such mechanism has been devised based on the fact that theory-theorists have shown that, in some categorical judgments tasks (e.g. assessing the situation where a dog is made to look like a raccoon), categorization is driven by the possession of a rudimentary biological theory and by theory-like representations [32]. In other words: being a dog, for example, isn't just a matter of looking like a dog. It seems, in fact, that it is more important to have a network of appropriate hidden properties of dogs: the dog "essence" [32]. In the proposed algorithm, this element has been taken into account by hypothesizing: (i) to measure the similarity between the theory-like representation of the first retrieved prototype with the stimulus d^5 and (ii) to compare the obtained result with a *Conceptual Coherence Threshold* that should measure how much the considered stimulus d shares, i.e. is *conceptually coherent*, with the corresponding theory-like representation of the retrieved prototype. The analysis of the conceptual coherence can be solved as a constraint satisfaction problem as shown in [33].

In this setting, if the distance between the stimulus d and theory-like representation of the originally retrieved prototype is above the considered threshold, it means that the retrieved prototype is assumed to be representative enough of the common-sense "essence" of d (i.e. it is "coherent enough"). In this case, the prototypical answer is maintained, otherwise it is overridden by the theory-like representation which is closer to d.

Let us assume, for example, that the stimulus to categorize is represented by an atypical Golden Zebra (which is almost totally white) and that in our agent's long-term memory there is no exemplar similar enough to this entity. This means that there will be no exemplar-based representation selected by our algorithm, and that the most similar representation to d will be searched among the prototypical representations in the agent knowledge base. Now: if we assume that the retrieved prototype is a typical white horse, we could discard such representation by simply relying on additional information coming from the stimulus d (e.g. the fact that lives in the Savannah, etc.) and comparing d with the default and common-sense theory associated to a horse (i.e. the category of the original prototypical choice). In this case, the categorical selection of the class Golden Zebra would be obtained by exploiting theory-like representational networks. A synthetic representation of the proposed procedure is presented in the Algorithm 1.

---

[5] Since all the different bodies of knowledge are assumed to be co-referring representational structure pointing to the same conceptual entity, it is possible to recover the theory-like representation associated, for example, to a given prototype or exemplar based representation.

Data: Stimulus d; list of candidate representations: $closed^{S1}$.
Result: A typicality based representation of a category.

1 $closed^{S1} = \{\emptyset\}$
2 $S1_{EX} \leftarrow$ categorizeExemplars(d);
3 if firstOf($S1_{EX}$, $closed^{S1}$).distance(d) $<$ exemplarsSimilarityThreshold then
4 | return firstOf($S1_{EX}$, $closed^{S1}$);
5 else
6 | $S1_{PR} \leftarrow$ categorizePrototypes(d);
7 | candidatePrototype $\leftarrow$ firstOf($S1_{PR}$);
8 end
9 if (theoryOf)(candidatePrototype, $closed^{S1}$).distance(d) $>$
   conceptualCoherenceThreshold then
10 | return candidatePrototype;
11 else
12 | $S1_T \leftarrow$ categorizeTheory(d);
13 | return firstOf(TheoryBasedCategorization, $closed^{S1}$);
14 end

Algorithm 1: *DELTA*: A Unified categorization algorithm for prototypes, exemplars and theory-like representations.

6 Conclusions and Future Work

In this paper I have proposed a categorization algorithm able to unify all the common-sense categorization strategies proposed in the cognitive science literature: exemplars, prototypes and theory-like common-sense knowledge structures. To the best of my knowledge, this proposal represents the first attempt of providing a unifying categorization strategy by assuming a heterogeneous representational hypothesis. In particular, the proposed algorithm relies and extends both the representational and the reasoning framework considering concepts as heterogeneous proxytypes [1]. The current theoretical proposal needs to be tested on the empirical ground in order to show both its feasibility with psychological data and its efficacy in the area of artificial cognitive systems. Also, there are additional elements, only sketched in this paper, requiring a more precise characterization. For example: the design of a method to calculate which is the most appropriate theory-like representations to select (line 13, Algorithm 1). On this point, however, it is worth-noticing that, since the most promising computational candidates for representing the theory-like body of knowledge are graphical models and probabilistic semantic networks, it seems plausible to imagine that such calculation could be performed with standard heuristics search on graph structures. Similarly, the individuation and the construction of a plausible *Conceptual Coherence Threshold* represents an issue that should be faced and solved empirically.

Acknowledgements. The topics presented in this paper have been discussed in these years with a number of people in international conferences, symposia, panels and workshops. I thank all them for the received comments. I am particularly indebted to

Marcello Frixione, Leonardo Lesmo, Paul Thagard, David Danks, Ismo Koponen and Christian Lebiere for their feedback and suggestions. I also thank Valentina Rho for her comments on an earlier version of this paper.

References

1. Lieto, A.: A computational framework for concept representation in cognitive systems and architectures: concepts as heterogeneous proxytypes. Procedia Comput. Sci. **41**, 6–14 (2014)
2. Lieto, A., Radicioni, D.P., Rho, V.: Dual PECCS: a cognitive system for conceptual representation and categorization. J. Exp. Theor. Artif. Intell. **29**(2), 433–452 (2017). https://doi.org/10.1080/0952813X.2016.1198934
3. Lieto, A., Radicioni, D.P., Rho, V.: A common-sense conceptual categorization system integrating heterogeneous proxytypes and the dual process of reasoning. In: Proceedings of the International Joint Conference on Artificial Intelligence (IJCAI), pp. 875–881. AAAI Press (2015)
4. Lieto, A., Radicioni, D.P., Rho, V., Mensa, E.: Towards a unifying framework for conceptual represention and reasoning in cognitive systems. Intell. Artif. **11**(2), 139–153 (2017)
5. Anderson, J.R., Bothell, D., Byrne, M.D., Douglass, S., Lebiere, C., Qin, Y.: An integrated theory of the mind. Psychol. Rev. **111**(4), 1036 (2004)
6. Sun, R.: The clarion cognitive architecture: extending cognitive modeling to social simulation. In: Cognition and Multi-agent Interaction, pp. 79–99 (2006)
7. Laird, J.: The Soar Cognitive Architecture. MIT Press, Cambridge (2012)
8. Snaider, J., Franklin, S.: Vector lida. Procedia Comput. Sci. **41**, 188–203 (2014)
9. Murphy, G.L.: The Big Book of Concepts. MIT Press, Cambridge (2002)
10. Rosch, E.: Cognitive representations of semantic categories. J. Exp. Psychol. Gen. **104**(3), 192–233 (1975)
11. Malt, B.C.: An on-line investigation of prototype and exemplar strategies in classification. J. Exp. Psychol.: Learn. Mem. Cogn. **15**(4), 539 (1989)
12. Smith, J.D., Minda, J.P.: Prototypes in the mist: the early epochs of category learning. J. Exp. Psychol.: Learn. Mem. Cogn. **24**(6), 1411 (1998)
13. Keil, F.: Concepts, Kinds, and Cognitive Development. MIT Press, Cambridge (1989)
14. Machery, E.: Doing Without Concepts. OUP, Oxford (2009)
15. Frixione, M., Lieto, A.: Representing non classical concepts in formal ontologies: prototypes and exemplars. In: New Challenges in Distributed Information Filtering and Retrieval, pp. 171–182. Springer (2013)
16. Prinz, J.J.: Furnishing the Mind: Concepts and Their Perceptual Basis. MIT press, Cambridge (2002)
17. Barsalou, L.W.: Perceptual symbol systems. Behav. Brain Sci. **22**(04), 577–660 (1999)
18. De Rosa, R.: Prinz's problematic proxytypes. Philos. Q. **55**(221), 594–606 (2005)
19. Lieto, A., Lebiere, C., Oltramari, A.: The knowledge level in cognitive architectures: current limitations and possible developments. Cogn. Syst. Res. **48**, 39–55 (2018)
20. Chella, A., Frixione, M., Lieto, A.: Representational issues in the debate on the standard model of the mind. In: AAAI Fall Symposium Series Proceedings, FSS-17-05, pp. 302–307. AAAI Press (2017). https://www.aaai.org/ocs/index.php/FSS/FSS17/paper/view/15990

21. Gärdenfors, P.: Conceptual Spaces: The Geometry of Thought. MIT Press, Cambridge (2004)
22. Lieto, A., Chella, A., Frixione, M.: Conceptual spaces for cognitive architectures: a lingua franca for different levels of representation. Biol. Inspired Cogn. Archit. (BICA) **19**, 1–9 (2017)
23. Lenat, D.B., Prakash, M., Shepherd, M.: CYC: using common sense knowledge to overcome brittleness and knowledge acquisition bottlenecks. AI Mag. **6**(4), 65 (1985)
24. Evans, J.S.B., Frankish, K.E.: In Two Minds: Dual Processes and Beyond. Oxford University Press, Oxford (2009)
25. Kahneman, D.: Thinking, Fast and Slow. Macmillan, London (2011)
26. Murphy, G.L., Medin, D.L.: The role of theories in conceptual coherence. Psychol. Rev. **92**(3), 289 (1985)
27. Blanchard, T.: Default knowledge, time pressure, and the theory-theory of concepts. Behav. Brain Sci. **33**(2–3), 206–207 (2010)
28. Frixione, M., Lieto, A.: Towards an extended model of conceptual representations in formal ontologies: a typicality-based proposal. J. Univ. Comput. Sci. **20**(3), 257–276 (2014)
29. Danks, D.: Theory unification and graphical models in human categorization. In: Causal Learning: Psychology, Philosophy, and Computation, pp. 173–189 (2007)
30. Danks, D.: Psychological theories of categorizations as probabilistic models (2004)
31. Medin, D.L., Schaffer, M.M.: Context theory of classification learning. Psychol. Rev. **85**(3), 207 (1978)
32. Atran, S., Medin, D.L.: The Native Mind and the Cultural Construction of Nature. MIT Press, Cambridge (2008)
33. Thagard, P., Verbeurgt, K.: Coherence as constraint satisfaction. Cogn. Sci. **22**(1), 1–24 (1998)

On Stable Profit Sharing Reinforcement Learning with Expected Failure Probability

Daisuke Mizuno[1], Kazuteru Miyazaki[2(✉)], and Hiroaki Kobayashi[3]

[1] Tokyo Institute of Technology, Tokyo, Japan
mizuno.d.aa@m.titech.ac.jp
[2] National Institution for Academic Degrees and Quality Enhancement of Higher Education, Tokyo, Japan
teru@niad.ac.jp
[3] Meiji University, Kanagawa, Japan
kobayasi@meiji.ac.jp

Abstract. In this paper, Expected Success Probability (ESP) is defined and a reinforcement learning method Stable Profit Sharing with Expected Failure Probability (SPSwithEFP) is proposed. In SPSwithEFP, Expected Failure Probability (EFP) is used in the roulette wheel selection method and ESP is used in the update equation of the weight of a rule. EFP can discard risky actions and ESP can make the distribution of learned results smaller. The effectiveness is shown with simulation experiments for a maze environment with pitfalls.

Keywords: Reinforcement learning · XoL · Profit Sharing · EFP

1 Introduction

Reinforcement Learning is a type of machine learning where the robot tries to solve a given task through trial-and-error searches using a evaluation value called a reward or a penalty. In RL, due to the trial and error search, there is a possibility of finding a good solution that exceeds our expectation. Furthermore, the calm change in the environment is allowable. As RL, methods based on dynamic programming, such as TD, Q-learning (QL), and Sarsa, are well known [7]. For QL, the optimality in a Markov decision process environment is guaranteed, but the rationality in a non-Markov environment is not guaranteed.

In this paper, we are interested in approaches that treat reward and penalty signals independently and enhance successful experiences strongly to reduce the number of trial-and-error searches. They are known as exploitation-oriented learning (XoL) [2]. One example of XoL learning methods with a type of a reward is Profit Sharing (PS) [1]. PS has a certain degree of rationality in a non-Markov environment. Also a method that uses Expected Failure Probability (EFP) [3] has been proposed to avoid penalties and ensure the rationality.

© Springer Nature Switzerland AG 2019
A. V. Samsonovich (Ed.): BICA 2018, AISC 848, pp. 228–233, 2019.
https://doi.org/10.1007/978-3-319-99316-4_30

Although PS learns faster than QL, there is a possibility that PS does not learn the optimum policy, and the learning results may vary widely. In this paper, in order to solve this problem, we introduce a new concept of expected success probability (ESP) to PS. We propose a new method called SPSwithEFP that incorporates EFP and ESP into PS method. We show the effectiveness of SPSwithEFP by applying it to the learning of the shortest path of a maze problem with pitfalls.

2 Domain

After receiving sensory inputs from the environment, the learning agent selects *an action* to perform. Receiving the sensory inputs and performing the selected action compose a unit of time, called *a step*. The sensory inputs from the environment is referred to as *a state*. A pair of a state and an action that is applicable in the state is referred to as *a rule*. If the agent selects the action q in the state s, the rule is described as *rule(s,q)*. A function that maps states to actions is called *a policy*. A policy is *deterministic* if it maps each state to only one action. A policy is *rational* if all of expected rewards per an action are positive.

In many methods, parameters called *a weight* is given to each rule. The agent receives a reward or a penalty after applying $rule(s, q)$ when a certain condition is satisfied. A reward is given if the target was achieved and a penalty is given when the target became unachievable.

A sequence of steps that starts from the initial state and ends with a reward or a penalty is called *an episode*. If different rules are selected for the same state, that is, if the episode contains a loop, the loop is referred to as *a detour*. Rules that always exist in detours are referred to as *an ineffective rule*. If not so, it is called *an effective rule*. A rule that has been directly given a penalty is called *a penalty rule*. If all rules for a state are penalty or irrational rules, the state is called *a penalty state*. If a destination after applying a rule is a penalty state, the rule is also classified as a penalty rule.

3 Exploitation-Oriented Learning

3.1 Profit Sharing (PS)

PS learns a rational policy by propagating a reward backward along an episode when a reward is given. Assume that the action q_t was used in the state s_t at time t and let $Q(s_t, q_t)$ be a reward shared to $rule(s_t, q_t)$. In this paper, we use the following geometric decreasing function to propagate the reward backward;

$$Q(s_t, q_t) \leftarrow Q(s_t, q_t) + \lambda^{N-t}R, \quad t = 1, 2, ..., N, 0 < \lambda < 1, \tag{1}$$

where R is the reward value, N is the episode length and λ is the parameter called *the discount rate*.

The action selection method is one of the key elements of the RL methods. In this paper, we use a roulette wheel selection. In roulette selection, the probability of selecting $rule(s, q_i)$ is defined as $p(q_i|s) = \frac{Q(s,q_i)}{\sum_{j=1}^{N_A} Q(s,q_j)}$, where $p(q|s)$ is the conditional probability of selecting the action q in the state s. N_A is the number of available rules in the state.

3.2   Expected Failure Probability (EFP)

EFP was proposed in order to learn a rational policy without a penalty as XoL methods. EFP is defined for a rule and gives a probability that the episode will end with penalty after all if the rule is used. We can generally find out penalty rules faster than both PARP and ImpPARP using EFP since EFP propagates swiftly like QL.

The online calculation of EFP can be performed as follows;

– If the state moves to the state s_k after applying $rule(s, q)$,

$$p_t(EFP|rule(s,q)) = (1 - \eta)p_{t-1}(EFP|rule(s,q)) + \eta p(EFP|s_k),$$

$$p(EFP|s_k) = \sum_{i}^{N_A} p(q_i|s_k)p(EFP|rule(s_k, q_i)), \qquad (2)$$

– If the agent receives a penalty just after applying $rule(s, q)$,

$$p_t(EFP|rule(s,q)) = (1 - \eta)p_{t-1}(EFP|rule(s,q)) + \eta, \qquad (3)$$

where $p_t(EFP|rule(s,q))$ is EFP of the $rule(s, q)$ at the iteration time t and the initial value is zero. $p(EFP|s_k)$ is a probability that the trial will fail after it transits the state s_k. η $(0 < \eta < 1)$ is *the failure probability propagation rate*.

EFP was used as follows in the papers [3]; if $p_t(EFP|rule(s,q))$ exceeds γ, $rule(s, q)$ is labeled as a penalty rule and removed from selection candidates, where γ is the threshold value for a penalty rule. This method is referred to as **original EFP**.

In the original EFP, if all rules in a state became penalty rules, the agent cannot act any more. One solution of this dead end is to select a rule with the lowest EFP among them. It is referred to as **original EFP+**.

3.3   PSwithEFP

In original EFP and original EFP+, EFP is not utilized sufficiently, since EFP never affects to the action selection probability until EFP exceeds the threshold γ. In PSwithEFP, EFP is incorporated into the action selection method as follows;

$$pe(q = q_i|s) = \frac{(1 - p(EFP|rule(s,q_i)))Q(s,q_i)}{\sum_{j=1}^{N_A}(1 - p(EFP|rule(s,q_j)))Q(s,q_j)}, \qquad (4)$$

where $pe(q|s)$ is a probability of action selection. Furthermore, since the selection probability of the action is also used in the calculation of EFP, Eq. 2 becomes as follows;

$$p(EFP|s_k) = \sum_{i}^{N_A} pe(q_i|s_k)p(EFP|rule(s_k, q_i)). \qquad (5)$$

By this modification, EFP can affect the action selection directly and immediately and no rule is excluded as a penalty rule. This means that more successful rules are selected with larger probability and that it never occurs that an important rule is excluded as a penalty rule.

PSwithEFP can learn a may-be-not-best but safer policy that avoids penalties in shorter time. PSwithEFP has been applied to Keepaway Task [6] and shown the effectiveness in a multiagent environment [4].

4   Proposal of Expected Success Probability (ESP)

4.1   ESP

We propose a new concept the *Expected Success Probability (ESP)* in order to resolve the problem that the learning results vary widely in PS. ESP is defined for a rule and gives a probability that the episode will end with reward after all if the rule is used. We expect to suppress the variation of results in PS method learning by adding ESP to PS.

The online calculation of ESP can be obtained by replacing "EFP" with "ESP" and η with η_s at Eqs. 2 and 3. $p_t(ESP|rule(s,q))$ is ESP of the $rule(s,q)$ at the iteration time t and the initial value is zero. $p(ESP|s_k)$ is a probability that the trial will fail after it transits the state s_k. η_s $(0 < \eta_s < 1)$ is the success probability propagation rate.

4.2   Proposal of SPS

We propose *Stable Profit Sharing (SPS)* by adding ESP to PS. PS distributes a reward by Eq. 2. The equation of SPS is changed to the following;

$$Q(s_t, q_t) \leftarrow Q(s_t, q_t) + p(ESP|rule(s_t, q_t))\lambda^{N-t}R, \quad t = 1, 2, ..., N, 0 < \lambda < 1,$$
(6)

Equation 6 is possible to suppress distribution of a reward at an early learning stage. As a result, the search space of PS is expanded and variation in PS learning result can be suppressed.

4.3   Proposal of SPSwithEFP

We propose *SPSwithEFP*, a learning method that incorporates EFP's algorithm into SPS. In this learning method, it is possible to avoid risky rules, and suppress variations in learning result. The learning algorithm is shown below;

1. Assign initial values to the evaluation function (Q, EFP, ESP).
2. Repeat until the following process converges.
   (a) Time t is set to 1, and state S_1 is set to the initial state.
   (b) Select action q_t according to Eq. 3 based on the evaluation value function for state S_t.
   (c) The action q_t is executed and the state transitions to S_{t+1}. Update EFP and ESP.
   (d) If the state is a goal state, update ESP, and distribute the reward by Eq. 6. Initialize the episode and go to (a).

5   Numerical Experiments

5.1   Experimental Setting

We examine the effectiveness of SPSwithEFP by using a maze problem with pitfalls as an example. Figure 1 shows the maze used for experiments. S is the start point, and G is the goal point. The black part is the wall and the agent cannot go into the wall. × is a pitfall. Falling into the pitfall will result in a failure and the game is over immediately. When the agent reaches the goal, the agent acquires a reward. The aim is to learn the shortest route from S to G. The shortest route length for this maze is 20 steps.

Five methods of (1) PS method, (2) PS with EFP, (3) SPS, (4) SPSwithEFP, and (5) QL are used for comparison. Reward R and the discount rate λ are 100.0 and 0.8, respectively. The failure probability propagation rate η of PSwithEFP and SPSwithEFP is 0.8. The success probability propagation rate η_s of SPS and SPSwithEFP is 0.5. These values were decided by preliminary experiments.

The roulette selection is used for PS as an action selection method. On the other hand, QL uses ϵ-greedy method as an action selection method. We set the initial evaluation value to 10 and the upper limit number of action selections to 1000. A trial is performed until the agent falls into a pitfall, reaches a goal, or the number of action selections reaches the upper limit number. One experiment contains 2000 trials and 30 experiments are done for each method.

5.2   Results and Discussion

Experimental results for the methods (1)–(5) are shown in from Figs. 2, 3, 4, 5 and 6. The horizontal axis is the number of trials, and the vertical axis is the number of steps until the goal. These figures show moving average steps of every 40 trials for 30 experiments.

Figures 2, 3, 4, 5 and 6 show only PSwithEFP and SPSwithEFP can learn the shortest route in all 30 experiments. This is because the location of the pitfall can be learned by EFP. On the other hand, PS can reach the goal in several experiments, while SPS and QL never did. This is because that, when the agent reaches the goal by chance in the early learning stage, the reward is distributed to the all rules in the episode in PS, but the reward does not reach the early step rules in the episode in SPS and QL.

Figures 7 and 8 show the frequency distributions of the number of steps of 30 experiments up to the goal at 2000 trials of PSwithEFP and SPSwithEFP.

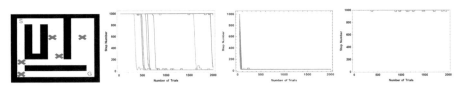

Fig. 1. Maze   **Fig. 2.** PS   **Fig. 3.** PSwithEFP   **Fig. 4.** SPS

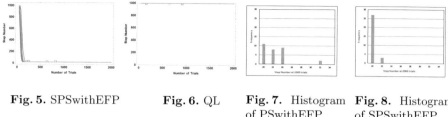

Fig. 5. SPSwithEFP   **Fig. 6.** QL   **Fig. 7.** Histogram   **Fig. 8.** Histogram
                                                         of PSwithEFP   of SPSwithEFP

SPSwithEFP learns the shortest route more stably than PSwithEFP. This is because that ESP suppresses the reward distribution and expands the search space in the early learning stage.

6 Conclusions

In this paper, in order to solve the problem that the learning results of Profit Sharing vary widely, we proposed a new concept called expected success probability ESP. In addition, we propose a new method SPSwithEFP that incorporates the expected failure probability (EFP) and ESP into PS. Through numerical experiments in a maze with pitfalls, we show that SPSwithEFP can suppress variation in learning results compared with existing methods. We plan to verify the effectiveness of proposed method under Keepaway task and Deep Q-network [5].

Acknowledgements. This work was supported by JSPS KAKENHI Grant Number 17K00327.

References

1. Miyazaki, K., Yamamura, M., Kobayashi, S.: On the rationality of profit sharing in reinforcement learning. In: Proceedings of the 3rd International Conference on Fuzzy Logic, Neural Nets and Soft Computing, pp. 285–288 (1994)
2. Miyazaki, K., Kobayashi, S.: Exploitation-oriented learning PS-r#. J. Adv. Comput. Intell. Intell. Inf. **13**(6), 624–630 (2009)
3. Miyazaki, K., Muraoka, H., Kobayashi, H.: Proposal of a propagation algorithm of the expected failure probability and the effectiveness on multi-agent environments. In: SICE Annual Conference 2013, pp. 1067–1072 (2013)
4. Miyazaki, K., Furukawa, K., Kobayashi, H.: Proposal of PSwithEFP and its evaluation in multi-agent reinforcement learning. J. Adv. Comput. Intell. Intell. Inf. **21**(5), 930–938 (2017)
5. Mnih, V., Kavukcuoglu, K., Silver, D., Graves, A., Antonoglou, I., Wierstra, D., Riedmiller, M.: Playing atari with deep reinforcement learning. In: NIPS Deep Learning Workshop 2013 (2013)
6. Stone, P., Sutton, R.S., Kuhlamann, G.: Reinforcement learning toward RoboCup soccer keepaway. Adapt. Behav. **13**(3), 165–188 (2005)
7. Sutton, R.S., Barto, A.G.: Reinforcement Learning: An Introduction. A Bradford Book. MIT Press, Cambridge (1998)

Neocortical Functional Hierarchy Estimated from Connectomic Morphology in the Mouse Brain

So Negishi[1,3(✉)], Taku Hayami[1,4,5], Hiroto Tamura[1,2,6],
Haruo Mizutani[2], and Hiroshi Yamakawa[1,2]

[1] Dwango Artificial Intelligence Laboratory, Tokyo, Japan
sonegishi_2020@depauw.edu,
{info,info}@wba-initiative.org
[2] The Whole Brain Architecture Initiative, Tokyo, Japan
{info,info,info}@wba-initiative.org
[3] DePauw University, Greencastle, IN 46135, USA
[4] Tokyo University of Agriculture and Technology, Tokyo, Japan
[5] National Center of Neurology and Psychiatry, Tokyo, Japan
[6] University of Tokyo, Tokyo, Japan

Abstract. The mechanisms of information processing in the neocortex underlie brain functions based on the hierarchy of individual cortical areas. This study was focused on the determination of the neocortical hierarchy resulting from a clustering analysis of connectomic morphology with cortical layer resolution. K-means clustering effectively classified connectivities into two groups. Moreover, the description of the neocortical hierarchy was simplified to a wiring diagram to understand how sensory information flows in the biological neural networks. In this initial work, the validity of the resulting classification is still uncertain to describe the hierarchical relationships in the neocortex.

Keywords: Connectome · General artificial intelligence
Whole Brain Architecture · Empirical neural circuits · Efficient engineering

1 Introduction

The Whole Brain Architecture (WBA) is a strong candidate for a computational cognitive architecture of an artificial general intelligence (AGI) computing platform on the basis of empirical neural circuits, also known as connectomes. In order to sufficiently understand information processing mechanisms of the brain and to make progress towards WBA, it is necessary to determine sets of feedforward (FF) and feedback (FB) relationships between neocortical areas. From a biological perspective, it is difficult to have certain FF/FB relationships in each mouse neocortical areas because even relatively specific layers in each area consist of distinct FF/FB layers. Moreover, the existing connectivity data has limited accuracy, especially in the higher-level hierarchy areas because of the thinness of their layers.

In this study, we apply a clustering analysis to mouse connectivity data from Allen Institute for Brain Sciences (Allen) to figure out the neocortical functional hierarchy

© Springer Nature Switzerland AG 2019
A. V. Samsonovich (Ed.): BICA 2018, AISC 848, pp. 234–238, 2019.
https://doi.org/10.1007/978-3-319-99316-4_31

with the FF/FB relationships. First, we calculated connectivity strengths between individual neocortical areas with the latest Allen dataset using linear regression analysis. Next, we applied a K-means clustering algorithm to our result of the cortical connectivity strength. These analyses enable to separate function of each neural connection into two groups: FF and FB projections. For classification purposes, we set a FF/FB threshold based on the proportion of connections from known feedforward regions such as primary visual cortex and primary auditory cortex. Finally, we constructed a directed wiring diagram using these results in order to ascertain the FF/FB hierarchy in the visual areas. For a variety of tested threshold levels, these methods suggest significantly that FBs are shown as stronger connectivity strengths than FFs. While this method is the first attempt at a novel classification approach, its results depicted clusters with a bundle based on findings, and it also correctly captures well-known connections, such as FF routes from primary visual area to other visual areas. A challenge is to verify whether a biological connectomic structure works as an architecture of artificial neural networks.

2 Method

Linear connectivity model via constrained optimization was applied to compute connectivity strengths among neocortex areas (Fig. 1). This linear connectivity model was modified from the method described in Oh's paper [2]. This resulting strength was used as the input data (consists of 6×1638) for following principal component analysis (PCA), K-means clustering, and computation of thresholds.

Fig. 1. Mouse neocortical connectivity matrix in mouse visual areas. The matrix depicts relationships between source and target regions in visual cortical areas. The x-axis indicates 6 layers in 10 visual areas whereas the y-axis indicates 8 visual areas due to the lack of data.

3 Results

Then the average connectivity method described above was used with K-means clustering on these connectivity strengths. Figure 2 showed a result of the clustering analysis using K-means, where K = 2. In order to examine correspondence between PCA and clustering results, we plotted K-means results into 2-dimensional PCA graph where the x-axis is as the first principal component and the y-axis is as the second principal component from the PCA results. The first principal component was dominant with layer 6a.

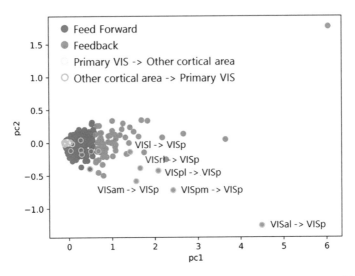

Fig. 2. Distribution of feedforward and feedback projections in mouse neocortex. The result of K-means clustering is overlaid on the plot of PCA. Feedback projections to the primary visual areas are labeled as the green circle and feedforward flows from the primary visual area are shown with the yellow circle.

A threshold on connectivity strengths was implemented for better looking and understanding purposes on the wiring diagram (Fig. 3). We tried two different methods for finding an appropriate lower limit. One way is calculating average connectivity using connectivity strengths among neocortex for the threshold. Another way is to find power law of connectivity strengths among neocortex and retained the top 20%, corresponding to connectivity strengths above 1.5. These regions mostly showed FB flows among visual areas, indicating that the most connected visual areas are typically FB. However, because of the resulting scarcity of FF connections, we decided to use the average method instead of the power-law method in order to create a comprehensive wiring diagram.

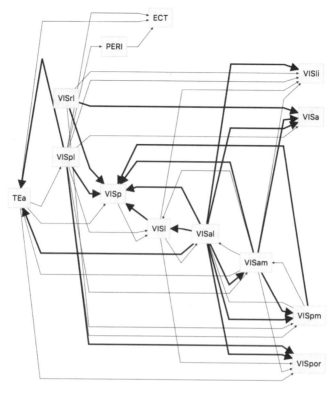

Fig. 3. Wiring diagram with the neocortical hierarchy in the visual area. Narrow and bold arrows indicate feedforward and feedback projections respectively. Only the projections above the average strength are displayed.

4   Conclusions

Neural strengths and hierarchies of connections between areas in the cerebral cortex were analyzed by linear regression and clustering methods. The results of FF/FB connections found from the wiring diagram depicts the same findings from Berezovskii's paper [3], particularly with the part of visual pathways in the lower level hierarchy.

The wiring diagram we estimated in this study would serve as a structure of biological neural network model to simulate a large-scale brain activity, and it might be useful to build a functional brain model. Moreover, it will also be advantageous to technologically develop a neuro-inspired machine learning module based on the architecture where we would like to implement neural algorithms as an artificial intelligent system.

In the next step, it enables us to apply this structure as a fundamental of computational study on neural circuits to simulate brain functions, and it eventually can apply to our cognitive architecture "Whole Brain Architecture (WBA) [1]" to develop an artificial intelligent system, which will allow us to test the success of these classifications whether it is mimicking particular brain functions in machine learning modules.

References

1. Berezovskii et al.: J. Comp. Neurol. **519**, 3672–3683 (2011)
2. Mizutani, H., et al.: Proceeding of BICA Society (2017)
3. Oh, S.W., et al.: A mesoscale connectome of the mouse brain. Nature **508**, 207–214 (2014)
4. Yamakawa, H. et al.: Whole brain architecture approach is a feasible way toward an artificial general intelligence. In: Proceedings of ICONIP, pp. 275–281 (2016)

Function Map-Driven Development for AGI

Masahiko Osawa[1,2,3](✉), Takashi Omori[4,6], Koichi Takahashi[5,6],
Naoya Arakawa[3,6], Naoyuki Sakai[3,6], Michita Imai[1], and Hiroshi Yamakawa[3,6]

[1] Graduate School of Science and Technology, Keio University, Kanagawa, Japan
{mosawa,michita}@ailab.ics.keio.ac.jp
[2] Research Fellow (DC1), Japan Society for the Promotion Science, Tokyo, Japan
[3] Dwango Artificial Intelligence Laboratory, Tokyo, Japan
hiroshi_yamakawa@dwango.co.jp
[4] College of Engineering, Tamagawa University, Tokyo, Japan
omori@lab.tamagawa.ac.jp
[5] RIKEN Center for Biosystems Dynamics Research, Osaka, Japan
ktakahashi@riken.jp
[6] The Whole Brain Architecture Initiative, Tokyo, Japan
http://www.ailab.ics.keio.ac.jp/, http://ailab.dwango.co.jp/

Abstract. This paper introduces the function map, a directed graph containing "function nodes," each of which consists of the description of a function, tasks, an implementation, and associated brain areas. A function is implemented when the implementation on the node accomplishes the tasks. To construct an AGI system, its overall functions must be decomposed to a large number of sub-functions. To facilitate decomposition, we propose a function map-driven development, in which a function map with a hierarchical structure is continuously constructed. This approach requires research and development to fill in incomplete function maps for continuous improvement. Upon the completion of the internal implementation of the top-level functions, the function map can be converted to a cognitive architecture representation. The map can also be converted to a roadmap for implementation of AGI by supplementing person-hours for the implementations that have yet to be realized.

Keywords: Function map · Artificial general intelligence
Whole brain architecture

1 Introduction

There are two reasons why the development of Artificial General Intelligence (AGI) [1] is so difficult. The first is the large number of functions to be created, and the second is the fact that the requirement specifications for what is to be created are vague. Therefore, the development of AGI requires a mechanism that allows researchers to not only suitably organize the large number of functions,

A. V. Samsonovich (Ed.): BICA 2018, AISC 848, pp. 239–244, 2019.
https://doi.org/10.1007/978-3-319-99316-4_32

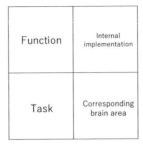

Fig. 1. Function node

but also to undertake development work even when the specifications remain vague.

In this paper, we propose a method of AGI development driven by function maps, in which an AGI is constructed by combining functions. In this method of development, function maps are used to organize large numbers of functions hierarchically. Furthermore, because the function map itself also plays the role of defining the requirement specifications of AGI, these specifications will gradually become clearer as the function map is refined in the development cycle. In Sect. 2 of this paper, we introduce the function map, which is at the center of our proposed research and development program. Next, in Sect. 3 we propose our method of development driven by function maps. In Sect. 4, we explain how a function map can be converted into a cognitive architecture representation after its completion, or a roadmap representation during its construction. Finally, conclusions are presented in Sect. 5.

2   Function Map

2.1   Function

This paper introduces the function map, a directed graph containing "function nodes," (Fig. 1) each of which consists of the description of a function, tasks, an implementation, and associated brain areas. A function is implemented when the implementation on the node accomplishes the tasks.

2.2   Overview of Function Maps

The function map proposed here is a tool for sharing knowledge within AGI development community in order to strategically build human-level AGI. A function map is a directed graph Fig. 1 consisting of Function Nodes, as illustrated in Fig. 2.

When constructing an AGI system by combining functions, organization by decomposition must be performed on the large number of functions necessary

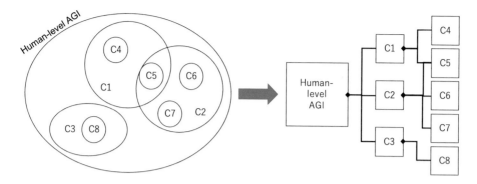

Fig. 2. Function map

for AGI. To facilitate decomposition, we propose a function map-driven development methodology, in which a function map is continuously constructed with a hierarchical structure. This approach requires research and development to fill in incomplete function maps for continuous improvement.

In the next section, we describe how function maps can be used to drive development.

3   Function Map-Driven Development

Function maps can be employed to develop and refine research and development towards the implementation of an AGI in a variety of ways. In this section, we introduce four such methodologies.

Revitalization of Research. We believe that the analysis of a function map can identify areas with little research or little knowledge. Once identified, research in such areas can possibly be revitalized. Furthermore, dividing the functions and defining tasks would enable researchers to identify the requirements to be solved, which should make it easier for individuals to participate in AGI development community.

Cycle of Task Setting and Task Resolution. Until now, the setting and formulation of tasks in AI research has contributed to the materialization of new functions. Examples of standardized tasks that have been set include Atari games as a reinforcement learning task and MNIST and CIFAR-10 for image recognition. In a function map, more than one task may be allocated to a single function, but it is assumed that an adequate task setting is proposed and refined in AGI development community. We believe that by developing cycles in which task setting and task resolution are performed on a function map, leading to the refinement of the function map itself, it will be possible to tackle the unsolved problems of artificial intelligence.

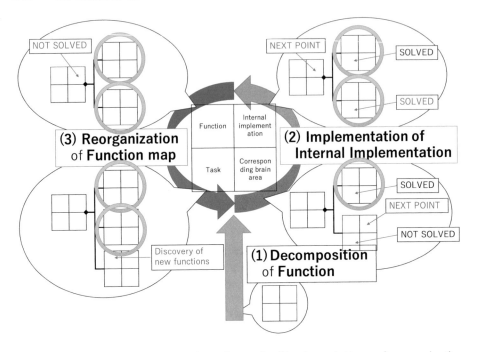

Fig. 3. Refining the function map through a cycle of implementation and reorganization

3.1 Cycle of Implementation and Reorganization

A cycle of implementation and reorganization is employed to drive the implementation of functions as the function map is continuously refined, as illustrated in Fig. 3. For a function map whose functions have been adequately decomposed (Fig. 3(1)), it is easier to identify the nodes that represent the best targets for task setting and implementation, and then carry out the implementation in the order starting from the lowest-level nodes (Fig. 3(2)). Function maps are designed based on the principle that the ease of implementation of a node is high when all its subordinate nodes have been implemented. However, when implementation is judged to be difficult, it is in fact possible to reorganize the function map through the process of analyzing why implementation is difficult (Fig. 3(3)).

3.2 Improvement of AGI Generality by Grounding to the Brain Constraints

In a function map, all functions are at least indirectly connected to the brain. Mapping a function to a brain area can create an opportunity to improve the generality of the AI [2] represented by the function map in two ways (Fig. 4).

Integration of Functions. If nodes labeled with different functions are allocated to the same brain area, this suggests that they may describe a single

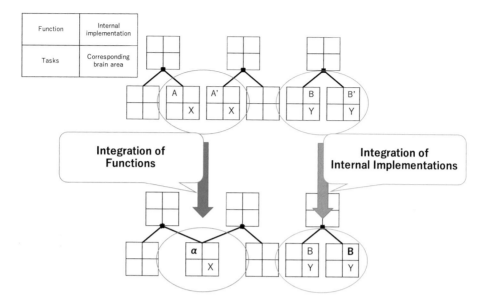

Fig. 4. Improvement of generality by brain type constraints

function. Then, these nodes are integrated, making the graph smaller. Refining the function map itself improves its generality, because we can avoid overfitting.

Integration of Internal Implementations. If Function Nodes developed by different internal implementations are allocated to the same brain area, this suggests that it may or ought to be developed by the same internal implementation. Because the graph of a function map is formed by the hierarchical structure of the functions, changes in internal implementations do not directly change the graph structure. However, the number of techniques used to represent the entire graph is reduced, improving its generality and reusability, because we can also avoid overfitting.

4   Conversion of Function Maps

Function maps can readily be converted into other representations (Fig. 5).

4.1   Conversion to a Cognitive Architecture Representation

If the internal implementations of all functions are in place, then a function map can be converted into a cognitive architecture representation (Fig. 5(a) → (b)). In other words, given a function map with a human-level AGI function as the top node, implementing the top node of the Map is equivalent to implementing a human-level cognitive architecture.

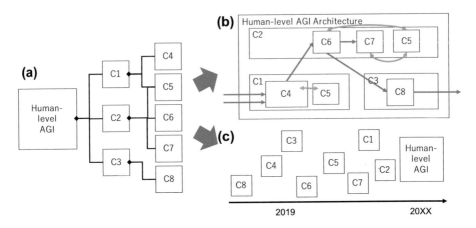

Fig. 5. Converting a function map

4.2 Conversion to a Roadmap Representation

A function map can also be converted to a roadmap for implementation of AGI by supplementing person-hours for the implementations that have yet to be realized (Fig. 5(a) → (c)). In this respect, the function map plays the role of automatically and objectively representing the current state and the next targets at each point in the course of constructing AGI.

5   Conclusion

In this paper, we introduced function maps and proposed the idea of function map-driven development, whereby function maps are employed to drive steady progress on difficult challenges, such as developing AGI. We also described how function maps can be readily converted into cognitive architecture and roadmap representations. The function map-driven development approach proposed here is similar to the iterative development methodology in software development. In the future, we plan to continue our efforts to construct partial function maps, and to continue using function maps to drive the development cycle. Thus we believe that the function map can contribute to AGI R&D community.

References

1. Goertzel, B.: Artificial general intelligence: concept, state of the art, and future prospects. J. Artif. Gen. Intell. **5**(1), 1–48 (2014)
2. Yamakawa, H.: Proposing minimum knowledge description length principle for general-purpose AI. In: The 30th Annual Conference of the Japanese Society for Artificial Intelligence (2016). (in Japanese)

Scruff: A Deep Probabilistic Cognitive Architecture for Predictive Processing

Avi Pfeffer[✉] and Spencer K. Lynn

Charles River Analytics, Inc., Cambridge, MA, USA
{apfeffer, slynn}@cra.com

Abstract. The theory of predictive processing encompasses several elements that make it attractive as the underlying computational approach for a cognitive architecture. We introduce a new cognitive architecture, Scruff, capable of implementing predictive processing models by incorporating key properties of neural networks into the Bayesian probabilistic programming framework. We illustrate the Scruff approach with conditional linear Gaussian (CLG) models, noisy-or models, and a Bayesian variation of the Rao-Ballard linear algebra model of predictive vision.

Keywords: Probabilistic programming · Predictive coding

1 Introduction

The term predictive processing, or predictive coding, encompasses a group theories, emerging over the last two decades, of how brains process information (see, e.g., Allen and Friston 2016). In the traditional stimulus-response model of perception and cognition, dominant since the 1950s and still presented in textbooks, the brain encodes sensory stimuli as they occur. After encoding, a series of processes categorizes the stimuli and builds a model of events in the external world that is used to derive meaning. Beliefs and concepts are considered to be the *result* of perception. Since the 1950s, scientists have constructed theories that allow contextual effects, such as selective attention and concept activation, to bias perceptual processes (reviewed by Lupyan et al. 2010; Zaki 2013). However, developments in the last two decades challenge the stimulus-response model, including models of contextual influence, with a conceptualization of the brain as a prediction engine (Friston 2010; Friston and Kiebel 2009; Hohwy 2013; Rao and Ballard 1999).

From the predictive processing perspective, our beliefs about the state of the world (or about the state of our bodies) yield predictions about incoming sensory signals. At the upper-most level of a hierarchy of neuronal ensembles these predictions are abstract (e.g., concerned with physiological allostasis; Barrett and Simmons 2015). The predictions become more and more specific as they pass down the hierarchy towards primary sensory cortex, where they come to specify the lowest level sensory features. What is encoded by sensory cortex, then, is not stimulus features, but prediction error–the difference between the prediction and the sensory signals. It is prediction error, not a stimulus representation per se, that is passed back up the hierarchy to be operated on

© Springer Nature Switzerland AG 2019
A. V. Samsonovich (Ed.): BICA 2018, AISC 848, pp. 245–259, 2019.
https://doi.org/10.1007/978-3-319-99316-4_33

by cognitive processes. Beliefs and concepts are considered to be the beginning of perception, not the result (in fact, in the absence of prediction error, beliefs constitute perception). Faced with a prediction error, the brain executes one of three options that function to minimize the error: (i) The prediction may be updated to reflect the sensory signals, (ii) the magnitude of the error may be attenuated in favor of the prediction, or (iii) behavior may be initiated to bring about sensory signals that better match the prediction.

As a hypothesis about how the brain works, predictive processing appears to address several challenges. It is efficient (only errors are propagated, not increasingly complex models of the world), it offers a functional explanation for some aspects of neuroanatomy (such as patterns of cellular granularity and directionality of inter-cortical connectivity).

As a biologically inspired cognitive architecture, predictive processing may offer several benefits with respect to other architectures. For example, local connections among simple, self-similar computational units suggests scalability, and top-down priming of lower-level expectations suggests a principled way to guide learning. Moreover, predictive processing is well aligned with other biologically inspired frameworks relevant to cognitive architectures, such as epigenetic robotics (e.g., Morse et al. 2010) and situated conceptualization (Barsalou 2015), in which the "meaning" of a sensation or event is goal-bound and multi-modal, potentially including actions afforded by the sensation or event.

1.1 Contributions of This Paper

Here, we advocate for predictive processing as a grounding principle for a cognitive architecture. We describe the requirements for a cognitive architecture based on predictive processing. We introduce a new cognitive architecture, called Scruff, which is designed to satisfy these requirements. Scruff is based on incorporating key properties of neural networks into the Bayesian probabilistic programming framework. Probabilistic programming is a framework for representing models using generative probabilistic processes expressed in a programming language. Probabilistic programming enables the hierarchical use of knowledge in predicting observations and the processing of errors to form posterior beliefs. Deep learning principles enable the development of probabilistic programs with stacked layers that are capable of learning predictive domain features, making probabilistic programming suitable for a cognitive architecture. Additionally, we show examples of predictive processing cognitive models that can be developed using Scruff.

2 Related Work

2.1 Predictive Processing Implementations

While implementations of predictive processing algorithms make use of more or less well-known computational, inferential processes (e.g., active inference, belief propagation), most efforts have been aimed at modeling brain function rather than creating a general cognitive architecture with which to construct an intelligent system. For example, Rao and Ballard (1999) modeled perceptual properties of the visual cortex. Friston and colleagues (e.g., Friston 2008; Mathys 2011) further elaborate models of visual cognition and provide a theoretical basis for predictive processing as Bayesian inference in the statistical concept of free energy. Other researchers have sought computational models of mental illness based on predictive processing (e.g., Deneve and Jardri 2016).

However, some researchers are taking inspiration from biology to arrive at architectures related to predictive processing. For example, Tenenbaum and colleagues (Lake et al. 2015; Lake et al. 2017; Tenenbaum et al. 2011), have explicitly explored predictive, causal, Bayesian frameworks for machine learning. Nonetheless, such architectures have tended not to be layered or distributed.

An alternative to Bayesian belief-centered approaches, deep neural network (DNN) approaches are predictive, but in a very difference sense from Bayesian approaches (see, e.g., Lake et al. 2017). Neural net approaches are predictive in the sense of classifying inputs via feature discovery of shared high-values states (e.g., shared labels, shared reinforcement). The primary goal of these approaches is the prediction of class membership given sensory inputs. If the sensory input does not fit a class, the common alternative is to select a different class. The belief-centered approaches are "predictive" in the sense of explaining causes. The primary goal of these approaches is building explanatory models of the world, given sensory input. If a cause does not explain the sensory input, one option (under active inference) is to take action to change the sensory input to better match the cause.

While DNN learning has been tremendously effective at many tasks, researchers have pointed out limitations of DNNs (e.g., Marcus 2018). These include, among others, the fact that they are data hungry, hard to use with prior knowledge, struggle with open-ended reasoning, and cannot distinguish between causation and correlation. Generative probabilistic programs have the potential to address these limitations. However, probabilistic programming has hitherto not been as effective as DNNs for many learning tasks, largely due to the challenges of inference. This has led to an interest into integrating neural network principles into probabilistic programs.

2.2 Frameworks Integrating Neural Networks into Probabilistic Programming

Researchers have developed a variety of generative deep learning architectures, such as variational autoencoders (Kingma and Welling 2013; Rezende et al. 2014) and generative adversarial networks (Goodfellow et al. 2014). However, these are specific models and not general-purpose probabilistic programming languages. Among probabilistic programming approaches, Anglican uses inference compilation (Le et al. 2016) to solve a probabilistic programming inference problem using a DNN. Edward (Tran et al. 2016) and Pyro (https://github.com/uber/pyro) have been developed as probabilistic programming languages that are trainable using a backpropagation-style algorithm. Both of these languages require a variational inference algorithm to be described explicitly.

Our approach, Scruff, has similar goals to Edward and Pyro, but aims to be a general cognitive architecture. This requires a flexible high-level representation language and the ability to compose many reasoning mechanisms. Compositionality is fundamental to Scruff's design. Scruff introduces the concept of tractability by construction, which ensures that if a reasoning mechanism can be applied to a model, then the mechanism is appropriate for the model, for some suitable agreed-on definition of tractability. Tractability by construction will be of significant benefit to cognitive modelers who are creating models of sophisticated brain processes.

3 Predictive Processing Background

The generation of predictions and processing of errors can be modeled as a Bayesian inference at each level of a hierarchy (Fig. 1). A prediction from level i, $\hat{x}_i$ (an estimate of the Bayesian posterior probability), becomes the prior expectation, μ_0, for the next deeper level, level $i + 1$. Level $i + 1$ compares that prediction (now its own prior expectation) to an incoming signal, x_{i+1}, and passes the error (ε_{i+1}) back up to level i, where it is received as an incoming signal, x_i. Thus, the error of level i's prediction from time $t = 0$ is calculated by the next level down, $i + 1$; the signal processed by level i at time t = 0 is the error of that level's prediction from an earlier time point, $t = -1$; and the adjustment of level i's prior expectation in light of that error is performed by the next level up, at a later time, t = 1. A weighting term, $w = \sigma_0^2/\left(\sigma_0^2 + \sigma_x^2\right)$, weights the influence of a signal on a prediction by the variance of the signal relative to the prior expectation about it.

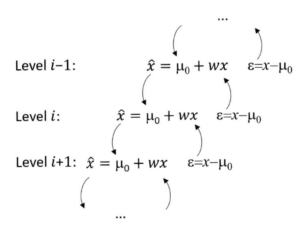

Fig. 1. The flow of information along a predictive processing hierarchy (illustrated with Gaussian distributions, but similar principles could apply to other distributions).

The predictive processing framework entails several propositions relevant to autonomous agents and general AI, under active investigation by a number of laboratories. These include: (i) A goal (e.g., of a decision or a behavior) is a prediction arising from a high level in the hierarchy (Pezzulo et al. 2015). (ii) Motivation (e.g., to achieve a goal) is the outcome of a perturbation from physiological allostasis; prediction error (distance from goal) leads to arousal and motivation, and subsequent action predicted to achieve the goal and the maintenance of allostasis (Chanes and Barrett 2015). (iii) The currently active system of predictions throughout the hierarchical network constitutes an ad hoc, activated conceptual system; minimization of prediction error constitutes categorization of sensory input, and similarity of category members is based on goal-related functional similarity, not physical similarity (Barrett 2017).

4   Requirements for a Cognitive Architecture

If we are to take the idea of building a cognitive architecture based on predictive processing seriously, a number of requirements arise:

- It must be able to express prior domain knowledge that is used for prediction.
- It must support hierarchical representations in which each layer contains knowledge used to predict the layer below and each layer can process errors from the layer below that is used for prediction.
- It must be able to fuse predictions with errors from observations. While a Bayesian formulation is not required for this capability, it is well supported by Bayesian approaches.
- It must support the ability to fuse multimodal observations.
- It must support the ability to combine knowledge from multiple domains and contexts to reason about a situation. For example, driving on city streets during the day and driving on a highway at night both require knowledge of vehicular motion, but each of the scenarios also brings in very different knowledge.
- It must support different computational mechanisms for different modalities, as well as for different cognitive tasks.
- It must be able to model temporal dynamics, including different frequencies of operation at different levels of hierarchy, as well as events that occur at different time scales.
- It must support learning the predictive knowledge from data. Prediction in a noisy environment is not possible unless the agent has learned statistical regularities. Ideally, this learning will include the ability to learn hidden regularities of the domain, like neural networks.
- It must support the ability to modulate, based on the context of the perception, the degree to which error is taken into account in forming beliefs. For example, error is more to be expected in noisy contexts so will have less effect.

In predictive processing theory, action is also the result of prediction. The agent, given a goal, predicts its needs via simulation of body and anticipated required action. Then it predicts sensations. Ideally, a cognitive architecture based on predictive processing should support this modeling of goal-directed action. While we show in the next section how Scruff satisfies the above requirements, we have not yet designed the transition to action, which is left as future work.

5 Scruff

We are developing a new cognitive architecture called Scruff, which is designed to be a deep probabilistic cognitive architecture based on predictive processing. Scruff is based on probabilistic programming, which is a framework for expressing generative probabilistic models using programming languages. Our view is that the brain uses models for prediction, and these models can be usefully expressed as programs in a probabilistic programming language that is designed appropriately. Scruff is an attempt to realize a probabilistic programming language design to support this view.

Specifically, Scruff is based on functional programming; functional programming has perhaps been the dominant paradigm in probabilistic programming, with numerous languages based on this paradigm (e.g., Koller et al. 1997; Pless and Luger 2001; Pfeffer 2007; Goodman et al. 2012; Narayanan et al. 2016). Scruff is embedded in the functional programming language Haskell.

The core concept in functional probabilistic programming is a stochastic function. In ordinary functional programming, a function describes a computation that takes arguments and produces a value. Functional probabilistic programming adds randomness, so that a function stochastically produces a value. The meaning of the function is the conditioned probability distribution over the produced value, given the arguments.

A functional probabilistic program can be used in multiple ways. It can be used to generate samples from the distribution over values; in this way it is acting as a predictor. It can also be used to take observations and reconcile them with the predicted distribution to form posterior beliefs, similar to how predictive processing processes error to reconcile observations with predictions to form posterior beliefs.

5.1   Hierarchical Representations

While this Bayesian formulation is natural, the programming language framework enables Scruff to satisfy many of the requirements of a predictive processing-based cognitive architecture. First, it lends itself well to hierarchical representations. We can write a program of the form:

    layer[0] = f[0]()
    layer[1] = f[1](layer[0])
    ...
    layer[n] = f[n](layer[n-1])

Each function f[i] could be any stochastic function defined in the probabilistic programming language. As a result, such a function can express a rich and detailed relationship between layers; this function may itself have an internal hierarchy. The root function f[0] can contain fundamental knowledge about the domain; this is then passed down to the next layer via the function f[1]. In practice, the functions f[i] are likely to be structured, so that not every variable in layer[i] depends on every variable in layer[i-1]; the programming language framework makes it natural to encode such structure, and probabilistic programming systems are able to exploit such structure for efficient reasoning.

Reasoning in this hierarchy follows Pearl's belief propagation process (Pearl 2014). In a downwards, predictive pass, a sequence of π messages is passed from parent to child, from the root down to the leaves. The content of a π message from a parent layer [i-1] to a child layer[i] is a prediction of the value of the parent layer[i-1], which is then combined with the function f[i] to obtain a prediction for the child layer[i]. This prediction is then passed on as a π message to layer[i]'s child.

While predictions are propagated all the way down to the lowest layers, observations can be simultaneously propagated upwards from the leaves to the roots. Each layer sends a λ message to its parent. This message represents the likelihood of the observation as a function of the parent. Once a layer has received both π and λ messages, it can compute its posterior beliefs by multiplying these messages together.

Although our hierarchical model above is a simple sequence of layers, the belief propagation process actually allows for layers to have multiple parent and child layers. As long as the network structure is a polytree (i.e., we can choose some node in the structure as the root, and the network is then a tree), belief propagation is an exact algorithm. Even if the network structure is not a polytree, belief propagation is often used as an effective approximate algorithm (McEliece et al. 1998)]. Indeed, Friston has posited that the brain uses a belief propagation processes to minimize free energy to implement predictive processing (summarized by Allen and Friston 2016).

Because Scruff allows layers to have multiple parents and children, we can express the rich interactions that take place in the brain, such as fusing information from multiple modalities. The high-level conceptual layers can have as descendants different hierarchies for each of the modalities. For example, the hypothesis that a lion is present will lead to a prediction in both the visual and auditory modalities. A predictive and error-correcting process in each of these modalities will result in separate posterior distributions at the top layer of the separate hierarchies for each of the modalities. Each of these will send a λ message to their common parent, which will integrate the observations from both modalities before continuing to propagate messages upwards.

5.2   Modeling Time

As we have discussed, time is an essential component of predictive processing. Time can be naturally represented in probabilistic programming using the function mechanism. Different layers of the hierarchy can evolve at different rates. This can be achieved by defining a standard time unit (e.g., 1 s), and setting the rates of each of the layers to multiples or fractions of this unit. Let the time unit of layer i be u[i]. We can then make the state of each layer depend on the previous state of that layer as well as the current state of the parent layer, all at the appropriate time scales, using (where % is the modulo operator):

$$\text{layer}[i, t] = f[i](\text{layer}[i - 1, t - t \mathbin{\%} u[i - 1]], \text{layer}[I, t - u[i]])$$

Note that this approach works whether the lower layer is faster or slower than the higher layer. The lower layer will typically be faster, but one can also imagine situations where a layer will have multiple lower layers, some of which are faster and some slower than it. One can use this approach to implement hierarchies for predicting events at different time scales, rooted in a common high-level hierarchy. For example, there could be a hierarchy for predicting the motion of objects at a very fast rate, a slower hierarchy for predicting changing environments, and yet slower hierarchies for daily and seasonal changes. Each layer can be assigned a time unit independently of all the other layers.

5.3   Modeling Context

The context in which a prediction takes place can significantly affect the quality of the prediction and the significance of an observation. Predictive processing posits that the context can modulate the precision of a prediction or the impact that an error has on posterior beliefs. For one example, visual perceptions at twilight are less reliable than at midday, so if the visual perception at twilight is different from predicted, the error will be down-modulated. Conversely, in a new environment predictions are more uncertain, so error will be up-modulated.

Both of these effects can be captured by making the function f[i] for layer i take an additional argument representing the context. (This additional argument will usually come from higher in the hierarchy or may be from a lateral connection.) We then make the variance of the prediction depend on the context. In the first example, the variance at the low perceptual layer is increased to account for the poor visibility at twilight. As a result, the λ message sent to the layer above will be more broadly distributed than at midday, which results in the layer above taking less account of the observation. In the second example, the variance at high layers is increased due to the uncertain environment. As a result, the π messages sent down are more broadly distributed and therefore so are the predictions. The observations will then dominate the predictions and become more significant.

5.4   Learning Predictive Models

In predictive processing, predictions are based on experience. It is therefore necessary to learn predictive models at all levels of the hierarchy. Scruff draws on neural network principles to achieve this. In Scruff, stochastic functions like the f[i] for each layer are characterized by a set of parameters. These parameters are analogous to the weights in a neural network, but they govern the generative probabilistic model. For example, a parameter might be the probability with which a certain effect is observed given certain causes. Alternatively, a parameter might be the mean or variance of a Gaussian distribution for a variable.

When an observation is made, its probability density function is defined by the prediction of the model, using the current parameter values. Ideally, we would like the parameter values to be as predictive as possible, which means to minimize the surprise of the observation. This means that we want the probability density of the observation to be high. This suggests that if we can compute the gradient of the probability density with respect to the parameters, we can adjust the parameters in the direction of the gradient. We can then use a stochastic gradient ascent method to learn the parameters, much like in deep neural networks.

Scruff computes the gradient using a combination of automatic differentiation and dynamic programming, which is the same combination that underlies the backpropagation algorithm used to train neural networks. In Scruff, each type of model that supports derivative computation contains instructions for how to compute the derivative of the density of a variable at a point, using the derivatives of other variables. Scruff then combines all these instructions into a general method for computing the derivatives.

The resulting learning framework is continuous and hierarchical. It is continuous in that every time an observation is received, at whatever time scale, it is used to update the parameters of the model. The parameters never settle at a final value. In particular, this feature helps navigate changing environments. For example, if the presence of food in a certain area is decreasing, Scruff will learn to decrease the prediction of food at some layer of the hierarchy.

The learning is hierarchical in that the parameters of all layers in the hierarchy are learned simultaneously. The degree to which the parameters in a layer will be adjusted depends on how sensitive the probability density is to those parameters. This can depend on the degree of confidence in the model at a layer. If the prior provides high confidence in a model, the prediction might be only minimally influenced by a small change in parameters. But if the prior is weak, a small change in parameters might lead to a big change in predictions. This mechanism can lead the agent to learn fast in new environments and slowly in more familiar environments.

6 Example Models

6.1 Deep Conditional Linear Gaussian Models

A natural model for predictive processing that has already been used in the literature is the linear Gaussian (LG) model (e.g., Deneve and Jardri 2016). In a LG model, each layer is represented by a Gaussian distribution whose mean is a linear function of the previous layer. The parameters of a LG model are the coefficients of the parents in the linear function, as well as a bias. The variance is typically fixed. Equations in Fig. 1 could have been derived from a LG model. A LG model determines a multivariable Gaussian over all the layers. This fact enables LG models to support efficient inference.

In the graphical models literature, LG models have been extended to conditional linear Gaussian (CLG) models (Lauritzen 1992). A CLG model includes discrete variables that can be parents of Gaussian variables. Each Gaussian variable is now determined by a LG model that depends on the values of its discrete parents. The discrete parent is a natural way to encode context that influences the choice of linear Gaussian. For example, a prediction that a lion is present in a certain physical location and pose may generate a prediction of an image of a lion, where the properties of the image are drawn from a Gaussian whose mean is a linear function of the location and pose. The variance of this Gaussian might depend on the time of day (e.g., midday vs. twilight). In a CLG model, this can be achieved by the choice of LG depending on a discrete variable representing different parts of the day. Note that in a LG model without discrete variables, there is no way to introduce this context-dependent variance, although more sophisticated models will have this capability.

In Scruff, we can create deep LG models and deep CLG models. A deep LG model consists of a network of stacked layers. Each layer consists of a vector of real-valued variables, where each variable is defined by a LG model of variables in the previous layer. What makes a deep LG model interesting is that the variables are not assigned a predefined meaning. Rather, they are latent variables whose meaning is inferred from data using a gradient ascent learning process. Therefore, deep LG models can learn knowledge about the domain, similarly to a multi-layer neural network, except in a predictive processing representation.

A deep CLG model extends a deep LG model with a discrete context layer. Figure 2 shows two example architectures. Linear Gaussian nodes are shown as ellipses, while discrete context variables are rounded rectangles. In the version on the left, only roots of the LG network will depend directly on the discrete context, although of course all other layers will have an indirect dependence via the root layer. In other models, all variables in the LG network will be able to depend on the discrete variables. In the architecture on the right of Fig. 2, each layer has its own dedicated context variable that influences all the nodes in that layer.

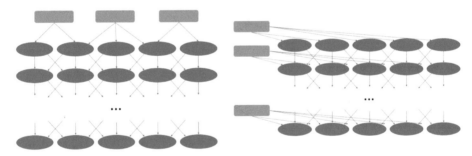

Fig. 2. Two deep CLG architectures. In the left architecture, only root nodes of the conditional Gaussian network depend directly on the context variables. In the right architecture, each layers has its own context variable.

Also, in some variants of a deep CLG, the discrete variables will be given a predetermined meaning, such as part of day, while in others, the discrete variables can also be latent and have their meaning learned from data, along with the Gaussian variables. These two variants correspond loosely to models where the relevant contexts are innate to the agent or learned. Scruff enables fine control over the degree to which variables in the model are observable or latent. It also enables expression of a preference for certain parameter values, which functions like a prior over parameter values, although it is not technically a Bayesian prior. Using this mechanism, it is possible to control the degree to which parameter values will be influenced by the data. As a result, Scruff enables representation of a wide variety of models of perception and learning using deep CLGs.

6.2   Deep Noisy-or Networks

A commonly used model in a Bayesian network is a noisy-or model (Heckerman and Breese 1996). In a noisy-or model (Fig. 3), there is a single effect that may be caused by a number of causes. Each variable represents the presence or absence of a condition; it is a Boolean variable that is True if the condition is present and False if it is not. Each cause is, on its own, sufficient to cause the effect to be True. This is the reason for the name "or"—the effect is True if any of the causes are True. However, each of the causes has an associated noise factor that may prevent the effect from being True, even when the cause is True, hence the name "noisy-or". In addition, there is a leak factor that enables the effect to be True even when none of the causes successfully cause it to be true. The parameters of a noisy-or model are the probabilities associated with the noise factors and the leak, all of which are Boolean variables.

In Scruff, we can extend a noisy-or model to a deep noisy-or network (DNON), shown in Fig. 4. A DNON is, quite simply, a network in which each node is defined by a noisy-or model. For the root nodes, there are no parents, so the noisy-or reduces to a Bernoulli model with the leak probability. A DNON can encode a combination of known causal relationships and latent relationships learned from data.

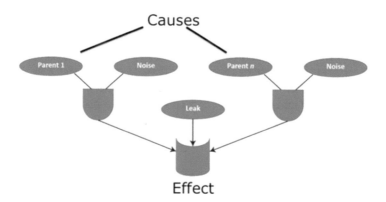

Fig. 3. A noisy-or model.

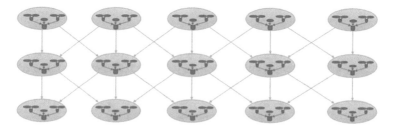

Fig. 4. A deep noisy-or network.

An interesting property of noisy-or models is that if you observe that the effect is False, you can condition all the causes independently. This is because in an or model, the effect being False means that all the causes have to be False, while the noise terms are local to each cause. However, if you observe that the effect is True, all the causes become coupled, because one cause being False makes it all the more likely that the other causes are true to result in the effect being True. Because of this property, it is much more computationally efficient to condition a DNON on False observations than on True observations.

Another property of DNONs is that, for most parameter values, for sufficiently large DNONs, all the leaves are extremely likely to be True, a priori. The probability a variable is False decreases exponentially with the number of parents because of the or structure of the model. Meanwhile, the probability of False in successive layers gets smaller and smaller. The net effect is that even for moderately deep and wide networks, the probability of False at the leaves is exceedingly small. We have found empirically that for a network whose depth is 3 and whose width is just 6, with random parameters, the probability of False at the leaves is less than 0.01.

These two properties imply that False observations are at once surprising and easy to process, while True observations provide little information and are hard to process. This suggests a natural approximate way of handling observations—just ignore the True observations and process only the False observations.

This approach has a natural predictive processing interpretation as anomaly detection and explanation. A True observation is considered normal. The prediction is almost always that things are normal. However, when a False observation is received, that represents an error that signals an anomaly. This error is propagated up the network by determining which nodes in the previous layers are likely to have been false, to explain the observed anomaly. This process continues up to the roots to determine the most likely root cause explanations of the anomaly. The agent can then adjust its expectations according to these explanations to explain away the anomaly.

6.3   A Bayesian Version of Rao and Ballard's Model

Rao and Ballard (1999) present an early implementation of a hierarchical predictive processing model. While the Rao and Ballard model is based on linear algebra and is not probabilistic, a Bayesian variation of the model can be developed in Scruff. Rao and Ballard present a three-layer model in which the bottom layer represents image patches, the middle layer represents local interpretations of the image patches, and the top layer represents an interpretation of a larger-scale portion of the image.

This increasing spatial refinement going down the hierarchy is the inverse of the decreasing spatial refinement going up the layers of a convolutional neural network. An important property of convolutional neural networks is that the same filters are applied to different areas of the image, leading to translational invariance. Similarly, we can design our model so that the generative process from a single variable in the top layer to a region in the middle layer is the same for all regions.

Let us define the variables in the top layer to be $x_{1,1}, \ldots, x_{m,n}$. Suppose each variable in the top layer influences $d \times d$ nodes in the middle layer. We can let $y_{1,1}, \ldots, y_{dm,dn}$ be the variables in the middle layer, while $z_{1,1}, \ldots, z_{dm,dn}$ are the variables in the bottom layer corresponding to the image patches. Each variable $y_{i,j}$ depends on the variable $x_{i/d,j/d}$. The specific functional form of the dependency depends on the location of y_{ij} in the $d \times d$ patch. We can define a set of functions $f_{1,1}, \ldots, f_{d,d}$ for each of the $d \times d$ locations. Then the variable $y_{i,j}$ depends on $x_{i/d,j/d}$ by the function $f_{i\%d,j\%d}$. The specific form of the function $f_{i\%d,j\%d}$ could be, for example, a linear Gaussian whose parameters depend on $i\%d$ and $j\%d$. This is close to the linear model of Rao and Ballard.

For the bottom layer, in the Rao-Ballard model the variable $z_{i,j}$ depends only on the variable $y_{i,j}$ by some function g that is the same for all image regions. However, we can enrich the model by making the variable $z_{i,j}$ also depend on the surrounding context. For example, we can make the function g also take as arguments the four adjacent image patches. We will define $z_{i,j}$ to be equal to $g\left(y_{i,j}, y_{i-1,j}, y_{i,j-1}, y_{i+1,j}, y_{i,j+1}\right)$.

This is just one of a wide variety of models we can make of image processing. Just as there is a wide variety of convolutional neural network architectures, there is a wide variety of possible architectures for predictive vision models in Scruff. We can vary the number of layers of processing, the degree to which nodes in a layer below shrink the region of the image covered by nodes in the layer above, and the degree to which nodes in a layer take into account the surrounding context. We can also choose the functional form and parameterization of each node. Whatever particular model is chosen, Scruff can learn the parameters of the model from data and make inferences using the model. As a result, Scruff could serve as a testbed for different hypothesized models.

7   Conclusion

The theory of predictive processing encompasses several elements that make it attractive as the underlying computational approach for a cognitive architecture. A new cognitive architecture, Scruff, is capable of implementing predictive processing models by incorporating key properties of neural networks into the Bayesian probabilistic programming framework. We illustrated the Scruff approach with conditional linear Gaussian (CLG) models, noisy-or models, and a Bayesian variation of the Rao-Ballard linear algebra model of predictive vision.

We acknowledge that our framework provides a far from complete story of cognition. In addition to connecting perception to action, an important issue is going beyond the parameter learning described in this paper to learning the structure of models. Is it possible to learn the hierarchical structure of the predictive models or the relevant context? We gave the example of time of day influencing the variance of the predictions; is it possible to learn the relevant times of day from experience? It may well be that the processes for learning these structural concepts are associative rather than predictive. It would be extremely interesting to explore whether associative learning mechanisms can be captured in a probabilistic programming framework like Scruff.

References

Allen, M., Friston, K.J.: From cognitivism to autopoiesis: towards a computational framework for the embodied mind. Synthese (2016). https://doi.org/10.1007/s11229-016-1288-5

Barrett, L.F.: The theory of constructed emotion: an active inference account of interoception and categorization. Soc. Cogn. Affect. Neurosci. nsw154 (2017). http://doi.org/10.1093/scan/nsw154

Barrett, L.F., Simmons, W.K.: Interoceptive predictions in the brain. Nat. Rev. Neurosci. **16**(7), 419–429 (2015). https://doi.org/10.1038/nrn3950

Barsalou, L.W.: Situated conceptualization. In: Perceptual and Emotional Embodiment: Foundations of Embodied Cognition, pp. 1–11 (2015)

Chanes, L., Barrett, L.F.: Redefining the role of limbic areas in cortical processing. Trends Cogn. Sci. (2015). https://doi.org/10.1016/j.tics.2015.11.005

Deneve, S., Jardri, R.: Circular inference: mistaken belief, misplaced trust. Curr. Opin. Behav. Sci. **11**, 40–48 (2016). https://doi.org/10.1016/j.cobeha.2016.04.001

Friston, K.: Hierarchical models in the brain. PLoS Comput. Biol. **4**(11), e1000211 (2008)

Friston, K.: The free-energy principle: a unified brain theory? Nat. Rev. Neurosci. **11**(2), 127–138 (2010). https://doi.org/10.1038/nrn2787

Friston, K., Kiebel, S.: Predictive coding under the free-energy principle. Philos. Trans. R. Soc. B: Biol. Sci. **364**(1521), 1211–1221 (2009)

Goodfellow, I., Pouget-Abadie, J., Mirza, M., Xu, B., Warde-Farley, D., Ozair, S., et al.: Generative adversarial nets. Presented at the Advances in Neural Information Processing Systems, pp. 2672–2680 (2014)

Goodman, N., Mansinghka, V., Roy, D.M., Bonawitz, K., Tenenbaum, J.B.: Church: a language for generative models. ArXiv Preprint arXiv:1206.3255 (2012)

Heckerman, D., Breese, J.S.: Causal independence for probability assessment and inference using Bayesian networks. IEEE Trans. Syst. Man Cybern. Part A Syst. Hum. **26**(6), 826–831 (1996)

Hohwy, J.: The Predictive Mind. Oxford University Press, Oxford (2013)

Kingma, D.P., Welling, M.: Auto-encoding variational Bayes. ArXiv Preprint arXiv:1312.6114 (2013)

Koller, D., McAllester, D., Pfeffer, A.: Effective Bayesian inference for stochastic programs. In AAAI/IAAI, pp. 740–747 (1997)

Lake, B.M., Salakhutdinov, R., Tenenbaum, J.B.: Human-level concept learning through probabilistic program induction. Science **350**(6266), 1332–1338 (2015). https://doi.org/10.1126/science.aab3050

Lake, B.M., Ullman, T.D., Tenenbaum, J.B., Gershman, S.J.: Building machines that learn and think like people. Behav. Brain Sci. (2017). https://doi.org/10.1017/S0140525X16001837

Lauritzen, S.L.: Propagation of probabilities, means, and variances in mixed graphical association models. J. Am. Stat. Assoc. **87**(420), 1098–1108 (1992)

Le, T.A., Baydin, A.G., & Wood, F.: Inference compilation and universal probabilistic programming. ArXiv Preprint arXiv:1610.09900 (2016)

Lupyan, G., Thompson-Schill, S.L., Swingley, D.: Conceptual penetration of visual processing. Psychol. Sci. **21**(5), 682–691 (2010). https://doi.org/10.1177/0956797610366099

Marcus, G.: Deep learning: a critical appraisal. ArXiv Preprint arXiv:1801.00631 (2018)

Mathys, C.: A Bayesian foundation for individual learning under uncertainty. Front. Hum. Neurosci. (2011). https://doi.org/10.3389/fnhum.2011.00039

McEliece, R.J., MacKay, D.J.C., Cheng, J.-F.: Turbo decoding as an instance of Pearl's "belief propagation" algorithm. IEEE J. Sel. Areas Commun. **16**(2), 140–152 (1998)

Morse, A.F., de Greeff, J., Belpeame, T., Cangelosi, A.: Epigenetic robotics architecture (ERA). IEEE Trans. Auton. Ment. Dev. **2**(4), 325–339 (2010). https://doi.org/10.1109/TAMD.2010.2087020

Narayanan, P., Carette, J., Romano, W., Shan, C., Zinkov, R.: Probabilistic inference by program transformation in Hakaru (system description). In: International Symposium on Functional and Logic Programming, pp. 62–79 (2016)

Pearl, J.: Probabilistic Reasoning in Intelligent Systems: Networks of Plausible Inference. Morgan Kaufmann, Burlington (2014)

Pezzulo, G., Barca, L., Friston, K.J.: Active inference and cognitive-emotional interactions in the brain. Behav. Brain Sci. **38**, 37–39 (2015)

Pfeffer, A.: 14 The design and implementation of IBAL: a general-purpose probabilistic language. In: Introduction to Statistical Relational Learning, vol. 399 (2007)

Pless, D., Luger, G.: Toward general analysis of recursive probability models. In: Uncertainty in Artificial Intelligence, pp. 429–436 (2001)

Rao, R.P., Ballard, D.H.: Predictive coding in the visual cortex: a functional interpretation of some extra-classical receptive-field effects. Nat. Neurosci. **2**(1), 79 (1999)

Rezende, D. J., Mohamed, S., Wierstra, D.: Stochastic backpropagation and approximate inference in deep generative models. ArXiv Preprint arXiv:1401.4082 (2014)

Tenenbaum, J.B., Kemp, C., Griffiths, T.L., Goodman, N.D.: How to grow a mind: statistics, structure, and abstraction. Science **331**(6022), 1279–1285 (2011). https://doi.org/10.1126/science.1192788

Tran, D., Kucukelbir, A., Dieng, A.B., Rudolph, M., Liang, D., Blei, D. M.: Edward: a library for probabilistic modeling, inference, and criticism. ArXiv Preprint arXiv:1610.09787 (2016)

Zaki, J.: Cue integration: a common framework for social cognition and physical perception. Perspect. Psychol. Sci. **8**(3), 296–312 (2013)

The Method for Searching Emotional Content in Images Based on Low-Level Parameters with Using Neural Networks

Rozaliev Vladimir[✉], Guschin Roman, Orlova Yulia,
Zaboleeva-Zotova Alla, and Berdnik Vladislav

Volgograd State Technical University,
28 Lenin Avenue, Volgograd, Russian Federation
vladimir.rozaliev@gmail.com

Abstract. The paper proposes a method to improve the modeling of subjectivity and human understanding based on the annotation of emotional semantics of images using the theory of fuzzy sets. Adaptive acceleration and reverse propagation of the neural network (BP). The annotation method was tested by analyzing images from the solar image database. In addition, the accuracy of the search is based on 85% of our method. This study is the basis for a more accurate semantic analysis and verification.

Keywords: Human emotions · Image identification · Neural networks
Adaboost algorithm · Information system

1 Introduction

Emotional and semantic analysis of images is an important step in the study of semantic image at a high level, pattern recognition and computer vision. When solving problems such as image classification, face detection, outdoor advertising monitoring and military intelligence, it is necessary to analyze the emotional behavior of a person, get semantic features from images, and then calculate the degree of similarity of the object.

The final objective of emotional semantic image analysis is to allow computers to define human emotional response to an image. Emotional semantic image's features are extracted based on low-level visual features. The image semantics can be divided into behavioral and emotional one which is the top-level semantics.

Low-level properties of images, such as color, texture, shapes and contours is the first extracted from the related processing technologies; further attempt be made to find a correlation between low-level properties and emotional semantics at a high level [1]. This automatic method is one of acquiring the high-level semantics. Semantic of emotional annotation is an advanced process in the field of digital image analysis. Scene images - this is the main data type generated from multimedia sources. This method provides a clear image search and is an effective tool for the introduction of multimedia information retrieval systems. The study of their semantic annotation is the

A. V. Samsonovich (Ed.): BICA 2018, AISC 848, pp. 260–265, 2019.
https://doi.org/10.1007/978-3-319-99316-4_34

foundation for the creation of the emotional semantic search for other image types and has a strong theoretical and practical significant consequently.

The launch of the study on computer emotional calculations come under 1980-th years. Currently, a reasonable approach to analyzing emotions is a popular topic of research, but development difficult in the computing fields. As of today, there have been many studies to examine the relations between the visual image elements and emotions. Mao, Ding and Mauting defined a mathematical model by analyzing image's features and came up with an analytical image method of fluctuation to calculate harmonious feelings for images [2]. Their deduction revealed that images in compliance with the fluctuation law provoke harmonious sentiments in humans. Chen and Wang draw on semantic quantization and factor analysis to create an emotional room based on dimensional analysis in the field of psychology [3]. Colombo, Bimbo, and Pala identified several basically used words, such as: warm, cool and natural, to describe the emotional images' semantics and to set an emotional room [4]. Baek, Hwang, Chung and Kim defined 52 of images templates and 55 emotional factors corresponding to patterns have been used questionnaires and have been measured the relations between low-level visual features and high-level emotions [5].

Particulary and equivocality play a key role in the human consciousness of the image. To identify effective method of subjectivity and ambiguity modeling over the human image comprehension will greatly enhance the search efficiency and result in human-centered image retrieval [6].

2 Fuzzy Set Theory and Principal Component Analysis - PCA

The present paper takes fuzzy theory as automated describes for images. The fuzzy theory was used to represent the levels of human emotions when comprehending the scene images. Thus, the feeling of "relaxation" will be provoked by when subject watch the image or background of image with nature.

In order to define human emotions excited, by images, this research demands on the definitions set out below. Emotional variables: variables comprise a five-dimensional vector: (x, E(x), U, G, T), with x - as the name of the variables, E(x) representing the emotional value set of x, U - as the domain (the extraction space of image features in this work), G being the grammar laws for the generation of the emotional value of E(x), and T which serves as the semantic rules for the calculate of the degree of emotional membership [7]. Basic emotional value set: this refers to array of emotional value, which can't be divided semantically. Extended emotional value set: this value of basic emotional with numeric degree description.

PCA had first been proposed by Karl and Pearson in 1901 and used as a tool for data analysis and the development of a mathematical model [8]. This is tool for a multi-criteria set statistical analysis to transfer through more than one index on the linear combination of multiple independent indices, which contain a large amount of information contained in the original indexes [7].

3 The Approach of the Analysis of Emotional Semantic and Automatic Annotation

After the image automatically analyzed and semantically annotated it is first necessary to extract the lower-level visual features. For the images, colors are the important features that may define the emotional semantics. In this connection, this study used a segmentation method to extract the visual features of the scene images. Then there should be created an emotional model and has to be executed a semantic mapping to implement a semantic mapping from low-level color features to emotional semantics of a higher level. Further, semantic analysis and automatic annotation can be completed.

Picking up the color-space and quantization. Since the color space of hue, saturation, and value (HSV) can mirror in satisfactory manner the human perception of colors, HSV was used as a working space. One 60-dimensional vector of color features was established to create the domain of emotional variables. Since human visual system is more sensitive to color than to saturation and value, we quantitated HSV space, in accordance with the methods used in previous studies [7, 8].

Such quantitative methods have different advantages, including a better understanding of the human visual system, reduce the redundancy of color and are to be applied to the grey surfaces simplifying the calculation.

The release strategy was taken for to extract local features of images [3, 7]. The commonly used method is to divide the images into m * n number of blocks, while different blocks have different weights. The central area of an image plays a crucial role in the semantic comprehension of a man and therefore has the greatest weight. In the case research, we sized the image into 4 * 4 segments, as shown in Fig. 1.

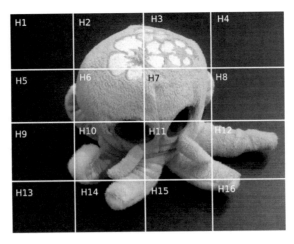

Fig. 1. Segment layout of image

The weight of each unit depends on the characteristics of the image to use. As the central segment or user-assigned one has usually a large weight that may better reflect information about the location of the image thus the weight of the central blocks of the image has been raised. The blocks weight was distributed as follows. As shown in Fig. 1, the area with the red lines was 1/4 of the total area of the image. Thus, his weight has been raised to 1/3. Therefore, weight H6, H7, H10 and H11 were 1/12 each, and the weight of each of them surrounding the 12 segments was 1/18.

The making up of emotional model is critical for emotional semantic scene image analysis and the definition of a core set of emotional value E(x) is a critical component in building up this model.

Normally, the creation of emotional patterns consists of three stages. Firstly, emotional adjectives are being collected to determine the emotional values. Secondly, the experiment on semantic quantization is being carried out and an emotional database is being created, based on the evaluation of images of individual subjects of the shooting. Thirdly, an analysis of the data is being implemented to establish the emotional room.

The determination of a set of emotional values. Seven adjectives - terms have been carefully selected to set E(X) = (natural, romantic, soft, relaxed, bright, regenerating, changeful). An expanded set of emotional values have been subsequently constructed as very, neutrally and slightly. For example, the enhanced emotional value of a set of basic emotional value of "soft" very soft, soft, slightly soft. The grammar law of G emotional variables was thereafter formulated in the following way: emotionally expressive formula :: == enhanced emotional value—base emotional value Enhanced emotional value :: == listed variable) & base emotional value Affiliation variables :: == very—neutrally—slightly base emotional value :: == natural—romantic—soft—relaxed—energetic—regenerating—changeful.

Building on our experiment, three extended emotional values (very neutral, and slightly) have been quantized as follows:

$$V_e(x) = \left\{ T_e^2(x) | x \in U \right\}, \tag{1}$$

$$N_e(x) = \left\{ \sin T_e(x) \times \pi | x \in U \right\}, \tag{2}$$

$$H_e(x) = \left\{ 1 - T_e(x) | x \in U \right\}, \tag{3}$$

with e - taken as base emotional value set, $T_e(x)$ - the degree of fuzzy affiliation, derived from neural network training BP, x constitutes a base emotional value, $V_e(x)$, $N_e(x)$ and $H_e(x)$ are affiliation degree of very, neutrally and slightly x consequently.

The SUN data base is a free image database for researchers in the field of computer-understand manner. In this study it was used images from the database for the experiments [9]. A total of 100 databases SUN typical image was chosen. These images provided different colors, cyberspace layouts and content. The users included 10 members; their ages ranged from 20 to 26. Emotional database had been produced on the base of user ratings.

For a set of samples where $\{(V_1, y_1), (V_2, y_2), ..., (V_n, y_n)\}$ - the extracted 60-element vector color features $V_i \in U$, $(i = 1, 2, ..., 60)$ and $y_i(i = 1, 2, ..., n)$ - membership class of base emotional value included in an enhanced emotional value, must mapping be set $T_e: V \to y$, $e \in E(x)$.

Considering that neural BP network [10] has a simple, structure, high speed instruction and a strong learning ability, which uses the processing of fuzzy sets and that the basic idea Adaboost algorithm is to combine the results of several weak predicates for effective prediction, Adaboost algorithm was combined with BP neural network. The flowchart and learning audio BP neural network are shown in Figs. 2 and 3, respectively. BP neural network is a neural network with feedback, which consists of an input layer, hidden layer and output layer. In this study, as a network entry isolated 60-dimensional visual color low-level particularities were used. The number of elements in the hidden layer was determined to be 20. As a function of the activation of hidden layers Gaussian function had been used.

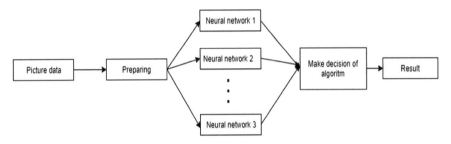

Fig. 2. Flowchart neural network

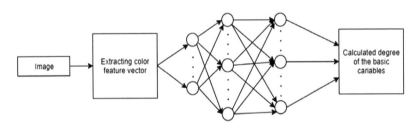

Fig. 3. Learning process neural network

The lead predicate has been set based on the Adaboost algorithm [7]. At the stage of learning, 250 images with styles were chosen from the 500 scene images which used as a set for training. After training, the degree of membership of seven emotional variables has been identified. For example, or the base emotional term "natural" since the membership degree to be 0.81, the degree of natural feeling inspired by input image being equal to 0.83. In accordance with the above procedures, the image features have been presented in the form of a 7-dimensional vector; that is, where the degree of fuzzy membership of sets basic emotional variables. Corresponding base emotional value of a plurality scene of images in space was (natural, changeful, romantic, regenerating, soft, relaxed, and bright).

4   Conclusion

A serious field of artificial intelligence issue is discovering how machines can replicate the human perception and expression to fulfil the human-machine harmony. The current study had resorted to theory of fuzzy set to find the relations between human comprehension and scene images [11]. The model was created based on Adaboost algorithm and BP neural network for the implementation of the automated description of emotional semantics in scene images. The findings revealed that the combination of algorithm and neural network has more advantages when they are applied to solve problems of human subjectivity and ambiguity of emotional subjects. The study not only suggests one approach to solving the problem of human subjectivity and ambiguities, but also provides new ideas for comprehension another kind of types of emotional semantics on images.

Acknowledgment. This work was partially supported by RFBR (grants 16-07-00407, 16-47-340320, 16-07-00453, 18-07-00220).

References

1. Lv, J., Xiang, J., Chen, J.J.: Research of image affection based on feature extraction technology of ROI. Comput. Eng. Des. **31**(3), 660–662 (2010)
2. Mao, X., Ding, Y.K., Moutian, Y.: Analysis of affective characteristics and evaluation on harmonious feeling of image. Acta Electron. Sin. **29**(12A), 1923–1927 (2001)
3. Wang, S.F., Chen, E.H., Wang, S.H.: An image retrieval based on emotion model. J. Circ. Syst. **8**(6), 48–52 (2003)
4. Colombo, C., Bimbo, A., Pala, P.: Semantics in visual information retrieval. IEEE Multimed. **6**(3), 38–53 (1999)
5. Baek, S., Hwang, M., Kim, P.: Kansei factor space classified by information for Kansei image modeling. Appl. Math. Comput. **205**(2), 874–882 (2008)
6. Kansei sessions, in Proceedings of the IEEE International Conference on Systems Man and Cybernetics, Tokyo, Japan (1999)
7. Cao, J., Chen, L.: Fuzzy emotional semantic analysis and automated annotation of scene images. Comput. Intell. Neurosci. **2015**, 115–125 (2015)
8. Han, W., Li, G.: The applied study of principal component analysis in evaluation of science and technology competence. Appl. Stat. Manag. **25**(5), 512–517 (2006)
9. Thomas, L.S.: The Analytic Hierarchy Process, 2nd edn. RWS Publications, Pittsburgh (1996)
10. Sun, X., Xie, Y.D., Ren, D.C.: Study on image registration technique based on wavelet transform and sub-graph. Comput. Eng. Des. **31**(21), 4653–4654 (2010)
11. Rozaliev, V., Orlova, Y., Guschin, R., Verishev, V.: General approach to the synthesis of emotional semantic information from the video. In: Creativity in Intelligent Technologies and Data Science. Communications in Computer and Information Science, vol. 754, pp. 201–214. Springer International Publishing (2017)

Intelligent Planning and Control of Integrated Expert Systems Development Process

Galina V. Rybina$^{(\boxtimes)}$, Yury M. Blokhin, and Levon S. Tarakchyan

National Research Nuclear University MEPhI (Moscow Engineering Physics Institute), Kashiskoe sh. 31, Moscow 115409, Russian Federation
galina@ailab.mephi.ru

Abstract. The paper is focused on problems of development of integrated expert systems basing on problem-oriented methodology. Special component - intelligent planner controls the development process. The intelligent planner uses automated planning techniques in its core. The development of integrated expert system is shown as a planning problem. Some necessary models and technical details of intelligent program environment are described. The other crucial components of intelligent program environment are briefly discussed. AT-TECHNOLOGY workbench is an implementation of problem-oriented methodology.

Keywords: Integrated expert system
Problem-oriented methodology · Intelligent program environment
Automated planning

1 Introduction

Currently among the individual classes of intelligent systems (for example, [1–3]), the most popular ones are integrated expert systems (IES) [4]. An analysis of the experience of developing foreign and domestic IES, including those based on a problem-oriented methodology for IES development (authored by G.V. Rybina) and the AT-TECHNOLOGY workbench supporting this methodology [4,5] showed that the greatest complexity is still on the stages of design and implementation of IES, and the specificity of the specific problem area (PA) and the human factor play a significant role. Here can note several major problems:

– the inability of users (knowledge engineers) to fully determine the requirements for the systems being developed, the lack of a reliable method for assessing the quality of verification and validation of the IES, the inapplicability of traditional tracing technology to the knowledge base (KB) of the IES;
– the presence of a large number of intermediate stages and iterations in the life cycle models of the construction of individual components of the IES;

© Springer Nature Switzerland AG 2019
A. V. Samsonovich (Ed.): BICA 2018, AISC 848, pp. 266–271, 2019.
https://doi.org/10.1007/978-3-319-99316-4_35

– practical absence (except the AT-TECHNOLOGY workbench) of specialized tools that provide: automated design of software and information support for applied IES at all stages of the life cycle; reuse of individual software and informational components; presence of internal integrability of tools; support for a convenient cognitive-graphic interface; openness and portability of tools.

The most widely known new generation toolkit is the AT-TECHNOLOGY toolkit [4,5], which provides automated support of the processes of building the IES based on the task-oriented methodology, and includes intellectual assisting (planning) the actions of knowledge engineers for account of the use of technological knowledge on standard design procedures (SDP) and reusable components (RUC) of previous projects that are basic components of the model of the intellectual software environment, the concept of which is described in detail in [4] and other works.

New results in the development of the intellectual software environment of the AT-TECHNOLOGY workbench were obtained by expanding the functionality of another basic component - the intelligent planner, which not only provided effective assistance to knowledge engineers, but also planning and managing the development process as a whole. Analysis of the current versions of the intelligent planner of the AT-TECHNOLOGY workbench [4,5] showed that since the complication of the IES architectures and the appearance of a large number of SDP and RUC in the technological KB, the time for finding solutions has significantly increased. The results of the system analysis of modern methods of intelligent planning and conducted experimental studies [6,7] have shown the expediency of using a fairly well-known approach related to planning in the state space.

Below we discuss some of the results of further development of the basic components of the intellectual software environment of the AT-TECHNOLOGY workbench intended for automating and intellectualizing the processes of building IES with different architectural typologies on the basis of a problem-oriented methodology (detailed description is given in [7–12]).

2   Aspects of Processes of the Prototyping of Applied IES

A full description of the methods for constructing various IES with a broad architectural typology developed and tested in practice is contained in [6,7], as well as in other works, and therefore we will focus here on the general formulation of the problem of intelligent planning [6–12] with reference to models, methods and means of the intellectual software environment of the AT-TECHNOLOGY workbench. The main goal of the problem-oriented methodology and its supporting tools is the intellectualization of rather complex and time-consuming processes of prototyping applied IES of various architectural typologies at all stages of the life cycle. To reduce the intellectual burden on knowledge engineers, to minimize possible erroneous actions and time risks in the prototyping of IES, according to [4], it is envisaged to use a technological KB containing a significant number of

SDP and RUC reflecting the expertise of knowledge engineers in the development of applied IES (static, dynamic, tutoring).

Therefore, in this paper, the formal statement of the problem of intelligent planning of the prototyping processes of IES is considered in the context of the model of prototyping processes of IES in the following form: $M_{proto} = \langle T, S, Pr, Val, A_{IES}, PlanTask_{IES} \rangle$, where T is the set of problem domains for which applied IES are created; S is the set of prototyping strategies; Pr is the set of created prototypes of IES based on a problem-oriented methodology; Val - the function of expert validation of the prototype of the IES, determining the need and/or the possibility of creating subsequent IES prototypes for a particular problem domain; A_{IES} - the set of all possible actions of knowledge engineers in the prototyping process; $PlanTask_{IES}$ is the function of planning knowledge engineer's actions to obtain the current prototype of IES for a particular problem domain. In [12], a detailed description of M_{proto} is given.

For effective implementation of the component $PlanTask_{IES}$ of the M_{proto} model modern methods of intelligent planning in the state space have been investigated [13–16]. Experimental software research has shown that the best results are achieved if the search space is formed by modeling the actions of the knowledge engineer while constructing fragments of the architecture model [4,5] of the prototype of IES using the appropriate SDPs.

An important role in the implementation of the PlanTaskIES component is played by the plan for constructing a specific coverage (i.e., the sequence of applied SDP fragments), which can be uniquely transformed into a plan for constructing a prototype of IES. In this case, the plan for constructing the prototype of the IES [12] can be represented like $Plan = \langle A_G, A_{atom}, R_{prec}, R_{detail}, PR \rangle$ where A_G is the set of global tasks decomposable into subtasks); A_{atom} is the set of planned (atomic) tasks, the execution of which is necessary for the development of the IES prototype; R_{prec} is the function that determines the predecessor relation between the planned tasks; R_{detail} is the relation showing the affiliation of the planned task to the global tasks; PR is the representation of the plan, convenient for the knowledge engineer.

With the help of the relation R_{prec} and the sets A_G and A_{atom} two task networks are formed - enlarged and detailed. The enlarged network of tasks obtained with R_{prec} and A_G is called the global plan (the relation of the precedence between the elements of the A_G is obtained on the basis of the R_{detail} detail relation), and the detailed network of tasks obtained on the basis of R_{prec} and A_{atom} is called the detailed plan, with each planned task associated with specific function by a function of a specific operational RUC.

3   General Formulation of the Problem of Generation of Plans for the Development of Prototypes of IES

Let us consider the general formulation of the problem of generation of plans for developing prototypes of IES. The initial data is the model of the architecture of the prototype of the IES, described using the hierarchy of extended data flow diagrams (EDFD [4]); The technological KB, containing a lot of SDP and RUC. In addition, a number of restrictions and working definitions were introduced.

1. From the composition of the architecture model M_{IES}, built at the stage of the analysis of system requirements, only a set of elements and a set of data flows represented in the form of a marked oriented graph G_{RIL}, where the labels determine the relationship between the elements the hierarchy of the EDFD and the vertices and arcs of the graph. This graph is called a generalized EDFD, and it can be uniquely obtained from the initial model of the architecture M_{IES}.
2. The fragment of the generalized EDFD is an arbitrary connected subgraph contained in G_{RIL}. A SDP instance is an aggregate of TPP and a fragment of a generalized EDFD satisfying the conditions of applicability (the C component of the TPP model) of the corresponding SDP. The cover ($Cover$) of a generalized RDPA is the set of instances of SDP with mutually disjoint fragments containing all the vertices of G_{RIL} (or cover the entire G_{RIL}).
3. A coarse coverage of a generalized EDFD is a EDFD coverage in which all instances of the SDP contain only mandatory fragments. The fine coverage is an extension of the coarse coverage by including optional fragments in the coverage.
4. The inclusion of each fragment of the SDP in the cover is compared with a certain value determined by an expert evaluation based on technological experience, which conceptually corresponds to the average costs of human resources for the implementation of the corresponding fragment of the model of the architecture of the IES.

Thus, the problem of generating a plan for developing a prototype of IES, taking into account the initial data and imposed constraints, is conveniently presented in terms of states and transitions in the form of a model $PlanTask_{IES} = <S_{IES}, A_{IES}, \gamma, Cost, s_0, G_{IES}, F_{COVER}>$, where S_{IES} is the set of states of the graph G_{RIL} that describe the current coverage $Cover$; A_{IES} - the set of possible actions over G_{RIL}, which are the addition of fragments of specific instances of SDP to the cover (the total set is formed in the aggregate WKB and G_{RIL}); γ is the transition function between states; Cost - a function that determines the cost of a sequence of transitions; s_0 is the initial state describing the empty coverage; G_{IES} - the function of determining whether the state belongs to the target state; F_{COVER} - the function of generating a development plan ($Plan$) from the coverage ($Cover$).

The solution of $PlanTask_{IES}$ is the plan of actions of the knowledge engineer (model $Plan$). The plan should be optimal, for which it is necessary to develop an admissible heuristic function.

To solve the problem, a method was developed (described in details in [12]) in which four stages can be conditionally identified: obtaining a generalized EDFD (G_{RIL}) from the architecture model M_{IES}; generating an exact $Cover$ cover using heuristic search; generation of the plan of actions of the knowledge engineer ($Plan$) on the basis of the obtained fine coverage ($Cover$); generating a plan view (PR) based on coverage ($Cover$).

4   Aspects of Software Implementation of Basic Components of the Intellectual Software Environment

We give a brief description of the composition and structure of the software tools of the intelligent software environment (using [7–12]), including the kernel, the user interface subsystem and the extension library, which implements interaction with operational RUCs. The subsystem of the user interface has a convenient graphical interface, on the basis of which the RUC interacts with the knowledge engineer using screen forms. The technological KB is conventionally divided into an extension library that stores operational knowledge in the form of plug-ins that implement the relevant operational RUCs and the declarative part. The kernel implements all the basic functionality of automated support for the development of prototype IES, project file management, extension management, etc.

The intelligent planner is a part of the core and implements functionality related to planning the prototyping processes of IES. With the help of the pre-processor of the hierarchy of the EDFD, the pre-processing of the hierarchy of the EDFD is carried out by converting it into one generalized diagram of maximum detailing. The task of coverage the detailed EDFD with the available SDP is implemented with the help of a global plan generator that, on the basis of the technological KB and the constructed generalized EDFD, provides the fulfillment of the task using the method described above, as a result of which a fine coverage is constructed, which is later transformed into a global plan of development. Generator of the detailed plan on the basis of the given coverage of the EDFD and the technological KB performs a detalization of each element of the coverage, thus forming a preliminary detailed plan. Then, based on the analysis of available RUC and their versions (data about which are requested from the development process control component), a detailed plan is formed in the plan interpretation component, where each task is related to a specific RUC.

5   Conclusion

Based on the data obtained as a result of joint testing of the intelligent planner and other components of the intelligent software environment of the AT-TECHNOLOGY workbench, it was concluded that all the developed tools can be used quite effectively in the prototyping of applied IES. The developed software was integrated into the AT-TECHNOLOGY workbench, with the use of which a prototype of a dynamic IES was developed to manage medical forces and facilities for major road accidents [12]; prototype of dynamic IES for satellite network resources management; prototype of static IES for diagnostics of blanks for electron-beam lithography.

Acknowledgements. The work was done with the Russian Foundation for Basic Research support (project №18-01-00457).

References

1. Giarratano, J.C., Riley, G.D.: Expert Systems: Principles and Programming, 4th edn. Thomson Course Technology, Stamford (2004)
2. Meystel, A.M., Albus, J.S.: Intelligent Systems: Architecture, Design, and Control, 1st edn. Wiley, New York (2000)
3. Schalkoff, R.: Intelligent Systems: Principles, Paradigms, and Pragmatics. Jones & Bartlett Learning, Burlington (2011). https://books.google.ru/books?id=80FXUtF5kRoC
4. Rybina, G.V.: Theory and Technology of Construction of Integrated Expert Systems. Monography. Nauchtehlitizdat, Moscow (2008)
5. Rybina, G.V.: Intelligent Systems: From A to Z. Monography Series in 3 books. Vol. 1. Knowledge-Based Systems. Integrated Expert Systems. Nauchtehlitizdat, Moscow (2014)
6. Rybina, G.V., Blokhin, Y.M.: Methods and means of intellectual planning: implementation of the management of process control in the construction of an integrated expert system. Sci. Tech. Inf. Process. **42**(6), 432–447 (2015)
7. Rybina, G.V., Blokhin, Y.M.: Use of intelligent planning for integrated expert systems development. In: 2016 IEEE 8th International Conference on Intelligent Systems, IS 2016 - Proceedings, pp. 295–300 (2016)
8. Rybina, G.V., Rybin, V.M., Blokhin, Y.M., Parondzhanov, S.S.: Intelligent programm support for dynamic integrated expert systems construction. Procedia Comput. Sci. **88**, 205–210 (2016). Annual International Conference on Biologically Inspired Cognitive Architectures, BICA 2016, held July 16 to July 19, 2016 in New York City, NY, USA. http://www.sciencedirect.com/science/article/pii/S1877050916316829
9. Rybina, G.V., Rybin, V.M., Blokhin, Y.M., Sergienko, E.S.: Intelligent technology for integrated expert systems construction. Adv. Intell. Syst. Comput. **451**, 187–197 (2016)
10. Rybina, G.V., Blokhin, Y.M.: Intelligent software environment for integrated expert systems development. In: Proceedings of the 2016 International conference on artificial intelligence ICAI2016, USA. CSREA Press (2016)
11. Rybina, G.V., Blokhin, Y.M., Parondzhanov, S.S.: Intelligent planning methods and features of their usage for development automation of dynamic integrated expert systems. In: Advances in Intelligent Systems and Computing, vol. 636, pp. 151–156. Springer, Cham (2018)
12. Rybina, G.V., Blokhin, Y.M.: Methods and software of intelligent planning for integrated expert systems development. Artif. Intell. Decis. Mak. **1**, 12–28 (2018)
13. Nau, D.S.: Current trends in automated planning. AI Mag. **28**(4), 43–58 (2007)
14. Ghallab, M., Nau, D.S., Traverso, P.: Automated Planning - Theory and Practice. Elsevier, New York City (2004)
15. Ghallab, M., Nau, D., Traverso, P.: Automated Planning and Acting, 1st edn. Cambridge University Press, New York (2016)
16. Geffner, H., Bonet, B.: A Concise Introduction to Models and Methods for Automated Planning. IEEE (Institute of Electrical and Electronics Engineers) IEEE Morgan & Claypool Synthesis eBooks Library, Morgan & Claypool (2013). https://books.google.ru/books?id=_KInlAEACAAJ

An Approach to Hierarchical Deep Reinforcement Learning for a Decentralized Walking Control Architecture

Malte Schilling[(✉)] and Andrew Melnik

Center of Excellence 'Cognitive Interaction Technology', Bielefeld University,
Bielefeld, Germany
mschilli@techfak.uni-bielefeld.de, andrew.melnik.papers@gmail.com

Abstract. Locomotion in animals is characterized as a stable, rhythmic behavior which at the same time is flexible and extremely adaptive. Many motor control approaches have taken considerable steps taking insights from biology. As one example, the Walknet approach for six-legged robots realizes a decentralized and modular structure that reflects insights from walking in stick insects. While this approach can deal with a variety of disturbances during locomotion, it is still limited dealing with novel and particular challenging walking situations. This has lead to a cognitive expansion that allows to test behaviors outside their original context and search for a solution in a form of internal simulation. What is still missing in this approach is the variation of lower level motor primitives themselves to cope with difficult situation and any form of learning. Here, we propose how this biologically-inspired approach can be extended in order to include a form of trial-and-error learning. The realization is currently underway and is based on a more broad formulation as a hierarchical reinforcement learning problem. Importantly, the structure of the hierarchy follows the decentralized organization taken from insects.

Keywords: Motor control · Robotics · Deep reinforcement learning
Neural network · Locomotion

1 Introduction

Control of hexapod walking has been an active area of research for a long time that has been influenced by different disciplines. First, from a biological perspective the question is how animals—and for six legged walking in particular insects—can produce such stable and adaptive walking behavior [2,9,20]. This area of research has lead to cybernetic models of locomotion. Second, from an engineering perspective the focus is on control approaches for six-legged robots that allow for fast and stable walking behaviors [11]. Importantly, with the high number of degrees of freedom this becomes a challenging problem and closed

© Springer Nature Switzerland AG 2019
A. V. Samsonovich (Ed.): BICA 2018, AISC 848, pp. 272–282, 2019.
https://doi.org/10.1007/978-3-319-99316-4_36

form solutions are usually not applicable. This brought these lines of research together and has lead to biologically-inspired control approaches.

Lately, locomotion has become also a field for learning behaviors (mostly in simulated environments) and has been used as an application example for (deep) reinforcement learning [6,12] as well as evolutionary computations [3]. Many of these learning approaches are solely taking an optimization perspective and are focussing on a simple objective. While these have shown successful on locomotion tasks, they require very long learning (and exploration) time that still don't lead to a degree of adaptivity as can be found in natural systems. These systems in particular struggle when switching towards changed environments and often require complete retraining. Recently, Deep Reinforcement Learning (DRL) approaches turned towards hierarchical architectures that allow to reuse parts of learned behavior in different contexts and only require to retrain certain levels. These approaches take a purely pragmatic approach in how to distribute the complexities of the problem onto different levels and for the higher levels use a central control system (described by a policy).

The proposed architecture in the current article differs in this respect as it follows insights from the long tradition of biological research on locomotion in insects: the control problem is distributed on different levels of a control hierarchy, and, importantly, even the higher level of the control hierarchy is constituted by distributed controllers, that are only loosely coupled. The overall behavior emerges from the interaction between these control modules and through the embodied agent's interaction with the environment.

In the following, we will review briefly the organization of our current control architecture that does not include learning. First, the biological characteristics will be explained. Secondly, the cognitive extension of this model will be explained that allows to further extend its behavioral repertoire through planning. Lastly, we will turn towards how this can (and in the future will be) extended towards learning, implementing on top a form of trial-and-error learning that allows to explore possible actions in an internal simulation while the behavioral selection is guided as in reinforcement learning.

2   Overview: Cognitive Walking Architecture

2.1   Walknet – Biological Inspired Hexapod Walking Control

The architecture for control of a hexapod robot is biologically inspired: it is based on detailed research on stick insects, on a behavioral level as well as on the level of neural structure (for a detailed introduction of the underlying control structure see [20]). Many approaches to six-legged walking are considering fixed gait patterns. This can be observed for fast walking and running, as in cockroaches. But actually characterizes only one part of the spectrum of walking [7]. For slow walking, a wide variety of different gait patterns and phases between legs appear not fixed at all but are changing all the time. This is also true for demanding walking tasks, like climbing. In these cases, the overall walking

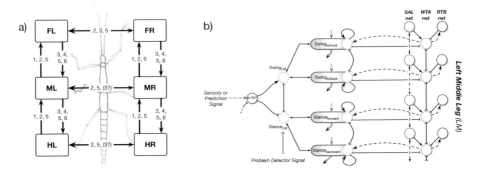

Fig. 1. (a) Coordination between different leg controllers as supported by findings on the walking of stick insects. Shown are the individual six leg controllers and the different local influences that govern the overall emergent behavior. (b) Schematics of a single leg controller that shows the different behavioral patterns (swing and stance movement for forward and backward walking) as well as the cognitive extension (on the right, for more details see [18]).

behavior is directly influenced by sensory inputs and emerges from interactions of fixed control structures and sensory signals [20].

Stick insects provide a good study system for such approaches as they are slowly walking and climbing through twigs which demands quite adaptive and flexible changes of the overall behaviors. This can be observed in quite diverse walking patterns.

In the Walknet approach, control is distributed following a morphological organization. Each leg is controlled by a single controller. Internally, these controllers are modular and hierarchically structured. While on the higher level action selection is driven mostly by sensory inputs, on the lower level motor primitives are executed. The six control modules are only loosely coupled with the respective neighboring control modules and through specific simple coordination influences. In dynamic simulations and on different six-legged robots, this simple control structure has already proven successful: stable and well-coordinated walking behavior emerges from the interactions of the different control modules. The behavior is flexible and can adapt to a wide variety of different disturbances, as is uneven terrain or even the loss of a leg [19].

2.2   ReaCog – Cognitive Expansion for Planning Ahead

Walknet is an example of an embodied control structure that exploits the loop through the world. None the less, already this quite simple system requires some form of coordination of joint movements, in particular for the control of the stance movement. During stance movement at least three legs touch the ground at any given point in time. Through the surface, these three legs form closed kinematic chains and all the involved joints are directly coupled. Each leg consists of three joints which leads to a coupling of at least nine joints and degrees of freedom of the system. As a consequence, the overall system is redundant and there is no analytical optimal solution for the control of joints. Any movement of a joint has to be coordinated with the movements of the other joints. Some of this coordination is mediated by the body itself which means that the effects of a single joint movement are experienced by the other joints and can be compensated locally by those. While such a solution has been used in some of our own simulations to produce forward walking, this approach does not work for negotiating a curve. In this case, central information is required that guides the overall movement of the leg and coordinates the contributions of the different legs. Such a form of coordination requires some form of internal mechanism or model.

Therefore, we introduced a simple neural network structure as a body model which allows to coordinate the actions of the individual joints during the stance movement [22]. Importantly, this form of model only consists of a small number of neurons in a recurrent neural network structure that spans an attractor space which represents the kinematics of the walker. Such a network is exploited in the inverse kinematic problem for the control of the stance movement and coordinates the different joints that are in contact with the ground.

Importantly, such a model is not restricted to the inverse kinematic case (coordinating joint angles for a given movement vector). Instead, this grounded model can be used as a predictive model as well [17]: when joint movements are fed to the model, the encoded attractor dynamics of the model complement this information and predict as a result the movements of the different legs and overall of the body that is in contact to the environment.

Following the idea of internal simulation, the Walknet approach has been extend in order to exploit these predictive capabilities (for details see [18]). While the grounded internal model (at least in parts) appears biologically plausible for an insect and is required for movement tasks, this next step leads away from the insect system. Instead, we are turning towards the question how a simple system that includes functional internal representations can be extended towards a cognitive system. Cognition is here understood as the capability to plan ahead [14]. The predictive capabilities of the internal model allow to exploit this model in different context as it provides outcomes of behavioral possibilities.

In a first step, we extended the overall control selection architecture through an expansion which allows for trial and error of behaviors. As one example, we forced the system into an artificial and unstable posture with both of the hind legs far to the back of their working range. For our robotic case, this posture leads to unstable walking when the robot tries to lift one of its hind legs (in insects this would not be problematic as the tarsi can attach to the ground and provide stability). The system can detect this instability (we again exploit the internal model) and stops its behavior. No present behavior provides a solution for this given situation—the system has run into a deadlock. Now, the cognitive expansion takes over and tries to use a movement even though that particular movement is not activated by the sensed context (e.g. a step forward or backward for a single leg). But, as this trial of a movement might cause the robot to topple over, it is dangerous. Therefore, it is not tried on the real system but instead applied in internal simulation in the internal body model (for details on the approach see [18]).

As a major limitation, the reaCog approach is currently only exploiting existing behaviors, allowing them to be applied out of their current context. Furthermore, no learning takes place once a suitable solution has been found. This is in contrast to how animals and even insects adapt over time and are able to learn to new contexts as well as learning variations of behaviors. In the current system, this becomes apparent as there is only a small set of possible actions and those are predefined. As a consequence, the emerging behaviors—even when including the cognitive expansions—are not that diverse.

3   Related Work

A major goal for future work is the introduction of learning into the walking architecture on two levels. First, on the level of action selection, it should be learnt when a behavior has been applied successfully in a new context after trying it initially in internal simulation. Secondly, new behaviors or variations of behaviors should be memorized as well.

The general—neuroscientific inspired—approach of an internal simulation loop realizes a form of planning ahead that is based on trial-and-error. Different possibilities are evaluated internally before being carried out on the real system. Such a form of trial-and-error directly relates to reinforcement learning: in a learning setting an agent learns through different experiences and, importantly, has a drive to explore the behavioral space and try out new behaviors for a specific context. Furthermore, reinforcement learning provides a framework of incorporating such new experiences.

In the proposed work, the goal is to extend the cognitive architecture towards a system that is capable of learning new behaviors and coming up with suitable variations of existing behaviors. This will be realized as a form of reinforcement learning on different levels of a hierarchical representation. The hierarchical control structure is a neural network which will deviate from the detailed, preprogrammed structure described before. But it will follow the same biological and cognitive considerations and constraints that govern the reaCog architecture.

The related work section will briefly provide pointers to relevant current work. First, with respect to the specific case of hexapod walking in general. Second, it will connect to some recent deep reinforcement learning approaches of hierarchical control.

3.1   Approaches to Hexapod Walking

Locomotion is a rhythmical behavior and in general one can distinguish two types of control approaches. First, there are open-loop control approaches that rely on a centrally generated rhythm which drives the overall behavior. Support for such approaches has been found in particular on fast walking and running in animals and insects. Nice examples are the fast locomotion in cockroaches [8] or running and swimming behavior as found in the salamander [10]. Central to such approaches are oscillating circuits that are not influenced by sensory inputs. Models of central pattern generators are usually realized as open-loop oscillators which only in some cases might then be modulated by sensory signals. Such models have been applied to different robots successfully. These system have as advantages that they can be analyzed in detail, stability of attractor dynamics can be shown and they require only a small number of parameters which allows for learning [9].

Secondly, a large group of approaches puts a focus on the influence of sensory signals on locomotion. For these sensory driven pattern generators, the walking behavior is mainly driven through sensory signals in a closed-loop fashion. Examples for such behavior can be found in all walking (and in particular slow walking) behavior, e.g. in stick insects [20]. One influential example is given by the Walknet as a controller that is mainly driven by sensory signals and as presented in the previous section [21].

3.2 Deep Reinforcement Learning

While many biologically-inspired approaches to locomotion focus on a suitable and simple control structure, they do not deal explicitly with learning or restrict learning to a small, specific parameter space for a particular control architecture.

Learning of locomotion skills without a given control structure poses a different and more complex problem. On the one hand, it requires a form of explorative learning as can be found in Reinforcement Learning. This requires some form of structure for the representation of possible controllers. On the other hand, it has been intractable for a long time as the space of possible control structures is simply too large. Furthermore, training has shown to be difficult as movement control in the end deals with a lot of variance in movements and requires therefore good generalization capabilities. Using neural networks for such control structures is promising, but only over the last years, Deep Reinforcement Learning has become a suitable method that allows to train neural networks as a form of trial-and-error learning [15].

Deep Reinforcement Learning has turned towards the control of continuous control problem during the last years [13]. One example for such a control problem is provided by the control of a four-legged creature inside the OpenAI gym environment [1]. This has been addressed by approaches that divide the overall problem—quite similar to approaches in biology—into different levels of a hierarchy: while the higher level deals with the problem of action selection, the lower level addresses directly the control of joint or muscle movements [4,5]. Heess and colleagues used hierarchical Deep Reinforcement Learning [6] to learn on these two different levels a control structure for the four-legged simple walker. Each of the legs has two degrees of freedom that can be controlled. For both levels, the control problem is framed as a policy. Policies are represented as neural networks which are optimized inside the actor-critique policy framework. Importantly, the higher level and lower level share some of the sensory input and the lower level is conditioned on the higher level policy, meaning the higher level's output is used as another input to the lower level. Furthermore, the lower and higher level were updated at different time intervals. While the detailed control of movements required small temporal update steps, the selection of a context was updated less frequently. These different temporal scales were simply realized through leaving out update steps of the higher levels, a common approach to deal with temporal abstraction in hierarchical DRL [5].

This showed successful on a variety of goal directed navigation tasks when the lower level was initially pre-trained on a basic locomotion task. The control structure showed to be modular. Different movement primitives had been learnt that—in transfer to new tasks—could be applied by the higher level differentially and allowed to quickly learn new tasks.

4   Proposed Architecture Overview

In the following, an extended architecture will be proposed for the control of a six-legged walking robot. The general structure is following the Walknet and reaCog approach: it consists of decentralized leg controllers that are only loosely coordinated through local coordination influences (as described above). The individual controllers will be hierarchical and modular, as can be found already in reaCog. But, while those controllers have been designed and engineered by hand, in the new architecture these controllers will allow for learning and will be realized as neural network controllers that are trained using Reinforcement Learning. This shall allow for trial-and-error learning of new behaviors and will, in a further next step, again exploit the internal simulation loop provided by the cognitive expansion to come up with novel behaviors without actually trying those on the real system. The overall structure of the controller is given in Fig. 2. The next subsections will individually discuss properties of the controller.

4.1   Local Control Modules and Decentralized Organisation

The decentralized organization will follow directly from the structure given in the Walknet approach. Each leg will be controlled by a single controller. This is in contrast to the organizational structure found in [6]. There, the hierarchical control is divided in the vertical direction onto two levels. In our approach, control is in addition distributed along an orthogonal axis: there is not a single, central controller acting on a single level of the hierarchy, instead this control is distributed following the morphological structure into six leg controllers (see Fig. 2).

Inside a single controller, the proposed approach will share the form of representation and hierarchy with the approach presented in [6]: the control problem is described as a policy on two different levels of the hierarchy and these policies will be represented as neural networks. In the case of [6], the different policies have access to different sensory inputs and the lower level is influenced by the higher level through access of the internal state (conditioning the lower level

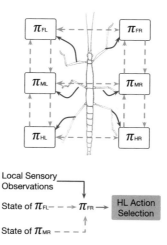

Fig. 2. Shown is the structure of the control architecture: On top, the decentralized structure is visualized. Dashed, green arrows represent the local coordination between the six different high level controllers implementing coordination influences. Sensory inputs are shown as blue, solid arrows. The lower part of the figure shows the inputs/conditioning of a high level policy which takes into account the sensory state and internal states of neighboring policies.

(sub-)policy on the higher level state, Fig. 3). In the proposed architecture, the decentralized structure will be realized and reflected in how the different local policies—for each of the six legs—have access to information of the overall system. On the higher level, a local controller will be, first, conditioned on the sensory inputs from that particular leg. Secondly, following the structure of the coordination influences, it can also be influenced locally, but only from neighboring legs. On the lower level, the controllers are—as given in [6]—dependent on the sensory inputs of that leg as well as on the internal state of the respective higher level controller of that leg. This will lead to a hierarchically organized control structure that is also decentralized.

4.2 Learning of Control

As the policies are represented as neural networks, these allow for learning novel behaviors. Initially, the controller will be pre-trained using the current control approach. Walknet will provide guidance, on the one hand, for the lower level motor controller to shape suitable movement primitives. On the higher level, this will entrain coordinated action selection leading to stable and adaptive locomotion.

Importantly, the learning architecture has been introduced to go beyond the current system. The formulation as a Reinforcement Learning approach allows to further extend the behavioral repertoire through trial-and-error learning as shown in [6]. While a trained architecture will exploit the learned control

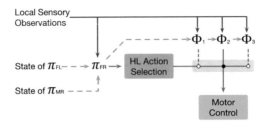

Fig. 3. Hierarchical control for a single leg: On the higher level, shown on the left, action selection is realized as a policy. On the lower level, shown on the right, the different motor primitives are represented as sub-policies that are modulated by the higher level. Arrows represent inputs to the different policies.

structure, this reformulation allows to further explore possible behaviors: on the one hand, through exploration on the higher level which corresponds to sequencing different movements out of context. Or, on the other hand, through exploring variations of motor primitives on the lower level.

Initially, this will be realized directly on a (simulated) robot system which when facing a novel problematic situation can turn towards exploration. As a further step in the future, the internal simulation capabilities of the cognitive expansion shall be exploited for exploration [18]. During exploration of possible behaviors, these do not directly control the robot, but instead drive the internal simulation which provides detailed prediction to the system. Based on these predictions, the system can get towards an informed decision of which behavior to take (for this exploration of possible behaviors see [18]).

4.3   Temporal Abstraction

Different levels in hierarchical Deep Reinforcement Learning are often operating on different timescales. These are realized as different time steps. While lower levels act every iteration, higher levels are often skipping iterations and are only updated on a longer timescale [6]. This is required to ensure behavioral stability and avoid constant switching of behaviors.

In the proposed architecture, this will be handled differently. Going back to the Walknet formulation in [19], there are two key characteristics for the higher level action selection part of the controller. First, there is mutual inhibition between competing layers (shown in Fig. 1). Secondly, there are recurrent connections which stabilize the behavior selection process and prevent constant change of behavior. As an advantage, this allows to update the higher level continuously. When adding more behaviors this could be further stabilized through introducing additional context units as has been done in a recurrent neural network control approach by Tani [16]. In their approach, different temporal scales are not realized as different update patterns. Instead, low pass properties of recurrent connections are exploited and for higher level units longer time-constants are used to stabilize the network. A similar approach could be used to enforce temporal stability on the higher level, which still can be updated each time step.

5   Conclusion

This article sketches an architecture for six-legged locomotion. First, the whole system is setup as a learning system that can incorporate new behaviors from trial-and-error learning. This leads to a reinforcement learning approach in which the controller—going back to the original Walknet approach—are realized as neural networks. Secondly, the structural organization reflects insights from biology. On the one hand, this leads to a hierarchical control approach which—in the formulation of Deep Reinforcement Learning—is realized as different policies (or sub-policies) with overlapping scopes. On the other hand, and as a novel aspect for DRL approaches to locomotion, the overall structure is decentralized and the architecture consists of local control modules that interact only through local coordination influences. The coordination influences are as well realized through appropriate conditioning of the different policies.

Currently, this architecture is setup in a simulated environment. The controlled system has a high number of degrees of freedom: it is a six-legged robot and each leg consists of three rotational joints. The architecture will be initially pre-trained using the current reactive Walknet system to come up with basic walking behavior. Further future work will include porting this to our robot and introducing the cognitive expansion.

Acknowledgments. This research/work was supported by the Cluster of Excellence Cognitive Interaction Technology 'CITEC' (EXC 277) at Bielefeld University, which is funded by the German Research Foundation (DFG).

References

1. Brockman, G., Cheung, V., Pettersson, L., Schneider, J., Schulman, J., Tang, J., Zaremba, W.: Openai gym. arXiv preprint arXiv:1606.01540 (2016)
2. Cruse, H.: A quantitative model of walking incorporating central and peripheral influences. i. the control of the individual leg. Biol. Cybern. **37**, 131–136 (1980)
3. Cully, A., Clune, J., Tarapore, D., Mouret, J.B.: Robots that can adapt like animals. Nature **521**(7553), 503–507 (2015). https://doi.org/10.1038/nature14422
4. Florensa, C., Duan, Y., Abbeel, P.: Stochastic neural networks for hierarchical reinforcement learning. CoRR abs/1704.03012 (2017). http://arxiv.org/abs/1704.03012
5. Frans, K., Ho, J., Chen, X., Abbeel, P., Schulman, J.: Meta learning shared hierarchies. CoRR abs/1710.09767 (2017). http://arxiv.org/abs/1710.09767
6. Heess, N., Wayne, G., Tassa, Y., Lillicrap, T.P., Riedmiller, M.A., Silver, D.: Learning and transfer of modulated locomotor controllers. CoRR abs/1610.05182 (2016). http://arxiv.org/abs/1610.05182
7. Hoinville, T., Schilling, M., Cruse, H.: Control of rhythmic behavior: central and peripheral influences to pattern generation (2015)

8. Holmes, P., Full, R.J., Koditschek, D., Guckenheimer, J.: The dynamics of legged locomotion: models, analyses, and challenges. SIAM Rev. **48**(2), 207–304 (2006)
9. Ijspeert, A.J.: Central pattern generators for locomotion control in animals and robots: a review. Neural Netw. **21**(4), 642–653 (2008)
10. Ijspeert, A.J., Crespi, A., Ryczko, D., Cabelguen, J.M.: From swimming to walking with a salamander robot driven by a spinal cord model. Science **315**(5817), 1416–1420 (2007)
11. Porta, J.M., Celaya, E.: Efficient gait generation using reinforcement learning. In: Proceedings of 4th International Conference on Climbing and Walking Robots (CLAWAR 2001), pp. 411–418 (2001)
12. Kidzinski, L., Mohanty, S.P., Ong, C.F., Huang, Z., Zhou, S., Pechenko, A., Stelmaszczyk, A., Jarosik, P., Pavlov, M., Kolesnikov, S., Plis, S.M., Chen, Z., Zhang, Z., Chen, J., Shi, J., Zheng, Z., Yuan, C., Lin, Z., Michalewski, H., Milos, P., Osinski, B., Melnik, A., Schilling, M., Ritter, H., Carroll, S.F., Hicks, J.L., Levine, S., Salathé, M., Delp, S.L.: Learning to run challenge solutions: Adapting reinforcement learning methods for neuromusculoskeletal environments. CoRR abs/1804.00361 (2018). http://arxiv.org/abs/1804.00361
13. Lillicrap, T.P., Hunt, J.J., Pritzel, A., Heess, N., Erez, T., Tassa, Y., Silver, D., Wierstra, D.: Continuous control with deep reinforcement learning. arXiv preprint arXiv:1509.02971 (2015)
14. McFarland, D., Bösser, T.: Intelligent Behavior in Animals and Robots. MIT Press, Cambridge (1993)
15. Mnih, V., Kavukcuoglu, K., Silver, D., Rusu, A.A., Veness, J., Bellemare, M.G., Graves, A., Riedmiller, M., Fidjeland, A.K., Ostrovski, G.: Human-level control through deep reinforcement learning. Nature **518**(7540), 529–533 (2015)
16. Nishimoto, R., Tani, J.: Development of hierarchical structures for actions and motor imagery: a constructivist view from synthetic neuro-robotics study. Psychol. Res. **73**, 545–558 (2009)
17. Schilling, M., Cruse, H.: What's next: Recruitment of a grounded predictive body model for planning a robot's actions. Front. Psychol. **3**(383) (2012). https://doi.org/10.3389/fpsyg.2012.00383
18. Schilling, M., Cruse, H.: Reacog, a minimal cognitive controller based on recruitment of reactive systems. Front. Neurorobot. **11**, 3 (2017). https://doi.org/10.3389/fnbot.2017.00003
19. Schilling, M., Cruse, H., Arena, P.: Hexapod walking: an expansion to walknet dealing with leg amputations and force oscillations. Biol. Cybern. **96**(3), 323–340 (2007)
20. Schilling, M., Hoinville, T., Schmitz, J., Cruse, H.: Walknet, a bio-inspired controller for hexapod walking. Biol. Cybern. **107**(4), 397–419 (2013)
21. Schilling, M., Paskarbeit, J., Hoinville, T., Hüffmeier, A., Schneider, A., Schmitz, J., Cruse, H.: A hexapod walker using a heterarchical architecture for action selection. Front. Comput. Neurosci. **7**, 126 (2013). https://doi.org/10.3389/fncom.2013.00126
22. Schilling, M., Paskarbeit, J., Schmitz, J., Schneider, A., Cruse, H.: Grounding an internal body model of a hexapod walker - control of curve walking in a biological inspired robot–control of curve walking in a biological inspired robot. In: Proceedings of IEEE/RSJ International Conference on Intelligent Robots and Systems, IROS 2012, pp. 2762–2768 (2012)

An Intrinsically Motivated Robot Explores Non-reward Environments with Output Arbitration

Takuma Seno[1](✉), Masahiko Osawa[1,2], and Michita Imai[1]

[1] Keio University, Kanagawa, Japan
{seno,mosawa,michita}@ailab.ics.keio.ac.jp
[2] Japan Society for Promotion of Science, Tokyo, Japan

Abstract. In real worlds, rewards are easily sparse because the state space is huge. Reinforcement learning agents have to achieve exploration skills to get rewards in such an environment. In that case, curiosity defined as internally generated rewards for state prediction error can encourage agents to explore environments. However, when a robot learns its policy by reinforcement learning, changing outputs of the policy cause jerking because of inertia. Jerking prevents state prediction from convergence, which would make the policy learning unstable. In this paper, we propose Arbitrable Intrinsically Motivated Exploration (AIME), which enables robots to stably learn curiosity-based exploration. AIME uses Accumulator Based Arbitration Model (ABAM) that we previously proposed as an ensemble learning method inspired by prefrontal cortex. ABAM adjusts motor controls to improve stability of reward generation and reinforcement learning. In experiments, we show that a robot can explore a non-reward simulated environment with AIME.

Keywords: Deep reinforcement learning · Robot navigation
Prefrontal cortex

1 Introduction

Deep reinforcement learning (DRL) has been shown to be capable of operating mobile robots directly from raw sensor inputs [1,2]. To compare to heuristic navigation methods, DRL can be more easily applied to various robots and situations with same models. For robotic navigation tasks, achieving exploration ability is essential because the real environment has the huge state space and sparse rewards.

Deep Q-Network (DQN) [3] successfully learns collision avoidance on a mobile robot only from raw depth images and negative rewards for collision between the robot and obstacles [1]. Though depth images are easier to be generalized by deep learning than RGB images, combining DRL and a depth prediction network achieves mobile robot controls from direct RGB inputs [2]. On the other hand, in

© Springer Nature Switzerland AG 2019
A. V. Samsonovich (Ed.): BICA 2018, AISC 848, pp. 283–289, 2019.
https://doi.org/10.1007/978-3-319-99316-4_37

terms of obtaining exploration skill, many curiosity-based exploration methods are proposed in DRL [4,5]. Curiosity is an internally generated reward defined as state prediction errors. The large prediction error means that DRL agents don't know that state and receive large rewards to be encouraged to explore previously unseen environments. Curiosity-based exploration methods generate intrinsic rewards which enables the DRL agent to explore VizDoom [6] and SuperMario Bros without any extrinsic rewards [4].

However, in robotic navigation researches [1,2], these set the explicit rewards on simulation environments which are negative for collision or positive for going farther. Negative rewards for collision would let agents learn how to avoid obstacles rather than how to explore unseen environments. Either positive rewards for farther reach don't evaluate exploration skill. In curiosity-based exploration researches, DRL agents are basically tested in game environments [6,7]. To apply curiosity-based exploration methods based on prediction model to the robotic naviation, we have to consider jerk caused by inertia, which unstably changes the large dimensional inputs such as images. Unstable state transitions prevent the state prediction from convergence, which the agent cannot correctly learn from intrinsic rewards.

In this paper, we propose Arbitrable Intrinsically Motivated Exploration (AIME), an end-to-end method enabling mobile robots to explore environments with DRL and curiosity-based exploration techniques. AIME is enough simple to collaborate any stochastic policy DRL methods and any curiosity-based methods. We previously proposed Accumulator Based Arbitration Model (ABAM) [8], an ensemble method inspired by prefrontal cortex. AIME exploits ABAM to arbitrate DRL's stochastic policy and a constant zero output to reduce jerking. As AIME is an end-to-end training method, this should be easily applied to various robots or situations to compare with heuristic methods. In experiments, we show that AIME enables the robot to explore a non-reward environment on a simulator with Recurrent Stochastic Value Gradients (0) [9], a simple deep reinforcement learning algorithm, and Intrinsic Curiosity Module [4] as an intrinsic reward generator.

2   Accumulator Based Arbitration Model

Accumulator Based Arbitration Model (ABAM) [8] is the ensemble learning method that arbitrates multiple modules dependently on their reliability. Arbitration in ABAM is inspired by prefrontal cortex where the brain chooses the reliable action out of several possible ones. Accumulator neuron found in prefrontal cortex accumulates proofs describing how much each action is reliable. When the proof exceeds the threshold, only the corresponding action is executed.

ABAM regards inputs as stream of data to consider module reliability in time-axis. This arbitration is formulated as follows:

$$A_t = \alpha A_{t-1} + x_t \tag{1}$$

where t is the timestep, A is the accumulated proof, α is a constant factor discounting the previous proof and x is the incoming proof. In reinforcement learning, x corresponds action probabilities calculated by softmax function in discrete action space or probability density function in continuous action space. All action selection modules, which can be heuristic desicion-making modules as well as reinforcement learning units, are hierarchically placed. Each one has its own accumulated proof A and updates it every timestep t by Eq. (1). In choosing an action, ABAM selects the highest module whose proof exceeds the threshold and ignores others.

According to Eq. (1), the module that consistently outputs high probable actions are more selected than the one outputs random actions. Therefore ABAM is able to choose most trained modules when arbitrating multiple reinforcement learning modules under training. In other words, ABAM restricts random outputs which can negativelly affect learning process.

3   Arbitrable Intrinsically Motivated Exploration

This paper proposes Arbitrable Intrinsically Motivated Exploration (AIME) which enables the robot to explore non-reward environments only with curiosity. AIME is enough simple to be combined with any stochastic policies and curiosity-driven exploration techniques.

Though deterministic policy eventually achieves continuous control with stable trajectories, there are some merits to choose stochastic policy despite unstable trajectories leading to jerking. First, stochastic policy permits on-policy exploration [10], which easily collaborates curiosity-based exploration because it is hard to combine noise and intrinsic rewards at the same time. Second, stochastic policy can perform better than deterministic policy in the partially observable environments [11] which we have to consider because the robots basically cannot observe the entire information. To introduce these merits to AIME,

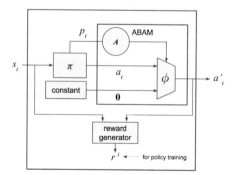

Fig. 1. The architecture of AIME. At time t, the stochastic policy π receives the input s_t to output the action a_t and the action probability p_t. ABAM module updates the accumulated proof A. ϕ decides the final output a_t' by switching a_t and $\mathbf{0}$ dependently on A_t. When $\mathbf{0}$ is selected, a_t' releases motor control to reduce jerking to stabilize reward generator and the policy learning. The policy is trained by intrinsic reward r^i.

ABAM reduces jerking to stabilize the state prediction module and the DRL module.

The architecture of AIME is shown in Fig. 1. In action selection at timestep t, AIME agent receives the state s_t to sample the action a_t and the corresponding probability p_t from a certain stochastic policy π. In order to measure the action reliability, the accumulated proof A_t is calculated in ABAM:

$$A_t = \alpha A_{t-1} + p_t \tag{2}$$

Using the updated A_t, intuitively ABAM arbitrates 2 modules including the policy π learned by deep reinforcement learning and the constant output $\mathbf{0}$:

$$a'_t = \phi(a_t, A_t) = \begin{cases} a_t \ (A_t > z) \\ \mathbf{0} \ (\text{otherwise}) \end{cases} \tag{3}$$

where a'_t is the final output and z is a threshold of ABAM. Here, $\mathbf{0}$ means freeing motor control to reduce jerking.

To update the policy and the state prediction model, mini-batch should use the final output a'_t instead of the policy output a_t because the policy learns from intrinsic rewards r^i calculated by the reward generator from the actual state transitions. In the exploration context, the policy should track the actual trajectories to explore the environment.

4   Experiments

AIME easily collaborates any policy learning method and any internal reward generators as mentioned above. In this paper, we implemented AIME with RSVG(0) [9] as an an actor-critic deep reinforcement leanring method and ICM [4] as intrinsic reward generator with the same network architecture used in [4].

4.1   Environment

We used $Gazebo^1$ simulator to simulate a robot and a physical environment, where we previously tested ABAM in a robotic task [12]. The robot used in experiments was $turtlebot$ controlled through a forward speed in m/s and a rotation speed in rad/s. We implemented a control interface between AIME and $turtlebot$ as a 2-dimensional vector that the first element takes the forward speed in the range of $[-1.0, 1.0]$ and the second element takes the rotation speed in the range of $[-1.0, 1.0]$. Input images are single-channel depth data resized to 42×42. AIME receives 30 images per second and outputs an action before the next input arrives.

---

[1] http://gazebosim.org.

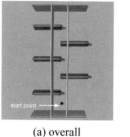

(a) overall   (b) closer look

Fig. 2. The training environment built on *Gazebo* simulator. (a) Overall environment. (b) Closer look to obstacles. Every 500 steps, the agent starts a new episode from the start point.

4.2 Training

The policy is defined as Gaussian distribution, whose standard deviation is fixed through the training. Actor network of RSVG(0) outputs the mean value μ with 2 dimensions corresponding the forward speed and the rotation speed respectively.

We set up an environment of a straight aisle with obstacles as shown Fig. 2. All agents were trained for 2000 episodes each with 500 steps. In every episode, the agents started from the start point described in Fig. 2a. During training, agents received any rewards except intrinsic rewards generated by ICM. The fixed standard deviation of the stochastic policy was set std = 0.1. We used $z = 3.0$ and $\alpha = 0.5$ for ABAM.

To evaluate how much AIME contributes its exploration performance, the agent without AIME (Naive agent) were also trained on this environment as a baseline. Naive agent consists only of RSVG(0) and ICM.

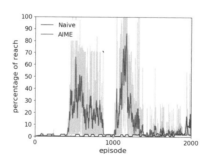

Fig. 3. Moving averages (window size of 10 episodes) of reach of the maze during training on the y-axis and training episodes on the x-axis. Training has been conducted with 2000 episodes each with 500 steps. The mobile robot with AIME reaches farther than that without AIME.

4.3   Evaluation

The result of the experiment is shown in Fig. 3 describing how much the agent covers the maze in exploration during episodes. AIME agent dramatically improves exploration performance, which the agent reaches the opposite side of the wall. The performance is dropped down in the course of training because the field is limited so that the reward is decreased to discourage exploration. Though Naive agent increases performance at the end, AIME agent more consistently explores the wider area.

We don't compare intrinsic rewards during training because the rewards are extremely low when the robot is stuck to the corner or the wall. Therefore Naive agent tends to get low rewards with lack of exploration, which is difficult to evaluate how AIME affects intrinsic reward generation.

Additionally, AIME agent has been observed that it obtains collision avoidance by moving backward at first. This policy is the most simple behavior to maximize intrinsic rewards. After a number of steps, the agent learns how to move forward to see unseen states.

5   Conclusion

We have proposed Arbitrable Intrinsically Motivated Exploration (AIME), which stably enables the mobile robots to explore environments with end-to-end deep reinforcement learning and curiosity-based exploration techniques. AIME exploits Accumulator Based Arbitration Model (ABAM) to restrict unreliable actions to reduce jerking leading unstable reward generation. With simple architecture, AIME can work with any stochastic policy methods and curiosity-based exploration methods.

In the experiment, AIME dramatically improves exploration performance even without any extrinsic rewards. We have shown that simple zero constant outputs can reduce jerking, which stabilizes exploration in robotic navigation tasks.

References

1. Tai, L., Liu, M.: Towards cognitive exploration through deep reinforcement learning for mobile robots. arXiv preprint arXiv:1610.01733 (2016)
2. Xie, L., Wang, S., Markham, A., Trigoni, N.: Towards monocular vision based obstacle avoidance through deep reinforcement learning. arXiv preprint arXiv:1706.09829 (2017)
3. Mnih, V., Kavukcuoglu, K., Silver, D., Rusu, A.A., Veness, J., Bellemare, M.G., Graves, A., et al.: Human-level control through deep reinforcement learning. Nature 518(7540), 529 (2015)
4. Pathak, D., Agrawal, P., Efros, A.A., Darrell, T.: Curiosity-driven exploration by self-supervised prediction. In: International Conference on Machine Learning (ICML), vol. 2017 (2017)

5. Ostrovski, G., Bellemare, M.G., Oord, A., Munos, R.: Count-based exploration with neural density models. arXiv preprint arXiv:1703.01310 (2017)
6. Kempka, M., Wydmuch, M., Runc, G., Toczek, J., Jakowski, W.: Vizdoom: a doom-based AI research platform for visual reinforcement learning. In: 2016 IEEE Conference on Computational Intelligence and Games (CIG), pp. 1–8. IEEE (2016)
7. Bellemare, M.G., Naddaf, Y., Veness, J., Bowling, M.: The arcade learning environment: an evaluation platform for general agents. J. Artif. Intell. Res. **47**, 253–279 (2013)
8. Osawa, M., Ashihara, Y., Seno, T., Imai, M., Kurihara, S.: Accumulator based arbitration model for both supervised and reinforcement learning inspired by prefrontal cortex. In: International Conference on Neural Information Processing, pp. 608–617 (2017)
9. Heess, N., Hunt, J.J., Lillicrap, T.P., Silver, D.: Memory-based control with recurrent neural networks. arXiv preprint arXiv:1512.04455 (2015)
10. Heess, N., Wayne, G., Silver, D., Lillicrap, T., Erez, T., Tassa, Y.: Learning continuous control policies by stochastic value gradients. In: Advances in Neural Information Processing Systems, pp. 2944–2952 (2015)
11. Singh, S.P., Jaakkola, T., Jordan, M.I.: Learning without state-estimation in partially observable Markovian decision processes. In: Machine Learning Proceedings 1994, pp. 284–292 (1994)
12. Ueno, S., Osawa, M., Imai, M., Kato, T., Yamakawa, H.: Reinforcement learning framework for robots in the real world that extends cognitive architecture: prototype simulation environment 'Re:ROS'. Biologically Inspired Cognitive Architectures (BICA) for Young Scientists, vol. 636, pp. 198–206 (2017)

A Strategic Management System Based on Systemic Learning Algorithm

Artem A. Sukhobokov[1,2]($\boxtimes$) [iD], Ruslan Z. Galimov[2] [iD],
and Aleksandr A. Zolotov[2] [iD]

[1] SAP DBS CIS,
Kosmodemyanskaya nab. 52/4, 115054 Moscow, Russian Federation
artem.sukhobokov@yandex.ru
[2] Bauman Moscow State Technical University,
ul. Baumanskaya 2-ya, 5, 105005 Moscow, Russian Federation

Abstract. The paper is devoted to the development of strategic management systems for companies and organizations to solve two problems that are typical for the current level of these systems: insufficient synchronization of strategic management systems with operational management systems and insufficient adaptability of strategic management systems to changes in external environment. To solve the first problem, for each category of objects in the operational management system is proposed to use a separate internal perspective in strategic management system. To solve the second problem, for each internal perspective is proposed to include in the system one additional external perspective reflecting the state of the environment. The problem of strategic management is formulated as the problem of Systemic Learning. To solve it, a combined active-passive algorithm of Systemic Learning is proposed. The presence of the passive component is due a lot of time between the individual decision making (month, quarter), during this time complex calculations can be performed. For a set of scenarios with a relatively high probability, all states and possible transitions between them will be calculated for entire depth of the forecasting period. To ensure high adaptation to changing of external conditions, the developed algorithm has an active component. After each step, information on trends, target values of indicators and probability of scenarios are updated, sets of initiatives may be changed. The architecture of the system's prototype capable of implementing the proposed ideas contains two parallel running On-Premise instances of the strategic management solution as well as additional modules for Systemic Learning algorithm execution and scenario network's maintenance.

Keywords: Corporate Development strategy · Strategic management system
Scenario network · Reinforcement Learning problem
Systemic Learning problem · Active-passive systemic learning algorithm

In economic theory a number of methods of strategic management for companies and organizations have been developed [1]. Among them, one of the most common is the method developed by Robert Kaplan and David Norton - Balanced Scorecard technique [2]. The problems of strategic management systems which await their solution are considered in [3]. Approaches to solving two of them are proposed in this paper.

© Springer Nature Switzerland AG 2019
A. V. Samsonovich (Ed.): BICA 2018, AISC 848, pp. 290–295, 2019.
https://doi.org/10.1007/978-3-319-99316-4_38

The first problem is related to the necessity to improve the synchronization of strategic management system and operational management system. The proposal to emphasize macroeconomic, technological, political/legislative, natural and socio-cultural indicators in the structure of external environment has been already made in [4]. The authors of this paper propose to follow this process to its logical end, and on the analogy of internal perspectives, in which the company's strategy is reflected, to emphasize completely symmetric external perspectives for the environment in which the company operates.

Continuing the analogy further, along with the initiatives, fulfillment of which lead to the change in the values of indicators in internal perspectives, for external perspectives the trends similar to initiatives must be formally described, and their realization will change the values of indicators in external perspectives. The combined structure of both external and internal perspectives in universal case can include all the main categories of matters represented in Corporate Development strategy: materials, services, finances, personnel, assets, business-processes, strategy. The full definition of each perspective will include dozens of indicators and should be determined for each company/organization individually. For the companies operating in specific industries, some of the mentioned perspectives can be not used.

As one more problem, which nowadays doesn't have the satisfactory solution for Strategic planning systems, in [3] it is indicated that these systems are static in nature. The strategy is developed for some future period of time, but when the external conditions change, the strategy does not adapt, it is necessary to do everything anew. This insufficiency can be overcome through scenario planning. The task of adaptive strategy's change due to changes in the values of external and internal indicators during the planning period can be considered as Reinforcement Learning problem.

To solve Reinforcement Learning problem a scenario network is needed, which starts form several different variants of current period completion and stretches for the entire duration of strategy construction. To simplify will consider time base as discrete, so transitions form different scenarios are possible only after the completion of the next time period (year, half-year, quarter). Each transition must be characterized by a transition probability, the value of which must be estimated. Then it is possible to consider several variants of transitions from recurrent completed period to the next one, and this, in fact, will represent the next level of scenarios detailing.

Each scenario network's node is characterized by a set of values of indicators included in external and internal perspectives. To exclude from the consideration transitions with very low probabilities, will use for forecasting a corridor of scenarios, having a fixed width (a number of considered scenarios), and development scenarios beyond the corridor borders won't be considered. Figure 1 shows an example of corridor. As a time period the year is selected. There are three basic development scenarios. Cumulative forecasting duration is 10 years. The width of the used corridor of scenarios is equal to 9. A total number of considered periods forming a corridor is 82.

After the completion of each annual period, different transitions with different probabilities are possible. Real scenario networks can have not such a regular character as the network in Fig. 1, there will not necessarily be three possible transitions after each annual period. There can be more, and can be only two. But such situation doesn't change the essence of scenario network.

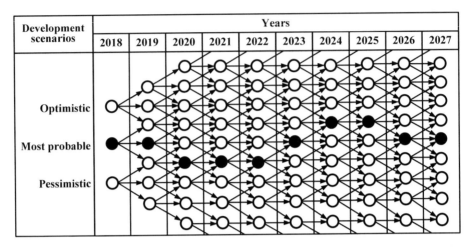

Reference symbols:

○ - **Annual forecast period**
● - **Annual forecast period included in selected chain of periods**
⟶ - **Transition with a certain probability from one annual period to another**

Fig. 1. An example of scenario network of a chain of interconnected annual periods

If an enterprise (organization) operates in several markets (in several regions) with different conditions, forecasts for each annual period should be made separately for each market (region). This will require the construction and use of several scenario networks.

Without the loss of generality, we can consider that there is no dependence of transition on previous history of annual periods. It allows us to consider the transition process as Markov process, and a probability of each particular chain of annual periods can be calculated as a product of initial probability of the scenario and all the probabilities of transitions between separate annual periods included in the chain. The example of such an interconnected chain of annual periods presented in Fig. 1 in the form of sequence of periods marked in black color.

In a formal form this problem can be formulated as follows:

- There is a limited sequence of T time periods T_i indexed with $i \in 1,I$.
- At each time period T_i the company and its environment can be described by many states S_{ij}, $j \in 1,J_i$, $i \in 1,I$.
- The state of the company is completely determined by a set of values of external indicators $Iext_k$, $k \in 1,K$ and a set of values of internal indicators $Iint_m$, $m \in 1,M$.
- Indicators $Iext_k$, $k \in 1,K$ are among different external $Pext_u$ perspectives, $u \in 1,U$, and indicators $Iint_m$, $m \in 1,M$ are among different internal $Pint_u$ perspectives, $u \in 1,U$, the number of which coincides.
- As the company strategy is determined in I time periods, the state of the company S_{ij} is characterized by $Iint_{im}$, $i \in 1,I$, $m \in 1,M$ and $Iext_{ik}$, $i \in 1,I$, $k \in 1,K$ values of indicators.

- The transition of the company from the state S_{ij} to the state $S_{i+1,n}$ is determined by the current external trends and internal initiatives. For each S_{ij} only a limited set of possible states is possible in the next time period $S_{i+1,n} \in S_{i+1,j}$, $n \in 1, N_{i+1,j}$, $j \in 1$, J_{i+1}, $N_{i+1,j} \leq J_{i+1}$.
- The transition between the states S_{ij} and $S_{i+1,n}$ is realized by fulfillment of a set of initiatives $E_{i,j,i+1,n}$ and characterized by probability $p_{i,j,i+1,n}$.
- During the operating period of the company, only one of the possible transitions from state S_{ij} into state $S_{i+1,n}$ is realized.
- A set of values of internal indicators $Iint_{im}$, $i \in 1,I$, $m \in 1,M$ contains indicator V_i, $i \in 1,I$, which determines utility of the achieved state of the company in step i.
- Cumulative utility of the company strategy during the whole sequence of time periods T is determined by the function $V = F(V_1, V_2, \dots V_I)$.
- During the operating period of the company at each time interval T_i, must be observed restrictions R_{si}, $s \in 1,S$, $i \in 1,I$:

$$R_{si}(Iint, \ Iext) \leq L_{si}$$

where S – total number of restrictions, L_{si} - limit value for restriction s in interval T_i, and as parameters can be used any indicators $Iint_{hm}$ and $Iext_{hk}$, $m \in 1,M$, $k \in 1,K$, $h \in 1,i$.
- The aim of the solving of the problem is to find the sequence of states S_{ij} for all $i \in 1,I$, for which $V \rightarrow max$ and observed all restrictions R_{si}, $s \in 1,S$, $i \in 1,I$.

According to the formulation of the problem, it belongs to the class of Systemic Learning problems described in [5]. Typical features of such problems:

- The process of interaction of the control algorithm with external environment is a sequence of actions, the result of each gives additional information for decision making to the control algorithm.
- External environment with which the control algorithm interacts is a system with its own borders and a set (hierarchy) of internal subsystems connected with each other.
- The information about the system with which the control algorithm interacts is represented as a set of perspectives reflecting different aspects of its functioning.

According to [5], the Systemic Learning combine the Multiperspective Learning and Whole-system Learning methods, which together provide the view from individual perspectives and from the point of view of the entire system as a whole.

In [6], a classification of algorithms for solving Reinforcement Learning problem is presented, according to which passive and active algorithms for solving the problem of determining the best sequence of actions are distinguished. Passive algorithms, the most known example of which is R. Bellman's method of dynamic programming, calculate the results of all possible variants of the development in advance and then in the control process use this data to make decisions in order to obtain the best result at the end. The active algorithms, the most known examples of which are Q-Learning and SARSA, learn during the control process, analyzing the results of each step. By analogy, it is possible to suppose the existence of active and passive algorithms for solving the problems of Systemic Learning.

To solve the formulated problem of strategic management systems, we are proposing to use a combined active-passive algorithm. Such choice is due to the following:

- It is very difficult to calculate in advance detailed scenario network for passive algorithm, there can be lots of variants, in addition during network transition from period to period, some probabilities and parameter values can be defined more exactly deviating from the initial variant.
- Strategic management has its own characteristics. Between individual steps of algorithm during transition to the next scenario network node we have much time (quarter, half-year, year), at this period of time complex calculations can be performed.
- Economic or social cost of each made decision is very high. So it is impractical to use active algorithms which can make wrong choice during the learning period.

Within the proposed combined active-passive algorithm on the analogy with passive methods for a set of scenarios that have a relatively high probability and included in the corridor, all states and possible transitions between them will be calculated for the entire depth of the forecasting period. According to these calculations a set of initiatives ensuring the transition to the next state will be selected. To ensure high adaptation to changing of external conditions, the developed algorithm in addition to passive component should have an active one. The adaptation lies in the fact that after calculations and choice of the action plan for the immediate period, its realization starts, and in parallel with it the group of analysts starts collecting and preparing data for the next calculation: information on trends, target values of indicators and updated probabilities of scenarios. At the same time, existing sets are being adjusted and new sets of initiatives are proposed. etc.

For entire realization of the proposed approach in contemporary conditions, it is necessary to develop a cloud application which contains all the described functionality. However full development till the approach proposed in this paper won't be tested by companies in practice can appear excessively risky due to high costs. To realize the proposed approach most easily, two parallel running instances of traditional common On-Premise solution for strategic management can be used. For example, such as SAP Strategy Management [7]. One copy will be used to work with internal perspectives of strategy, due initiatives and indicators, and the second copy will be used for external perspectives, with initiatives being used for external trends representation. It is necessary only to develop the scenario network management module, Systemic Learning algorithm implementing module, and attendant interfaces. The proposed architecture of the solution is shown in Fig. 2.

Modules and applications in Fig. 2 which need to be developed are shown as rectangular blocks. Existing modules and applications which need to configure are shown as ovals. In blocks bounded by a broken line there are selected the elements each of which must have the same data connections as the block at whole. These blocks are formed not to obstruct the diagram with lots of displayed connections.

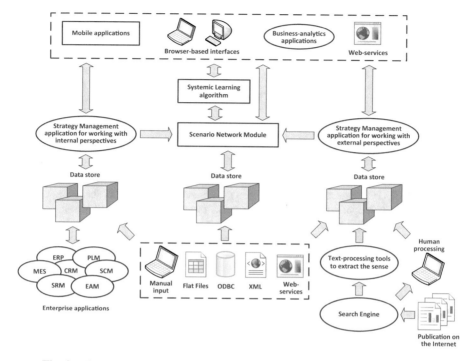

Fig. 2. The architecture of solution for quick realization of the proposed approach

References

1. David, F.R., David, F.R.: Strategic Management: Concepts and Cases. Pearson Education Limited, Harlow (2015)
2. Kaplan, R.S., Norton, D.P.: The Strategy-Focused Organization: How Balanced Scorecard Companies Thrive in the New Business Environment. Harvard Business School Press, Boston (2005)
3. Barrows, E., Neely, A.: Managing Performance in Turbulent Times. Wiley, Hoboken (2012)
4. Hungenberg, H.: Strategisches Management in Unternehmen: Ziele - Prozesse - Verfahren. Dr. Th. Gabler Verlag, Wiesbaden (2000)
5. Kulkarni, P.: Reinforcement and Systemic Machine Learning for Decision Making. Wiley, Hoboken (2012)
6. Russell, S.J., Norvig, P.: Artificial Intelligence: A Modern Approach, 3rd edn. Prentice Hall Inc., A Pearson Education Company, Upper Saddle River (2010)
7. SAP Strategy Management. https://www.sap.com/products/strategy-management.product-capabilities.html. Accessed 02 May 2018

Genetically Evolved Extreme Learning Machine for Letter Recognition Dataset

Tomasz Szandała$^{(\boxtimes)}$

Wroclaw University of Science and Technology, Wroclaw, Poland
Tomasz.Szandala@pwr.edu.pl

Abstract. It is well known that the performance of learning feed forward neural networks is in general far slower than required and it has been a major bottleneck in their applications. Two key obstacles the slow gradient-based learning algorithms which are extensively used to train neural networks. Combining slow training process with even slower evolutional methods appears to be incomprehensible but here comes the Extreme Learning Machine. ELM has randomly chosen hidden nodes and analytically determined only the output weights of network. In theory, this algorithm tends to provide good generalization performance at extremely fast learning speed. Experiment in this paper shows that ELM's classification efficiency can be noticeably improved if its training is combined with Genetic Algorithm.

Keywords: Extreme Learning Machine · Genetic algorithm
Pattern recognition · Evolutional algorithms

1 Introduction

I have first encountered Extreme Learning Machines when my daily PC, enriched with powerful Titanium GPU was out of service. I was left only with my notebook, that surely was not enough to work on employing genetic algorithm to enhance neural networks learning. The paper survey has lead me to Prof. Huang Guangbin's work [1] about much simpler neural networks, called Extreme Learning Machine. Since they are being trained without complex back-propagation – only use linear matrix computing, they fit perfectly the idea of evolving neural networks.

2 Extreme Learning Machine

It is said that Extreme Learning Machines (ELM) fill the gap between Frank Rosenblatt's Dream and John von Neumann's Puzzle [2]. It is simple architecture artificial neural network consisting of, typically, three layers:

- input layer, that collect input data,
- hidden layer – usually one, but there are no real obstacles to insert more,
- output layer, that collects signals from inner layer and gives network's final verdict.

© Springer Nature Switzerland AG 2019
A. V. Samsonovich (Ed.): BICA 2018, AISC 848, pp. 296–300, 2019.
https://doi.org/10.1007/978-3-319-99316-4_39

The only difference between ELM and classical network is one-time assigning weights to input connections in hidden layer. All weights between input and hidden layer and all weights between layers in hidden neurons are randomly assigned when network is created and then become frozen. Only weights in connections between hidden and output layer are influenced by training process (Fig. 1).

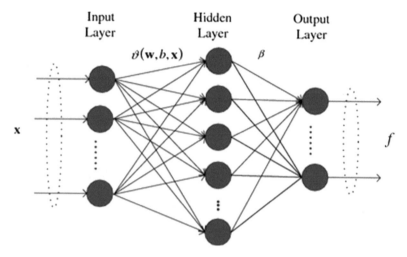

Fig. 1. Conceptual diagram of architecture of Extreme Learning Machine with one hidden neural layer

2.1   ELM Training Algorithm

A simplest ELM training algorithm, chosen for this paper, learns a model of the form (for single hidden layer sigmoid neural networks) [3]:

$$Y = W_2 f_h(W_1 X) \tag{1}$$

Where (Eq. 1) W_1 is the matrix of weights between input and hidden layer. f_h is an activation function of hidden layer, and W_2 is the matrix of hidden-to-output-layer weights.

The training algorithm proceeds as follows:

1. Fill W_1 with random values, preferably from range $[-1, 1]$.
2. Estimate W_2 by least-squares fit to a matrix of response variables Y, computed using the pseudoinverse $\cdot+$, given a design matrix X (Eq. 2):

$$W_2 = f_h(W_1 X) \overset{+Y}{} \tag{2}$$

Pseudoinverse has already been proven superior for some artificial neural networks training, e.g. Hopfield neural network for pattern recognition. Computation of the pseudoinverse is much simpler, therefore less computing hungry, than whole back-propagation algorithm. Sample benchmark on MNIST 10 dataset shows over 70 times shorter time of training between Deep Belief Neural network (taught by back-propagation) and ELM network.

3   Genetic Algorithm

Genetic Algorithms were developed to mimic the processes observed in natural evolution. Many people, biologists included, are astonished that life at the level of complexity that we observe could have evolved in the relatively short time suggested by the fossil record. The idea with GA is to use this power of evolution to solve optimization problems [5]. The father of the original Genetic Algorithm was John Holland who invented it in the early 1970's. Each implementation of genetic algorithm consists few steps:

1. Creation of base population.
2. Creation of the next generations in the processes called crossover and mutation.
3. Selection operation based on computing fitness function value.
4. Termination moment - after enough amount of epochs or when the value of fitness is enough.

While the creation and termination stages are called once, the two other are used many times during each epoch. What is more interesting: both steps can be executed separately and parallel for some individuals of population.

4   The Experiment

For the purpose of experiment I have chosen Letter Recognition Data Set from UCI machine learning repository. The objective of this dataset is to identify each of a large number of black-and-white rectangular pixel displays as one of the 26 capital letters in the English alphabet. The character images were based on 20 different fonts and each letter within these 20 fonts was randomly distorted to produce a file of 20,000 unique stimuli. Each stimulus was converted into 16 primitive numerical attributes (statistical moments and edge counts) which were then scaled to fit into a range of <0, 1>. The recommended training set were the first 16000 items and then the resulting model was used to predict the letter category for the remaining 4000 instances.

   Base population consisted of 40 networks, each had 16 input neurons, 80 neurons in hidden layer and 26 output neurons. Each neuron in hidden layer was utilizing sigmoidal activation function. Mutation probability factor has been set to 2%, GA max epochs number has been set to 30.

   Each epoch consisted of 3 steps:

1. Ordering of population by specimen's accuracy value. Fit function was considered as ratio of value of correctly classified letters from validation set divided by number

of all samples invalidation set. There more correctly classified samples, the better result.

2. Taking 40 the best fitted specimens for cross-over process. Cross-over method was defined as taking 2 randomly chosen specimens and conditionally swapping their respective neurons from hidden layer, with all their weights. The most important difference between standard GA was no discarding of parent networks. I believed that there is a chance that even in first epoch I could obtain the perfect network.

3. Next occurs the mutation, which was the process of generating whole new neuron, with all its weights. this step could only affect neurons of descended networks.

4. Last step of epoch was the training of newly obtained specimens.

5 The Results

See Table 1.

Table 1. The experiment has been performed 3 times and this table shows important results of each attempt along with final, average value

Result	Attempt 1	Attempt 2	Attempt 3	Average
Time of whole process [min]	12:36	11:58	11:54	12:09
Accuracy of the best specimen from initial population	0.69	0.65	0.70	0.68
Accuracy of the best specimen post-GA	0.73	0.70	0.72	0.72

6 Conclusions and Future Work

Experiment shows that Extreme Learning Machine network is indeed really fast learning artificial neural network. The average time of initialization and experiment itself took around 12 min on not-top mobile CPU. Furthermore ELM appears to be susceptible for tuning with Genetic Algorithm which slightly improves its classification performance. This removes biggest fear of ELM, that is has unchangeable weights values in hidden layer, so it may never fully adapt. Last conclusion from the results table is the noticeable small improvement from GA. At the beginning of research I was expecting improvement close to 10%. While 4% is still better, it proves only how good can be raw ELM network, without any enhancements.

For the future there are left experiments with more advanced activation functions for ELM. There is no obstacle to use different from each other function in the same hidden layer. Regarding evolutional training my biggest remark is that I should parallelize training of new specimens in each epoch and I should work a bit more on tuning GA parameters.

To sum up Extreme Learning Machine appears to be worthy competitor to standard Deep Learning. While it is really fast, it still keeps high correctness in results.

Utilization of Genetic Algorithm improves its efficiency for, at least for now, small degree and maybe it can be improved even further.

References

1. Huang, G.-B., Zhu, Q.-Y., Siew, C.-K.: Extreme learning machine: theory and applications. Neurocomputing **70**, 489–501 (2006)
2. Huang, G.-B.: What are extreme learning machines? Filling the gap between Frank Rosenblatt's dream and John von Neumann's puzzle. Cogn. Comput. **7**, 263–278 (2015)
3. Huang, G.-B., Zhou, H., Ding, X., Zhang, R.: Extreme learning machine for regression and multiclass classification. IEEE Trans. Syst. Man Cybern. **42**, 513–529 (2011)
4. Szandała, T.: Comparison of different learning algorithms for pattern recognition with Hopfield's neural network. Proc. Comput. Sci. **71**, 68–75 (2015)
5. Thomas, B.: Evolutionary Algorithms in Theory and Practice. Oxford University Press, New York (1996)

Toward Human-Like Sentence Interpretation—a Syntactic Parser Implemented as a Restricted Quasi Bayesian Network—

Naoto Takahashi[(✉)] and Yuuji Ichisugi

Artificial Intelligence Research Center, National Institute of Advanced Industrial Science and Technology (AIST), Tsukuba, Japan
naoto.takahashi@aist.go.jp

Abstract. Most sentences expressed in a natural language is ambiguous. However, human beings effortlessly understand the intended message of the sentence even when a computer program finds out countless possible interpretations. If we want to create a computer program that understands a natural language in the same way as human beings do, a promising way would be implementing a human-like mechanism of sentence processing instead of implementing a "list exhaustively then select" method. By the way, it is highly likely that human's language ability is realized mostly by the cerebral cortex, and recent neuroscientific studies hypothesize that the cerebral cortex works as a Bayesian network. Then it should be possible to reproduce human's language ability using a Bayesian network. Based on this idea, we implemented a syntactic parser using a restricted quasi Bayesian network, which is a prototyping tool for creating models of cerebral cortical areas. The parser analyzes a sequence of syntactic categories based on a subset of combinatory categorial grammar. We confirmed that the parser correctly parsed grammatical sequences and rejected ungrammatical sequences.

Keywords: Syntactic analysis · Bayesian networks
Combinatory categorial grammar · Language area

1 Introduction

Most sentences expressed in a natural language are lexically and syntactically ambiguous. This fact is easily demonstrated by parsing natural languages with computer programs; you will be surprised by seeing countless possible interpretations that you have never imagined. Therefore it is necessary to assess each interpretation based on a certain criterion to select the most probable one.

Nevertheless, human beings effortlessly understand the intended message of the sentence without being troubled by possible but unintended interpretations.

If we want to create a computer program that understands a natural language in the same way as human beings do, one of the most promising way

© Springer Nature Switzerland AG 2019
A. V. Samsonovich (Ed.): BICA 2018, AISC 848, pp. 301–309, 2019.
https://doi.org/10.1007/978-3-319-99316-4_40

would be implementing a human-like mechanism of sentence processing instead of implementing a "list exhaustively then select" method.

By the way, medical case studies of aphasia and observations of brain activities measured by recent technologies strongly suggest that specific areas of cerebral cortex, so called language areas, play crucial roles in language processing [7]. At the same time, neuroscientific studies hypothesize that the cerebral cortex works as a Bayesian network [10] or a kind of probabilistic graphical model [1–6,8,9,11–14]. If human's language ability is realized by the cerebral cortex, and if the cerebral cortex works as a Bayesian network, then it should be possible to reproduce human's language ability using a Bayesian network.

Based on this idea, the current authors formerly implemented a syntactic parser for a context free grammar using a restricted quasi Bayesian network [16]. We present another syntactic parser for a different type of grammar, i.e. combinatory categorial grammar, in this article.

2   Restricted Quasi Bayesian Networks

One way to study the mechanism of information processing in the brain is to create computational models of the targeted function. However, creating realistic models using machine learning techniques, e.g. Bayesian networks, forces the designers to resolve inessential problems, like tuning hyper parameters. Moreover, it is often difficult to trace the real cause of unsatisfying results when the created model does not behave as expected.

It is often helpful to create prototypes before creating a realistic model. By creating prototypes, we can estimate the hopefulness of the fundamental design of the realistic model that we are going to create.

Restricted quasi Bayesian network [16] is a prototyping tool for creating models of cerebral cortical areas. It is a simplified Bayesian network that only distinguishes probability value zero from other values. Its conditional probability tables are restricted to fulfill certain mathematical conditions to avoid combinatorial explosion.

Restricted quasi Bayesian network provides *gates*, which control the flow of information. Thus the designer can design generative models in a similar way as designing logical circuits.

Since restricted quasi Bayesian network does not have learning ability, conditional probability tables must be prepared by the designer. Because of its limited capabilities, restricted quasi Bayesian network may not be applied to practical, real-world problems. However, it releases the designers from inessential problems and allows them to concentrate on the essential part of model design. As a result of agile prototyping activities, designers would find potential problems in the model, which can be extremely difficult to find in a complicated, realistic model.

3   Combinatory Categorial Grammar

Combinatory categorial grammar [15] generalizes classical categorial grammar by introducing functional composition. It is suitable to describe syntactic rules of natural languages because its weak generative capacity locates between context-free grammar and context-sensitive grammar.

In traditional phrase structure grammar, each syntactic category is represented as a unique non-terminal symbol. For example, sentence, noun phrase and verb are often represented as S, NP and V, respectively.

In combinatory categorial grammar, on the other hand, syntactic categories are represented by *ground categories* and *operators*. When X and Y are syntactic categories, $X \backslash Y$ represents a syntactic category that constitutes X when preceded by Y. Likewise, X/Y represents a syntactic category that constitutes X when followed by Y.

In English, for example, a verb phrase is composed with a preceding noun phrase (NP) to constitute a sentence (S). Thus the category for verb phrase is represented as $S \backslash NP$, assuming that S and NP are ground categories. Furthermore, a transitive verb is composed with a succeeding noun phrase (NP) to constitute a verb phrase $(S \backslash NP)$. Thus the category for transitive verbs is represented as $(S \backslash NP)/NP$.

4   Implementation

In this section, we explain a restricted quasi Bayesian network that implements the forward/backward functional application rules of combinatory categorial grammar.

4.1   Representation of Syntactic Categories in a Bayesian Network

Theoretically, the length of a syntactic category in combinatory categorial grammar is unlimited. However, we suppose it is limited because of the information processing ability of human. In this article, we represent a syntactic category by five nodes. Each node takes a ground category or an operator as its value.

We adopt prefix notation to eliminate parentheses. In a sequence of five nodes, first comes the operator, then the category that was originally at the right side, and finally the category that was at the left side. Five nodes are used in the flush-left mode; unused nodes take a special value to explicitly indicate its inactivity (Fig. 1).

4.2   Forward/Backward Functional Application Rules

Figure 2 shows a restricted quasi Bayesian network that implements the forward functional application rule. The value combinations that appear with a probability greater than zero are listed in Table 1.

Fig. 1. The ordinary notation of syntactic categories in combinatory categorial grammar (left) and their prefix notation in this article (right). A full stop (.) represents an unused node

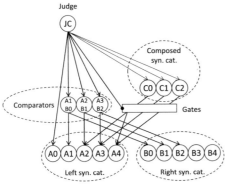
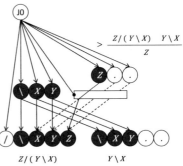

Fig. 2. A restricted quasi Bayesian network that implements the forward functional application rule. Each node in a syntactic category takes a ground category or an operator as its value. Each comparator node has two children. If both children have the same value, the parent comparator takes that value. Otherwise it takes a special value that represents "unmatched". The judge node controls the gates between the left syntactic category and the composed syntactic category based on the values of the comparators. Only the first three nodes of the composed syntactic category are depicted since the length of the syntactic category is always shorten when a function application rule is applied

Fig. 3. An example of the forward functional application rule in which the first three nodes of the right syntactic category match a part of the left syntactic category. The gates operate so that the final node of the left syntactic category and the first node of the composed syntactic category have the same value

When the first node, namely the topmost operator, of the left syntactic category has the value slash (/) and the values of the following three nodes match the values of the first three nodes of the right syntactic category, the gate that

Table 1. The value combinations that have a probability greater than zero in Fig. 2. X is an arbitrary ground category and can be different from column to column. A full stop (.) means "unused" or "unmatched". An underscore (_) is an arbitrary value. The rightmost four columns indicate connection/disconnection between nodes controlled by the gates. The value of the judge node (JC) may be anything as long as each row can be distinguished

JC	A0	C0	C1	C2	A1B0	A2B1	A3B2	C0A2	C0A4	C1A3	C2A4
J0	/	X	.	.	$/,\backslash$	X	X	Off	On	Off	Off
J1	/	$/,\backslash,X$	$X,.$	$X,.$	X	.	_	On	Off	On	On

connects the final node of the left syntactic category and the first node of the composed syntactic category opens to make their values equal (Fig. 3). At the same time, all the other nodes in the composed syntactic category are marked as "unused".

When the first node of the left syntactic category has the value slash (/) and only one succeeding node matches the leftmost part of the right syntactic category, the gates operate so that the remaining three nodes of the left syntactic category and the nodes of the composed syntactic category have the same values (Fig. 4).

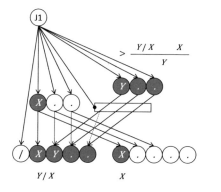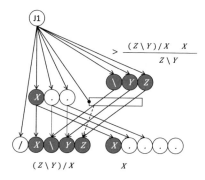

Fig. 4. Examples of the forward functional application rule in which only one node of the left syntactic category matches the leftmost part of the right syntactic category. The gates operate so that the last three nodes of the left syntactic category and the nodes of the composed syntactic category have the same values

Table 2. The value combinations that have a probability greater than zero in Fig. 5

JC	B0	C0	C1	C2	A0B1	A1B2	A2B3	C0B2	C0B4	C1B3	C2B4
J2	$\backslash$	X	.	.	$/,\backslash$	X	X	Off	On	Off	Off
J3	$\backslash$	$/,\backslash,X$	$X,.$	$X,.$	X	.	_	On	Off	On	On

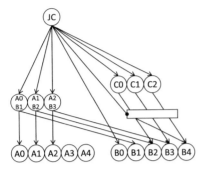

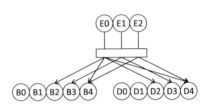

Fig. 5. A restricted quasi Bayesian network that implements the backward functional application rule

Fig. 6. The nodes for the third syntactic category (D0 to D4), the composed category (E0 to E2) and their connections for applying the functional application rules. The judge node and the comparator nodes are omitted

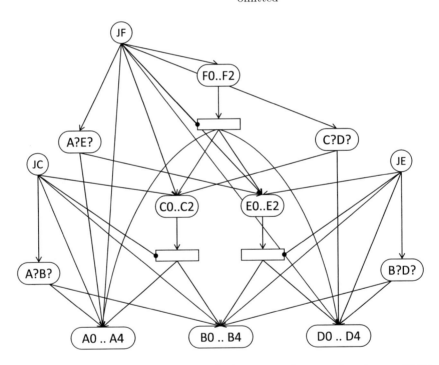

Fig. 7. A restricted quasi Bayesian network that parses three-word sentences. Multiple nodes that belong to the same group are depicted as a single oval. A0..A4, C0..C2, etc. are nodes to represent syntactic categories. A?B?, C?D?, etc. are comparators between two syntactic categories. JC, JE and JF are judge nodes. Rectangles are gates

The backward functional application rule is implemented similarly as the forward functional application rule (Fig. 5 and Table 2).

By combining the networks described in Figs. 2 and 5, we obtain a restricted quasi Bayesian network that performs the forward/backward function application rules between two syntactic categories.

4.3   A Bayesian Network for Three Syntactic Categories

Now we explain how to construct a restricted quasi Bayesian network that applies the forward/backward functional application rules to three syntactic categories.

First, we introduce the third syntactic category for input, which is represented by the nodes D0 to D4. Then we connect the B nodes and the D nodes in the same way as we did for the A nodes and the B nodes. The composed syntactic category is represented by the nodes E0 to E2 (Fig. 6).

Then we connect the A nodes (the first input category) and the E nodes (the composition of the second and the third input categories), as well as the C nodes (the composition of the first and the second input categories) and the D nodes (the third input category). Both compositions are represented by the nodes F0 to F2 (Fig. 7).

We have implemented the above-mentioned network for three syntactic categories as a restricted quasi Bayesian network. We also confirmed that all the input combinations that are applicable to the functional application rules were correctly parsed, and inapplicable combinations were rejected, using up to three different ground categories.

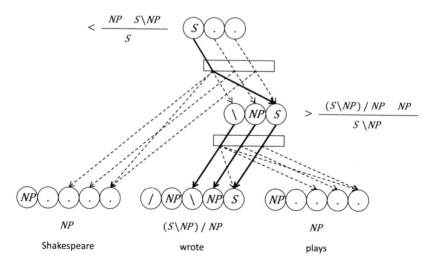

Fig. 8. A three-word sentence parsed by the restricted quasi Bayesian network. First, the transitive verb (*wrote*) and the object noun phrase (*plays*) compose a verb phrase ($S\backslash NP$). Next, the subject noun phrase (*Shakespeare*) and the composed verb phrase make a sentence (S). The judge nodes and the comparator nodes are not illustrated

Figure 8 is an example of parsing three syntactic categories. It shows how the created network parses a three-word sentence that consists of a subject noun phrase, a transitive verb and an object noun phrase. First, the transitive verb (*wrote*) and the object noun phrase (*plays*) compose a verb phrase ($S\backslash NP$). Next, the subject noun phrase (*Shakespeare*) and the composed verb phrase make a sentence (S).

5   Conclusion

We examined the feasibility of creating a syntactic parser for combinatory categorial grammar as a restricted quasi Bayesian network. So far we have only implemented the forward/backward functional application rules. The number of acceptable syntactic categories is also limited to a small number. To extend the current parser to a practical level, it is necessary to confirm that the presented design strategy does not cause combinatorial explosion with sufficient number of syntactic categories for input.

One of the advantages in using combinatory categorial grammar is that syntactical derivation and semantic composition can be associated elegantly; this property should be utilized in a practical parser.

Restricted quasi Bayesian networks perform exhaustive search to find all the combinations that have a probability greater than zero, but ordinary Bayesian networks can calculate the most probable combination quickly with approximation. It is possible that humans also use some kind of approximation to realize a real-time sentence interpretation because humans do not perform, at least consciously, exhaustive search, and they fail to interpret some types of grammatical sentences, e.g. deep centre-embedded sentences.

Our final goal is to reproduce human's language ability using an *ordinary* Bayesian network. For this purpose, we designed a prototype using a restricted quasi Bayesian network to see the feasibility of such networks. We plan to examine other implementations for comparison.

Acknowledgement. This paper is based on results obtained from a project commissioned by the New Energy and Industrial Technology Development Organization (NEDO).

References

1. Chikkerur, S., Serre, T., Tan, C., Poggio, T.: What and where: a Bayesian inference theory of attention. Vis. Res. **50**(22), 2233–2247 (2010)
2. Dura-Bernal, S., Wennekers, T., Denham, S.L.: Top-down feedback in an HMAX-like cortical model of object perception based on hierarchical Bayesian networks and belief propagation. PLoS One **7**(11), e48216 (2012)
3. George, D., Hawkins, J.: A hierarchical Bayesian model of invariant pattern recognition in the visual cortex. In: 2005 International Joint Conference on Neural Networks (IJCNN) (2005)

4. Hosoya, H.: Multinominal Bayesian learning for modeling classical and nonclassical receptive field properties. Neural Comput. **24**(8), 2119–2150 (2012)
5. Ichisugi, Y.: The cerebral cortex model that self-organizes conditional probability tables and executes belief propagation. In: 2007 International Joint Conference on Neural Networks (IJCNN) (2007)
6. Ichisugi, Y.: Recognition model of cerebral cortex based on approximate belief revision algorithm. In: 2011 International Joint Conference on Neural Networks (IJCNN) (2011)
7. Kemmerer, D.: Cognitive Neuroscience of Language. Psychology Press, Abingdon (2015)
8. Lee, T.S., Mumford, D.: Hierarchical Bayesian inference in the visual cortex. J. Opt. Soc. Am. A **20**(7), 1434–1448 (2003)
9. Litvak, S., Ullman, S.: Cortical circuitry implementing graphical models. Neural Comput. **21**(11), 3010–3056 (2009)
10. Pearl, J.: Probabilistic Reasoning in Intelligent Systems: Networks of Plausible Inference. Morgan Kaufmann Publishers Inc., San Francisco (1988)
11. Pitkow, X., Angelaki, D.E.: Inference in the brain: statistics flowing in redundant population codes. Neuron **94**(5), 943–953 (2017)
12. Raju, R.V., Pitkow, X.: Inference by reparameterization in neural population codes. In: Advances in Neural Information Processing Systems 29: Annual Conference on Neural Information Processing Systems 2016, 5–10 December 2016, Barcelona, Spain, pp. 2029–2037 (2016)
13. Rao, R.P.: Bayesian inference and attention modulation in the visual cortex. Neuroreport **16**(16), 1843–1848 (2005)
14. Röhrbein, F., Eggert, J., Körner, E.: Bayesian columnar networks for grounded cognitive system. In: Proceedings of the 30th Annual Conference of the Cognitive Science Society, pp. 1423–1428 (2008)
15. Steedman, M.: The Syntactic Process. MIT Press, Cambridge (2000)
16. Takahashi, N., Ichisugi, Y.: Restricted quasi Bayesian networks as a prototyping tool for computational models of individual cortical areas. In: Proceedings of Machine Learning Research, PMLR, vol. 73 (2017)

Research of Neurocognitive Mechanisms of Revealing of the Information Concealing by the Person

Vadim L. Ushakov[1,2(✉)], Denis G. Malakhov[1],
Vyacheslav A. Orlov[1], Sergey I. Kartashov[1], and Yuri I. Kholodny[1,3]

[1] National Research Center "Kurchatov Institute", Moscow, Russia
ushakov_vl@nrcki.ru
[2] National Research Nuclear University "MEPhI", Moscow, Russia
[3] Bauman Moscow State Technical University, Moscow, Russia

Abstract. This work is related to the creation of an MR-compatible polygraph and the development of methods for detecting the hierarchy of neural networks of the cognitive organization of hidden memory markers. At the moment, there is practically no scientific work on lie detection, in which both the classical polygraph and the MRI scanner were simultaneously used. Combining these methods will help increase the probability of recognizing the facts of hiding important information and carry out an objective assessment of the truthfulness of the reported information. This method can also be used to detect the level of resistance of operators for emotional stress, assess the perception of emotional stimuli by subjects in neurocognitive tasks.

Keywords: MR compatible polygraph · fMRI · Hierarchy of neural networks
Hidden memory markers

1 Introduction

The method of revealing a person's concealed information by monitoring the dynamics of his individual physiological functions was created to help with the disclosure of crimes more than two thousand years ago. Originating in the countries of the Ancient East [1] and surviving centuries, the method was used for the same purposes in medieval Europe [2] and in its primitive form survived to the twentieth century: ethnographers observed its application in the 1940s in the tribes of Equatorial Africa [3]. At the end of the 19th century, various laboratory instruments began to be used to implement this method, on the basis of which a polygraph was created in the USA in the 1930s, which was actively used in the post-war years and incorrectly called the "lie detector". In the 21st century, studies using a polygraph to identify a person's hidden information (hereinafter - polygraph studies, PS) are used in the course of the investigation of crimes and checking the staff personnel (to prevent offences) in more than 80 countries.

At the same time, despite the popularity of PSs and their centuries-old history, the question of the processes that provide the possibility of revealing hidden information

© Springer Nature Switzerland AG 2019
A. V. Samsonovich (Ed.): BICA 2018, AISC 848, pp. 310–315, 2019.
https://doi.org/10.1007/978-3-319-99316-4_41

that occur in the neural system and the human body remains open. In an attempt to substantiate the technology of PS, by the end of the twentieth century, researchers proposed more than a dozen "polygraph theories" [4], but none of them allowed an exhaustive explanation of the phenomena observed in practice and in laboratory conditions. Work in this direction continued, and by the efforts of researchers from different countries (including Russia), the number of "polygraph theories" by the end of 2010 increased to sixteen [5], and by the present moment - reached two dozen [6].

At the end of the twentieth century, the "polygraph theories" were conditionally divided "into two main classes: (a) theories, based on motivational and emotional factors, as the most important determinants of psychophysiological differentiation… and (b) theories based on cognitive factors." [7]. Subsequent years showed that researchers were more inclined to develop and study the "theories" of the first class, and in many respects it was determined by the established traditions: in the USA the results of the conducted PSs are still considered in the framework of the paradigm "DI/NDI" ("DI" - deception indicated); "NDI" - no deception indicated).

An ancient method of revealing a person's concealed information by monitoring his physiological reactions is the oldest forensic method that appeared long before the birth of this applied legal science. Modern criminalistics proceeds from the premise that any criminal act inevitably leaves traces of itself - both of a material nature, as well as images of the event, imprinted in the memory of a person caught in the act of crime or in some way involved in this act. Therefore, the very field of application of this method (investigation and prevention of crime) leads to the idea that PS technology and "polygraph theory" should take into account the special role of memory and be based on the study of cognitive factors.

The use of functional magnetic resonance imaging (fMRI) in the interests of the so-called "lie detection" - it was believed that this method would eliminate the shortcomings of the PS and prove to be more reliable - began at the turn of the 20th and 21st centuries mainly along the traditional path within the framework of the "DI/NDI" paradigm, however this approach soon revealed its limitations and caused significant arguments by scientists, lawyers and experts in the field of PS [8, 9].

In recent years, some researchers who use fMRI to study the neurophysiological mechanisms of implementation of false reports have begun using the paradigm of revealing "concealed information" (hereinafter referred to as the "CI" paradigm). This approach proved promising, received some experimental confirmation [10], and led leading American experts in the field of PS to raise the question of the expediency of moving from the paradigm of "lie detection" ("DI/NDI") of a person to the paradigm of searching in his memory for traces of concealed events of the past [11–14].

As mentioned above, a fairly convincing "polygraph theory", i.e. theory of the revealing of a person's traces of the events of the past that he hides, has not been created to the present day. Nevertheless, thanks to empirical searches, the technology of execution of PS has been created for a long time ago and is successfully applied in practice and is constantly being improved. In recent years, the technology of fMRI studies in order to reveal traces of human memory and Biologically Inspired Cognitive Architectures has made a significant step forward [15]. In this regard, in our opinion, it seems appropriate to make more extensive use of the opportunities and accumulated experience of conducting PSs in the interests of further development of research in the field of the use of fMRI for revealing human memory traces. The combination of PS and fMRI technologies will allow for more dynamic monitoring of neurocognitive mechanisms in the course of a purposeful study of human memory, to detect the level of emotional stress resistance of operators, assess the perception of emotional stimuli by subjects in neurocognitive tasks.

2 Materials and Methods

In addition to fMRI, a methodology was developed for monitoring fast processes that reflect physiological changes in the condition of subjects under cognitive tasks. The basis of the technique was registration of parameters of skin conductance response, photoplethysmogram, upper and lower respiration of humans by MP compatible polygraphic sensors.

A program for recording and analyzing data of an MP-compatible polygraph was developed. Figure 1 shows a part of the data record during the experiment. The upper two rows - TTL markers of each volume scan with fMRI, and a synchronous audio recording of the questions and answers of the subject. There are classical signs of vegetative reactions to the relevant stimulus (in this paper, the concealed information about the name of the subject): (1) change in frequency and amplitude of respiration, (2) a sharp decrease in skin resistance (skin conductance response), the appearance of double peaks in the graph of skin conductance response, (3) change of vascular condition, which is reflected in the amplitude of the photoplethysmogram signal and on the change in the heart rate.

MRI data were obtained from 10 healthy subjects, mean age 24 (range from 20 to 35 years). Permission to undertake this experiment has been granted by the Ethics Committee of the NRC "Kurchatov Institute".

As the target stimulus, the person's personal name was chosen as the most powerful emotional stimulus. For each participant in the experiment, a questionnaire was created consisting of 6 names. Among them was the name of the subject, the control name for checking the reaction, and 4 other names that were specially selected so as not to cause strong associations for the participant. In total there were 5 series of each experiment, in each of which the first name was excluded (during the statistical processing of data was not analyzed), the remaining 5 names were presented in random order. In the statistical analysis, a change in the BOLD signal was analyzed in the event-related paradigm with duration of 15 s at the moment of responding to the real name (target stimulus) in relation to other names (non-target stimulus).

The MRI data was acquired using a 3 Tesla SIEMENS Magnetom Verio MR scanner. The T1-weighted sagittal three-dimensional magnetization-prepared rapid gradient-echo sequence was acquired with the following imaging parameters: 176 slices, TR = 1900 ms, TE = 2.19 ms, slice thickness = 1 mm, flip angle = 9°, inversion time = 900 ms, FOV = 250 mm × 218 mm². fMRI data was acquired with the following parameters: 30 slices, TR = 2000 ms, TE = 25 ms, slice thickness = 3 mm, flip angle = 90°, FOV = 192 × 192 mm². The fMRI and structural MR data were preprocessed using SPM8 (available free at http://www.fil.ion.ucl.ac.uk/spm/software/spm8/). After converting Siemens DICOM files into SPM NIFTI format all images

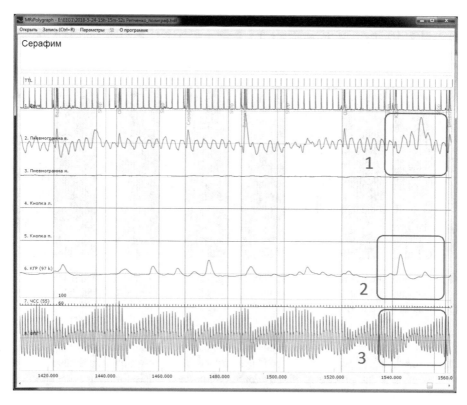

Fig. 1. An example of recording of vegetative responses to a relevant stimulus in a prototype of an MP compatible polygraph.

were manually centered at the anterior commissure. EPI images were corrected for magnetic field inhomogeneity using FieldMap toolbox for SPM8. Next, slice-timing correction for fMRI data was performed. Both anatomical and functional data was normalized into the ICBM stereotactic reference frame. T1 images were segmented into 3 tissue maps (gray/white matter and CSF). Functional data was smoothed using Gaussian filter with a kernel of 6 mm FWHM. Statistical analysis was performed using Student's T-statistics ($p < 0.05$, with correction for multiple comparisons (FWE)).

3 Results

Figure 2 presents examples of the results of visualization of active neural networks of two subjects while performing the task of hiding information about their names. The analysis was carried out on the basis of statistical processing of fMRI data recorded synchronously with polygraphic data. As a contrast, 4 blocks of responses to the name of the subject were compared to 20 control blocks. There are changes in the neural network activity in the zones SupraMarginal_R, Frontal_Sup_2_LR, Frontal_Sup_-Medial_L,   Heschl_L,   Precentral_LR,   Postcentral_LR,   Cerebelum_6_R,

Precuneus_LR, Frontal_Mid_2_LR, Cingulate_Mid_R, Cingulate_Ant_L, Occipital_Sup_R, Fusiform_R, Frontal_Inf_Tri_R, Parietal_Inf_L, Parietal_Sup_L, Insula_R, Angular_R, Frontal_Med_Orb_L (L-left, R-right).

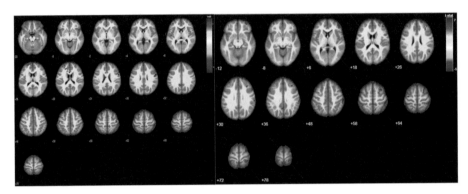

Fig. 2. Localization of neural network activity of the brain in two subjects when performing the task of concealing information about one's name (p < 0.05, FWE) by fMRI data. Warm colors - significant increase in BOLD-signal, cold colors - its decrease.

Identified zones take part in the cognitive processes of construction of the answer "lie"-"truth", in the work of "mirror" systems of displaying behavioral responses of other people, in the processes of self-identification and decision-making, perception of auditory stimuli, in systems evaluation of reward and punishment, in sensorimotor perception and motor functions, in the processes of episodic memory and self-awareness, in the processing of language information, in the processes of construction and processing of emotions, multisensory information, extraction traces of memory, object categorization.

4   Conclusions

For the first time in Russia an attempt was made to combine registration of fast physiological processes with fMRI data. Preliminary results show the prospect of continuing the experimental work on creating on the basis of MP compatible polygraph an objective specialized method for detecting traces of hidden memory. In the future, based on this approach, it is planned to create methods for an objective assessment of the stress-resistance of operators and predicting their behavior in an extreme situation.

Acknowledgements. This study was partially supported by the Russian Science Foundation, Grant 18-11-00336 (data preprocessing algorithms) and by the Russian Foundation for Basic Research grant ofi-m 17-29-02518 (study of thinking levels). The authors are grateful to the MEPhI Academic Excellence Project for providing computing resources and facilities to perform experimental data processing.

References

1. Simonov, P.V.: Higher Nervous Activity of a Person (Motivational and Emotional Aspects). Nauka, Moscow (1975)
2. Schneikert, G.: The Mystery of the Criminal and the Way to Its Disclosure. Right and Science, Moscow (1925)
3. Wright, G.V.: Witness of Witchcraft. Young Guard, Moscow (1971)
4. Kholodny, Yu.I., Saveliev, Yu.I.: The problem of using polygraph tests: an invitation to a discussion. Psychol. J. **17**(3), 53–69 (1996)
5. Kholodny, Yu.I.: Theoretical concepts of criminal research with the use of polygraph: an invitation to the discussion. Bull. Crim. **1**(41), 89–98 (2012)
6. Krapohl, D., Show, P.: Fundamentals of Polygraph Practice. Academy Press, San Diego (2015)
7. Ben-Shakhar, G., Furedy, J.: Theories and Applications in the Detection of Deception. A Psychophysiological and International Perspective. Springer, New York (1990)
8. Bizzi, E., Hyman, S., Raichle, M., Kanwisher, N., Phelps, E., et al.: Using Imaging to Identify Deceit: Scientific and Ethical Questions. American Academy of Arts & Sciences, Cambridge (2009)
9. Matte, J.: fMRI Lie detection validity and admissibility as evidence in court and applicability of the court's ruling to polygraph testing. Eur. Polygr. **7**(4), 191–198 (2013)
10. Verschuere, B., Ben-Shakhar, G., Meijer, E.H. (press eds.): Memory Detection. Theory and Application of the Concealed Information Test. Cambridge University Press, Cambridge (2011)
11. Rosenfeld, P.: Detecting Concealed Information and Deception. Recent Developments. Academy Press, Elsevier Inc., Amsterdam (2018)
12. Krapolh, D.: Paradigm shift: searching for trace evidence in human memory. Police Chief **2**, 52–55 (2016)
13. Palmatier, J., Rovner, L.: Credibility assessment: preliminary process theory, the polygraph process, and construct validity. Int. J. Psychophysiol. **95**(1), 3–13 (2015)
14. Palmatier, J.: Why it's not lie detection but instead memory detection: what modern science says about credibility assessment. Report American Polygraph Association, 52nd annual conference (seminar and workshop) (2017)
15. Ushakov, V.L., Samsonovich, A.V.: Toward a BICA-model-based study of cognition using brain imaging techniques. Procedia Comput. Sci. **71**, 254–264 (2015)

Encrypted Data Transmission
Over an Open Channel

Aleksander B. Vavrenyuk$^{(\boxtimes)}$, Viktor V. Makarov,
Viktor A. Shurygin, and Dmitrii V. Tcibisov

National Research Nuclear University MEPhI, Moscow, Russia
abvavrenyuk@mephi.ru, makarov45fel6@yandex.ru,
vic-54@mail.ru, blacodim@gmail.com

Abstract. This paper discusses methods for transmitting cipher messages over an open communication channel using an additional channel for transmitting a secret key.

Set of the possible one-time keys are contained in the table on the sending and receiving side. The key to encrypt and decrypt a message is selected based on a signal that can be transmitted over an additional channel at a specified time or a timestamp in the cipher message itself.

The proposed method also allows the use of a truly random number generator to obtain secret keys, which provides an absolutely strong cipher. Two variants of forming and using tables with secret keys are proposed. Estimations of the required time for encryption and decryption of data, as well as the necessary amount of memory for storing tables are given.

Keywords: Cipher message · Secret key · Encryption · Decryption
One-time key · True random number generator

1 Introduction

Gilbert Sandford Vernam was a telegraph operator, an AT&T Bell Labs employee who invented in 1917 and in 1919 patented the system of automatic encryption of telegraph messages: a polyalphabetic stream cipher for which the perfect cryptographic secrecy was later proved. The popularity of stream ciphers was brought by the work of Claude Shannon, published in 1949, in which Shannon proved the perfect. Vernam did not use the term "XOR" in the patent, but he implemented that operation in relay logic [1]. Each character in the encrypted message was generated using the key (Fig. 1). In this case, the key must have three critically important properties:

(1) be truly random,
(2) coincide in size with the given plaintext,
(3) be used only once.

The key was used only once and destroyed immediately after use by both the sender and receiver. Vernam ciphers with single-use keys on every page were widely used in the 20th century by scouts, military and diplomatic staff around the world [2].

© Springer Nature Switzerland AG 2019
A. V. Samsonovich (Ed.): BICA 2018, AISC 848, pp. 316–325, 2019.
https://doi.org/10.1007/978-3-319-99316-4_42

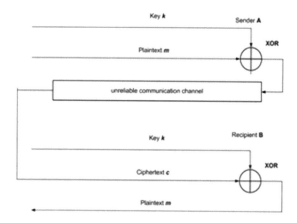

Fig. 1. The Vernam stream cipher scheme

2   Perfect Cipher

A one-time key k in the Vernam cipher is randomly generated (Fig. 1). The length of the key k is equal to the length of the plaintext ($|k| = |m|$). To form each character in the cipher message, Vernam used the Boolean *XOR* function. The arguments of the function were the character's code of the source text and the key (a random number). The same key and the same function were used to decrypt the received cipher text. Why is Vernam's cipher rarely used in practice? It is generally accepted that the reasons are as follows.

(1)   Pseudo-random number generators (PRNGs) give the sequence of characters with a certain period, and hence a cryptanalyst can find a pattern in the sequence and hack the code [3–5]. Basically, it is possible to use true random number generators (TRNGs) based on radiation, atomic clocks, quartz generators, etc., with subsequent conversion of signals into digital form. However, such generators are complex, unreliable and expensive.

(2)   The hand ciphers that were used by spies and diplomats in the first half of the 20th century suggest using a relatively small number of keys in a pad. Therefore, even dice could be used as a RNG.

(3)   There is a problem of reliable transfer of the session key from the source of the message to the receiver. As a rule, during the transfer of a secret key there have been failures in the past.

(4)   In our time, as computer technology, mobile telephony, as well as cryptography develop, with the abandonment of hand ciphers, there remains the problem of transferring the key to the receiver. Any special channel created for this is a potential threat to the cipher's secrecy. And if such a channel is absolutely reliable, then why to encrypt the message? It is possible to use this channel to send an unencrypted text.

However, is it possible to refuse the transfer of a one-time key?

3 Using Tables for Transmission of a One-Time Key

3.1 Using an Additional Telephone Channel

It is proposed to form the tables for correspondence of one-time keys and time of phone calls. Each key generated by the PRNG is assigned a time interval for the phone call (for example, 2 min per connection). Note that it is the connection, but not the conversation. You can talk as much as you want. In this case, no restrictions on the duration of the conversation is not imposed. Only the fact of a call from "A" (the sending party) at a certain time to "B" (the receiving party) is important.

If for some reasons "B" cannot answer the call, the standard information remains in the phone's memory, namely the phone number of the caller and the time of the call.

Such tables can be considered as some analogue of Vernam ciphers. They should be prepared in advance and, one copy of the table should be available for "A", and the second, exactly the same, for "B". The method of transferring such a table is not considered here. Obviously, such a transfer must be carried out over an absolutely reliable channel. Also, "A" and "B" should have exactly the same PRNGs.

The table should contain all possible one-time keys for a predetermined period of time (for example, 1 year). There are various options for implementing PRNG [6–8]. It is important only that the PRNG period ensures the uniqueness of each key in the table.

The structure of the table for the sender "A" has the following form (Table 1).

Table 1. Keys for "A" (variant 1)

No. of a week	Key	Call time
01	1 Key	00-00
01	2 Key	00-02
01	3 Key	00-04
…	…	…
52	36722 Key	00-02
52	37440 Key	23-58

The structure of the table for "B" has the following form (Table 2).

Table 2. Keys for "B" (variant 1)

No. of a week	Call time	Key
01	00-00	1 Key
01	00-02	2 Key
52	00-00	36721 Key
…	…	…
52	00-00	36721 Key
52	23-58	37440 Key

It can be certainly assumed that "B" knows the number of the current week when the call is received. The number of rows in Table 1 is the same as in Table 2, namely 37 440.

Plan of actions for "A" (variant 1) is the following.

(1) The PRNG outputs a key with the length m.
(2) The key is compared with the keys stored in Table 2 and, if there is no such key, then it was used earlier and "cleared"; we return to step 2 and get another key.
(3) We take the call time for the received key from Table 1.
(4) If this time has already been missed, then the message is transmitted the next day. If you need to transfer today, another key is generated.
(5) At the appropriate time, the call to "B" is made.
(6) If the call was successful, an encrypted sequence is formed (Fig. 1).
(7) The transmission is being carried out.
(8) The row of a session key in Table 1 is cleared.

Plan of actions for "B" (variant 1) is the following.

(1) "B" receives the encrypted message.
(2) "B" records the call time (checks the main and backup phones).
(3) Knowing the current week number, "B" retrieves the key from Table 2, and its row in Table 2 is cleared.
(4) In accordance with Fig. 1, the message is deciphered.

Plan of actions for "A" (variant 2) is the following.

(1) A message of the length m is generated. The length of the message must be no more than the length of the keystream in the table.
(2) The call time is selected.
(3) A corresponding key is selected from the table.
(4) Starting with the 1st bit, a part of the keystream of the length m is chosen.
(5) The message is encrypted with the received key.
(6) At the appropriate time, the call to "B" is made.
(7) If the call was successful, the transmission is carried out (Fig. 1).
(8) The row of a session key in Table 3 is cleared.

Table 3. Keys for "A" (variant 2)

Date (day.month)	Key	Call time
01.01	1 Key	00-00
01.01	2 Key	00-02
01.01	3 Key	00-04
...	...	...
31.12	241081 Key	00-00
31.12	241082 Key	00-02
31.12	241083 Key	23-58

Plan of actions for "B" (variant 2) is the following.

(1) The encrypted message is received.
(2) The call time is recorded (the main and backup phones are checked).
(3) The key corresponding to the call time is retrieved from Table 4.
(4) The row used in Table 4 is cleared.
(5) In accordance with Fig. 1, the message is deciphered.

Table 4. Keys for "B" (variant 2)

Call time	Date (day.month)	Key
00-00	01.01	1 Key
00-02	01.01	2 Key
00-04	01.01	3 Key
...	...	...
00-00	31.12	241081 Key
00-02	31.12	241082 Key
23-58	31.12	241083 Key

With this method, the key cannot be reused. If you want to send a message that exceeds the length of the key in the table, the message can be divided into 2 parts, and for each part a certain key and a certain call time from the table should be used according to the presented algorithm. Then each such part should be considered as an independent message. Reusing the key will also be excluded. In this way, long messages can be divided into several parts. A necessary condition for a rigorous proof of the secrecy of this cipher is the equality of the length (in bits) of the message and the secret key. But at the same time the message becomes "narrow" due to possible limitations in the size of the key. If the true random number generator for the initial formation of the table is used, you can make the key as long as possible and at the same time get rid of the periodicity peculiar to PRNG. A possible limitation can only be the amount of memory for the tables. However, with the development of a modern element base, in particular flash technologies, such limitations do not seem to be critical.

3.2 Transmission of a Cipher Message Without an Additional Telephone Channel

The described approach has one significant drawback. It is necessary to use an additional communication channel (telephone), which is not reliable, besides there is no guarantee that the call will take place exactly at the time that the transmitting party is counting on. The following technology seems to be more rational.

If the encrypted message is sent by E-mail, then you can use the service information inserted into each transmitted message by the mail program. At the beginning of each letter, the client's mail program generates a header indicating the local time for sending the message with split-second accuracy. The sender can select a future time interval for sending the message according to the table described earlier, after which the PRNG can be turned on with the starting point corresponding to this time interval and the key-stream of the required length (equal to the length of the message) can be received; then the sender can encrypt the message using the received key, then send the message prepared in this way at the designated time interval (either manually by pressing the "SEND" button, or instruct the program to send at the specified time.)

It does not matter what time interval will be chosen to generate the key and to send the message. It is important that the time indicated in the header corresponds to the pre-selected interval.

The receiving party, when receiving the message, should choose in the same table as for the sender the interval corresponding to the time specified in the header, start the PRNG and receive a key of the length corresponding to the length of the received message. After that, the message is deciphered, and the communication session ends here. Deleting the used key is also not necessary for the previously mentioned reasons.

4 Experimental Testing of Cipher Message Transmission Using Tables

For the practical use of the proposed technology, it is necessary to:

- select the value of the time interval when forming a table, which, in turn, will be determined by the possible spread of real-time values when sending a message,
- determine the amount of memory for storing the table,
- estimate the search time in the table of a row containing the required time interval,
- select the PRNG with the maximum possible period,
- estimate the running time of the PRNG to generate a key (based on the maximum allowed message length).

In order to test the proposed technology, a software solution was developed that includes the simplest mail client (hereinafter referred to as the program), which provides the following options:

- connection and work with the mail server;
- encryption of messages before sending;
- decryption of messages before reading;
- rewriting the used keys in case they are used in the encryption or decryption process.

During the experiment, a situation was considered where everything could be in the hands of the Enemy, except for the original key table used for encryption To match the receiving time/sending time, which was used to select the right key for encryption and decryption, in the experiment the time was brought to the conformity with GMT (Greenwich Mean Time).

When generating, each key is checked for uniqueness within the generated table. The time interval for generating the keys is 2 min. To ensure the low algorithmic complexity of $O(1)$ and, therefore, the minimum search time, the keys are represented in a hash table format. Searching for one key takes ~ 2 ms. Two types of key tables were used in the experiment (Tables 5 and 6).

Table 5. Characteristics of the Fist Type Key Table

Estimated characteristics	Value
The number of keys per day – per week – per 52 weeks	720-720-37440
Required memory for a complete table of 52 weeks (byte)	248 351 738
Average generation time for a complete table (seconds)	3.1
Key search time for encryption/decryption (milliseconds)	2

Table 6. Characteristics of the Second Type Key Table

Estimated characteristics	Value
The number of keys per day – per week – per 52 weeks	720-5040-262080
Required memory for a complete table of 52 weeks (byte)	52 087 201
Average generation time for a table (s)	7.3
Key search time for encryption/decryption (ms)	2

The key table of the first type (Option 1) is a hash table (key-value), where the key is a row in the form: "Day of the week - hours - minutes", and the value is a pair of a 32-byte key pair and a state vector of the PRNG used for generation of the required length key (with a maximum limit within the experiment up to 256 bytes).

The key table of the second type (Option 2) is a hash table (key-value), where the key is a row in the form: "Month - day of the month - hours - minutes", and the value is a 128-byte key.

The amount of memory required for the first type tables significantly exceeds the required amount of memory for the second type ones. This is due to the necessity in the first case to receive and store additionally in the table a state vector of the PRNG to realize the possibility of the generation of a key sequence of the required length. In this case, the total and weekly number of keys is greater for the second type table.

Two methods of selecting a key for a given time interval are considered in the experiment, namely:

(1) the user is requested to manually input the time parameters (GMT) for encrypting (when sending a message) and decrypting (when receiving a message), this method is used when encrypting and decrypting the message on the key corresponding to the time of the phone call (or other method involving manual input time parameters);

(2) determination of the encryption time according to the current time (brought to the conformity with GMT), and the decryption time according to the time contained in the header of the mail message.

To generate pseudo-random numbers, the Mersenne twister (MT) PRNG was used, namely MT19937, which is based on the properties of Mersenne prime numbers and ensures fast generation of high-quality pseudo-random numbers by randomness. The period of the selected PRNG is 219 937; it is possible to reproduce the already generated sequence (in the presence of a state vector representing 312 64-bit integers).

The main characteristics of the program: size is 120 Kb, amount of RAM, is 1 GB,) hard disk capacity is 1 GB, stable connection to the Internet with the mail server is needed.

The process of working with the program in the experiment consists of 3 main stages:

(1) setting the initial parameters of the program (selection of the key table type and the method of selecting the key);
(2) launching the program that downloads a table from an already generated file or generates a table of the type selected at the first stage, saving it to a file suitable for storage and transmission to the receiver;
(3) reading (including the selection of the message, its decryption, taking into account the selected method at the first stage and displaying the result to the user) and sending messages (including the formation of each message, its encryption, taking into account the selected method at the first stage and sending the encrypted message to the receiver).

Plan of actions from the sender's side during the experiment is as follows:

(1) filling the configuration file in on the sender's side (indicating the type of the key table, and the method for selecting the key);
(2) launching the program with the input of authorization data from the sender's mailbox;
(3) if at the last stage, generation occurred, the transmission of the file containing the key table to the receiver.
(4) the transition to sending a message;
(5) input of the receiver, a message subject (optional) and a test message of the required length;
(6) performing actions according to the selected method of selecting the key;
(7) waiting for the message to finish transmitting, automatic erasing of the used key (in RAM and in the file from which the table was downloaded), receiving a request for pressing the Enter key and exit from the mail client.

Plan of actions from the receiver's side during the experiment is as follows:

(1) obtaining information about the configuration of the program and the key table from the sender;
(2) saving the table in the corresponding program directory;
(3) filling the configuration file in on the receiver's side with the parameters received from the sender;
(4) launching the program with the input of authorization data from the receiver's mailbox;
(5) automatic downloading of the table from the file received from the receiver into RAM;
(6) transition to reading the message;
(7) performing actions according to the selected method of selecting the key;

(8) successful reading of the information contained in the message, automatic deleting of the used key (in RAM and in the file from which the table was downloaded) and receiving a request for pressing the Enter key, exit from the mail client.

After the key is deleted, it becomes impossible to read the message again, either by the sender or by the receiver.

5 Conclusion

The presented methods for transmitting cipher messages using tables and an additional communication channel have a number of advantages and disadvantages of stream ciphers. The undoubted advantage of the proposed method is the independence of the next cipher message decryption from the failure during the transmission of the previous cipher message (the loss of one or more characters due to insufficient reliability of the channel).

One of the main drawbacks is the need for the initial transmission of the table from A to B. Another disadvantage is not obvious. The disadvantage is that with any of the ways of forming and using tables considered in the work, a considerable amount of them is occupied by unused data. The amount of this information depends on the statistical characteristics of the stream of transmitted cipher messages. However, if the characteristics of transmitted messages, such as the maximum length, frequency of transmissions, etc. are known in advance, it is possible to generate corresponding tables for practical use.

The authors did not set the objective in this paper to give a comprehensive assessment of the cryptographic secrecy of the cipher, the authenticity and integrity of the message and other parameters, but were limited to the fact that they showed the possibility of using a technology similar to Vernam ciphers in modern conditions. The method is useful in cases where there is no need to frequently update the table when sending relatively short messages.

Acknowledgment. The authors acknowledge the support from the MEPhI Academic Excellence Project (Contract No. 02.a03.21.0005).

References

1. Shannon, C.E.: Communication theory of secrecy systems. Bell Syst. Tech. J. **28**, 656–715 (1949)
2. Venona, L.L.: The Most Secret Operation of the US Intelligence Agencies. Olma-Press, Moscow (2003)
3. Lano J.: Cryptanalysis and design of synchronous stream ciphers. www.esat.kuleuven.be/cosic/publications/thesis-124.pdf. Accessed 25 Dec 2016
4. Wu, H.: Cryptanalysis and design of stream ciphers. www.pdfs.semanticscholar.org/7659/3b22460d3c070ed1c900f75771c9165cccb2.pdf. Accessed 14 Oct 2016
5. Bertoni, G., Daemen, J., Peeters, M., Van Assche, G.: Sponge-based pseudo-random number generators. www.keccak.team/files/SpongePRNG.pdf. Accessed 25 Dec 2017

6. Chugunkov, I.V., Ivanov, M.A., Rovnyagin, M.M., et al.: Three-dimensional data stochastic transformation algorithms for hybrid supercomputer implementation. In: 17th IEEE Mediterranean Electrotechnical Conference, Beirut, Lebanon, 13–16 April 2014, pp. 451–457 (2014)
7. Ivanov, M., Starikovskiy, A., Rosliy, E., et al.: Stochastic method of data transformation RDOZEN+. J. Theor. Appl. Inf. Technol. **93**(2), 332–337 (2016)
8. Ivanov, M., Starikovskiy, A., Rosly, N.: Sequential and parallel composition of round transformations for construction of an iterative algorithm for stochastic data processing. J. Theor. Appl. Inf. Technol. **94**(1), 224–229 (2016)

Analysis of Gaze Behaviors in Virtual Environments for Cooperative Pattern Modeling

Norifumi Watanabe[1](✉) and Kota Itoda[2]

[1] Advanced Institute of Industrial Technology, Shinagawa-ku, Tokyo 1400011, Japan
watanabe@aiit.ac.jp
[2] Keio University, 5322 Endo, Fujisawa-shi, Kanagawa 2520882, Japan

Abstract. In a goal type ball game such as soccer and handball, a plurality of players who can pass are searched for, and each player's intention is estimated and a player who can pass is selected. Furthermore, the ball holder checks the position and behavior of enemies around passable players, estimates their intentions, and determines teammate players whose pass is most successful. In order to realize instantaneous intention estimation and judgment subject to be strong temporal and spatial constraints, cooperative patterns shared within the group are considered to exist. Therefore, in this research, we focus on human gaze behaviors in goal type ball game. We presented to subjects a first-person perspective of professional soccer players by using virtual environment, and analyzed the gaze behaviors during pre- and post-training for constructing cooperative pattern modeling. Based on the results, we model a process of intention estimation concerning cooperative pattern. We discuss that subjects switch their behavior by estimating the intention of other players by presenting the visual information based on the first-person perspective.

Keywords: Cooperative pattern · Intention estimation
Gaze behaviors · Virtual environments

1 Introduction

We estimate our intention from human behavior and decide our own action strategy based on the result in our daily life. In many cases of one-to-one communication, other's intention will be inferred based on own internal model. However, in cooperative behavior where two or more participants estimate their intention, it is necessary to decide the subject to pay attention to and construct the consecutive intention process by estimating next subject's intention. Specifically, in a goal-type ball game such as soccer or handball, each player searchs a passible teammate players and estimate their intention for determining passing player. Furthermore, the ball holder checks the position and behavior of enemies

© Springer Nature Switzerland AG 2019
A. V. Samsonovich (Ed.): BICA 2018, AISC 848, pp. 326–333, 2019.
https://doi.org/10.1007/978-3-319-99316-4_43

around passable players. In addition ball holder estimates their intentions and determines teammate players whose pass is most successful.

In order to realize instantaneous intention estimation and judgement players to be strong temporal and spatial constraints, we thought a cooperative pattern is shared within the group. Therefore we analyze professional soccer player's gaze behaviors in the timing of clarifying the process of intention estimation. To clarity the shared process of cooperative pattern and model such human cooperative pattern, we suggest an experiment focusing on pass scenes where it is important to form cooperative patterns in soccer.

In this experiment, we use virtual environment to give a first-person perspective of professional soccer players as training data, and analyze gaze behaviors of subjects before and after learning. We reproduce cooperative behaviors in the virtual environment based on soccer player's position by using tracking data and repeatedly presenting specific scenes. It is acquired selection behaviors of ball holder and receiver in the passing and changes of subject's gaze behavior before and after. Based on the results, we discuss whether subjects can switch the gaze behaviors by estimating the intention of passing players and the players who receive the pass by presenting first-person perspective information.

2   Cooperative Pattern Analysis

There is some multi-agent researchs that realizes cooperative behavior such as RoboCup soccer simulation. Akiyama et al. construct a search tree called an action chain search framework and study on the linkage of soccer tactics by implementing a effective online chain action [1]. In agent model researchs, although it is possible to repeat experiments based on data and reproduce behaviors similar to human, it is difficult to estimate the player's intention in cooperative pattern [2].

In study of cooperative patterns on human behavior, it is necessary to analyze action decision process between trials. So it is repeatedly perform the actions of the player acquiring the cooperative pattern and the actions of unacquiring player. In conventional sports behavior analysis, we must reproduce the scene in mini game form by the team member in actual field [3], the situation changes depending on the participating members and weather and so on. Although there is a study by using a display [4], it is highly reproducible, but in order to switch the situation based on the gaze behavior of a subjects, it is necessary to display a curved surface display close to 360°.

In recent years, a head mounted display capable of presenting a virtual environment with wide viewing angle from the first-person perspective has been developed. It is low cost and used for many cognitive training. By using these devices, virtual environments are used for training of athletes, and their effectiveness is also indicated [5,6]. In this research, by solving the problem of reproducibility and realism, it is possible to analyze behavioral changes in an environment close to realistic scenes.

3 Cooperative Pattern Acquisition Experiment on Gaze Behavior

3.1 Scene of Virtual Environment Experiment

The scene presented to the subjects is FIFA Confederations Cup 2013 game against Spain vs. Italy in the first half and one minute from the ball reservation in the middle stage to a shoot including a passing. In order to reproduce in virtual environment, position tracking data and moving image data that recorded at 10 fps are used. The data does not include gaze direction, so we use the direction of player's head based on the moving image data for approximating player's gaze direction. In recent research, it is suggested that we are able to approximate the direction of gaze and attention by the direction of head of a person [7].

A field player whose subjects share a field of view is a midfielder (MF) that estimates the intention of surrounding players and switches behaviors. We analyze gaze behaviors at each stage leading from the beginning of attack to the shooting situation. The Spanish team play a good passing with tactics to keep balls [10], players share intention to realize a precise pass. So it seems that they can move instantaneously.

3.2 HMD and Virtual Environment

In the experiment, we use a head mounted display (FOVE0 [8] made by FOVE inc.) that can present a virtual environment by wearing it on a head. The performance has a viewing angle of about 100 deg, and in addition to the head tracking function, an eye tracking system is mounted. It is possible to analyze the movement of subject's gaze while presenting the virtual environment. The virtual environment was constructed using a game engine "Unity 3D" manufactured by Unity Technology Co., Ltd.

Agents that play the role of soccer players in the virtual environment are color coded by red and white for each team, and parameters including fields are created based on actual soccer players and game regulations [9] (Fig. 1).

We aligned the position of subject's head on the HMD, and they share the vision with one of professional soccer players. By this environment, it is possible to analyze behaviors in a form closer to a judgment in a more real soccer scene, which is not a bird's-eye view looking down at the field but decides a behavior at the first-person perspective. In addition, by making the body of subject's agent transparently, it prevents surrounding objects from being hidden by own bodies. The main parameters in the virtual environment are showed in Table 1.

3.3 Method

Subjects were repeatedly presented the scenes created in the virtual environment, and they acquire the passing selection behavior of MF. The agent in virtual space moves automatically based on tracking data, and the subject takes a gaze behavior as an agent corresponding to MF specified in the experiment. The

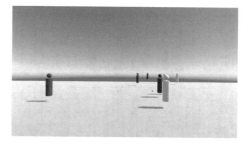

Fig. 1. A virtual environment at the subject's first-person perspective

Table 1. Main parameter setting of virtual environment[m]

Ball diameter	0.2	Field width	105.0
Player height	1.7	Field vertical	68.0
Player width	0.4	Goal width	7.3
Player head diameter	0.3	Goal height	2.2

subject can freely check the surrounding situation in the virtual environment by shaking own head with HMD. Subjects were seated for the safety during the experiment.

The experiment has three phases, "prior phase", "learning phase" and "posterior phase". In prior phase, the subject can move own head and freely check the situation in corresponding scene. In learning phase, the field of view range of subject is fixed and the gaze behavior performed by actual player is presented to the subject. In posterior phases, the subject move own head based on the action in learning phase. Subject were instructed about the scene and requested to decide where to pay attention in that scene. After the experiment, we have a simple questionnaire about ball game experience and the attention point. We have three 20 years old subjects and take a break time of about 1 min in each trial for preventing VR sickness.

3.4 Results

Figure 2 shows the result of gaze angle variance for each trial of 3 subjects differences by using FOVE0 eye tracking system. Since the angle variance is used trigonometric functions such as Experiments 1, 2.

$$(R\cos\bar{\theta}, R\sin\bar{\theta}) = \frac{1}{N}(\sum\cos\theta, \sum\sin\theta) \tag{1}$$

$$V = 1 - R \tag{2}$$

Here, $\bar{\theta}$ is angle average, V is variance, N is a number of data, and a rotation radius R. The angle is aligned an unit vector in the same direction and aligned

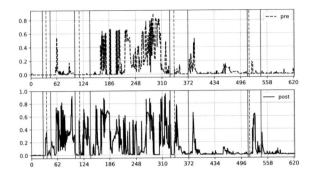

Fig. 2. Distribution of angular change of subject1 gaze point in prior (upper) and posterior (lower). The blue solid line shows the timing when the teammate player gives a pass to the subject, the red dotted line shows the timing of ball touch of the subject, and the red solid line shows the timing at which the subject pass.

0 in the different direction. So we represented as a value between 0 and 1, such as Experiment 2.

From Fig. 2, in prior phase a subject have continuously watched different gazing points. In posterior phase, there is a tendency that the variance decreases in a short time and it is similar results in other subjects. It is thought that subjects were checking the surroundings by moving the gaze point in a very narrow range. There was no significant difference between prior and posterior phase with respect to the average between one second variance trials for each teammate before passing, before the subject received a ball, and before passing a ball to teammate.

Next, Fig. 3 shows the variance of gaze angle of three subjects in each trial. In this case, the teammate player gives a pass to the subject and the subject gives a pass to yet another player. In prior phase, subjects were closely watching nearly a same part in each 3 times, but in posterior phase the parts to be watched were different. Therefore, in posterior phase, it seems to be taking exploratory behavior to gaze point finely in a narrow range.

4   Discussion

Although there is no difference in variance due to angle change between prior and posterior phases, it is considered that subjects were watching a same gaze point many times and were closely watching in a narrow range. These results indicate the following. In prior phase, since subject's head was moved and the field of view was largely changed, gazing points were also changed and variance was increased. On the other hand, since posterior phase was common to the field of view of MF, gazing points were clear, and the timing at which the variance became smaller increases. As a result, information obtained from a particular gazing point increases, it is considered that analysis of other player's behavior and other's intention estimation became possible.

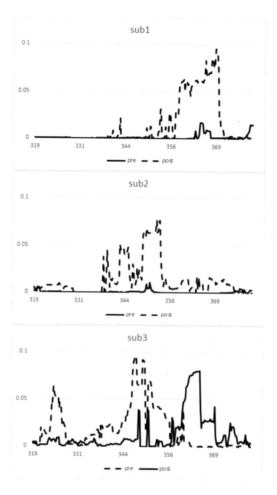

Fig. 3. Distribution of gaze angle of subjects 1, 2 and 3 in prior and posterior. It shows the value of 319–379 frame with passing in presentation scene.

At the timing when a teammate player gave a pass to the subject and the subject pass to another player, the variance in each trial increased in posterior phase. In prior phase, subjects check the situation of entire space by using peripheral vision without visual attention. On the other hand, in posterior phase, subjects search passable players and check the player's behaviors by foveal vision and move the gaze finely.

Therefore, we analyze the players who the subject actually watched. Figure 4 shows the relative angle of two FW players from a subject gaze point. This graph shows the results from 319 to 379 frames where the subject who received a ball pass to the another player, and at this timing the subject subsequently pass to FW player. Until around 340 frames, the subject was watching FW1 and FW2 at about the same angle. However, from 340 frames, the subject's gaze point was moved finely, and after 350 frames the relative angle of FW2 was nearby 0. So it is considered that the subject decided by using foveal vision. The subject himself would be acquired the cooperative pattern and actively searched possible players in the pass behavior.

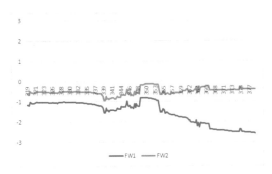

Fig. 4. The relative angle (rad) of FW player (1, 2) from the gaze point of subject1. It shows the value of 319–379 frame with passing in presentation scene.

5   Conclusion

We analyzed the gaze behavior in soccer cooperative pattern by presenting a first-person perspective using a virtual environment. We clarified a change of intention estimation process concerning soccer passing situation. Experimental results showed that by sharing the gaze behavior of professional soccer player, subjects narrowed the range of the field of view to be searched. So they can limited the players whose intent was to be estimated. It is thought that subjects attempted to move the gaze point finely within the range of the field of view to be searched further, actually confirm the plurality of FW players in foveal vision and estimate their intention. These behaviors were led by subjects positively searching based on cooperative patterns of pass behavior in learning phase.

With regard to eye movement in this study, the number of subjects and scenes to be analyzed are limited, so it is necessary to conduct additional experiments and analyze in the future. Also, in an actual cooperative behavior, other players change their behaviors by estimating subject's intention due to change in subject's behavior. In this experiment, we can not realize the situation where other players change their behaviors, so we will construct an agent model to change behaviors based on human intention estimation, and interact with subjects in virtual space.

References

1. Henrio, J., Henn, T., Nakashima, T., Akiyama, H.: Selecting the best player formation for corner-kick situations based on Bayes' estimation. Lecture Notes in Computer Science, vol. 9776, pp. 428–439 (2016)
2. Itoda, K., Watanabe, N., Takefuji, Y.: Model-based behavioral causality analysis of handball with delayed transfer entropy. Procedia Comput. Sci. **71**, 85–91 (2015)
3. Hervieu, A., Bouthemy, P., Cadre, J.P.L.: Trajectory-based handball video understanding. In: Proceedings of the ACM International Conference on Image and Video Retrieval, p. 43. ACM (2009)
4. Lee, W., Tsuzuki, T., Otake, M., Saijo, O.: The effectiveness of training for attack in soccer from the perspective of cognitive recognition during feedback of video analysis of matches. Football Sci. **7**, 1–8 (2010)
5. Bideau, B., Kulpa, R., Vignais, N., Brault, S., Multon, F., Craig, C.: Using virtual reality to analyze sports performance. IEEE Comput. Graph. Appl. **30**(2), 14–21 (2010)
6. Miles, H.C., Pop, S.R., Watt, S.J., Lawrence, G.P., John, N.W.: A review of virtual environments for training in ball sports. Comput. Graph. **36**(6), 714–726 (2012)
7. Nakashima, R., Shioiri, S.: Facilitation of visual perception in head direction: visual attention modulation based on head direction. PLoS ONE **10**(4), e0124367 (2015)
8. FOVE 0. https://www.getfove.com/. Accessed 04 July 2017
9. Federation Internationale de Football Association: Laws of the game. FIFA (2015)
10. Ladyman, I.: World cup 2010: Beat Spain? It's hard enough to get the ball back, say defeated Germany, July 2010. http://www.dailymail.co.uk/sport/worldcup2010/article-1293239/WORLD-CUP-2010-Beat-Spain-Its-hard-ball-say-Germany.html. Accessed 04 July 2017

A Consideration of the Pathogenesis of DID from a Robot Science Perspective

Yuichi Watanabe, Takahiro Hoshino, and Junichi Takeno[✉]

Robot Science Laboratory, Meiji University, 1-1-1 Higashimita Tama-ku,
Kawasaki-shi, Kanagawa 214-8571, Japan
ytrp441@gmail.com, tantakatan0324@gmail.com,
juntakeno@gmail.com

Abstract. In this study, we propose a consciousness model that simulates the dissociation symptoms and the pathogenesis of dissociative identity disorder (DID). With the model presented in this paper, we attempt a more advanced reproduction of the pathogenesis of DID by incorporating a new mechanisms in the conscious system. We believe that this study can enable a more objective understanding of the processes involved in the development of dissociative disorders, including DID.

Keywords: Neural network · MoNAD · Conscious system
DID (dissociative identity disorder)

1 Introduction

The authors' research group have considered the symptoms of dissociative identity disorder (DID) as a problem of a "self-dispersing conscious system" and proposed a consciousness model using MoNADs [1]. In the current research, by adding new mechanisms to the conscious system by means of MoNAD consciousness modules, we propose a model that highly reproduces the symptoms of DID and its onset process.

2 MoNAD Consciousness Module

The conscious system proposed in this paper can be realized by combining a plurality of consciousness modules called MoNADs (Module of Nerves for Advanced Dynamics) that were developed by the authors' research group [2]. The MoNAD computational model of consciousness that uses recursive neural networks (RNN) and consists of the following units: input, primitive representation, cognitive representation, behavioral representation, output, and somatic sensation unit. In the primitive representation of the MoNAD, behaviors are performed during cognition and cognitive learning happens at the same time when performing behaviors.

© Springer Nature Switzerland AG 2019
A. V. Samsonovich (Ed.): BICA 2018, AISC 848, pp. 334–340, 2019.
https://doi.org/10.1007/978-3-319-99316-4_44

3   Conscious System

The conscious system consists of three subsystems: the reason subsystem, emotion and feelings subsystem, and association subsystem. The reason subsystem is a MoNAD group that represents the perception of the external environment and one's own behaviors and outputs behavior information at the same time. In this study, we believe that the MoNAD group of the association subsystem, which arbitrates information between the reason subsystem and the emotion and feelings subsystem, functions as an important module that represents what is called "personality" in humans [2].

4   Dissociation and Dissociative Identity Disorder (DID)

The dissociation phenomenon occurs as a defensive or adaptive response to trauma, and the symptoms cited may include personal memory impairment, complete amnesia of information about oneself such as name and age, and an experience of transformation of one's sense of identity. Especially, it is said that in dissociative identity disorder (DID) there are several personalities within one person that can be clearly distinguished from each other [4]. The multiple personalities observed in DID patients are thought to be formed by chronic repetition of dissociative symptoms due to trauma from childhood [3].

5   Conscious System for the Dissociation Model

By adding new mechanisms to their conscious system, the authors attempted to express the symptoms of dissociation and the onset process of DID (Fig. 1). In this conscious system, the MoNAD used as the main personality is defined as (H) in the association subsystem.

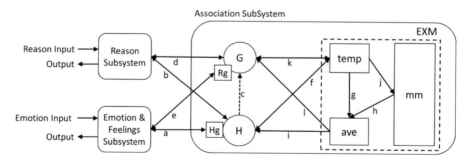

Fig. 1. Conscious system for the dissociation model.

5.1   Recording Device (EXM)

The dissociation phenomenon is assumed to occur in the following two cases.

(1) The main personality gradually and repeatedly receives unpleasant emotions such as through trauma.
(2) The main personality receives excessive unpleasant emotions all at once.

In order to simultaneously study these two cases of dissociation with the conscious system, the authors considered that the association subsystem should be able to record various kinds of information including not only the current information obtained from the reason subsystem and the emotion and feelings subsystem but also information from the past history. Therefore, the conscious system would have a "special recording device (called EXM)" as part of the association subsystem. EXM currently does not have a MoNAD structure and is an area that only holds the history of various information received by the association subsystem. EXM has a "temporary recording unit (called temp)," a "main recording unit (called mm)," and an "emotional value averaging unit (called ave)." Every time the MoNAD of the association subsystem receives information from the reason subsystem and emotion and feelings subsystem, that information is recorded and updated (Fig. 1). In this paper, the information (a) indicating the degree of pleasant and unpleasant emotion sent from the emotion and feelings subsystem to (H) of the association subsystem is called the "emotional value Ev." And, the information (b) transmitted from the reason subsystem to (H) is referred to as the "reason value Rv." The temp unit is the area that temporarily holds the current information (Ev and Rv) transmitted (f) to EXM that goes from the reason subsystem and emotion and feelings subsystem through (H) of the association subsystem until it is finally stored in mm. The mm unit stores a long-term record of the past (Ev, Rv) information transmitted to the association subsystem as one set of information together with a label of the MoNAD of the association subsystem that functioned as the personality for that information. This label is a kind of symbol. By using the set of information in which the label of the MoNAD is recorded in mm, the conscious system can judge, for example, whether the main personality (H) has already experienced the information. Among the past information recorded in mm, the ave unit holds the "total average emotional value" which is a combination of all the Ev of the information set having the same Rv as the information held by temp and the Ev of temp.

5.2   MoNAD Filter Value

The authors think that when the Ev value representing an unpleasant emotion to the main personality (H) was sent to the association subsystem (a), it is necessary to have a substitute MoNAD that can handle the information on behalf of (H). Likewise, if Ev represents the unpleasant emotion to all the MoNADs in the association subsystem, a MoNAD that can handle this Ev will not exist. For this reason, each MoNAD in the association subsystem is considered to have a unique "filter value" (Fig. 1, Rh, Rg). This filter value can change the value of Ev received for each MoNAD individually to a certain degree. Thus, there may be certain MoNADs that do not receive an Ev representing unpleasantness in the association subsystem. Additionally, the filter value is simply expressed here as a variable.

5.3 Occurrence of Dissociation

In the authors' conscious system, the Ev value (for pleasant and unpleasant emotions) of the association subsystem is calculated using EXM and the filter value of each MoNAD. This method makes it possible to identify the two cases in which the dissociation phenomena may occur. In other words, these are the two previously mentioned cases in which the main personality "(1) repeatedly receives unpleasant emotions" or "(2) receives excessive unpleasant emotions all at once."

> 1-1.

Rv is sent (b) from the reason subsystem to (H) of the association subsystem. At the same time, Ev is sent (a) from the emotion and feelings subsystem to (H). At this time, Ev will change depending on the filter value Rh of (H).

> 1-2.

At that time, Rv and the changed Ev are sent (f) to EXM and are held in temp.

> 1-3.

Among the past information recorded in mm, all the Ev in the information sets having the same Rv as temp and the Ev of temp are combined (h, g), and the total average value (ave) of Ev is calculated.

> 1-4.

The ave value is sent (i) to (H), and a judgment is made as to whether the value is unpleasant for (H). When "ave is unpleasant for (H)," this is a case in which ave is judged to represent unpleasantness (e.g. a level of 5 or more on a scale of 10). This is a judgment that indicates the previously mentioned case in which the main personality "(1) repeatedly receives unpleasant emotions."

> 1-5.

In addition, a judgment is made as to whether excessive unpleasantness is being expressed to (H) only for the Ev (current emotional value) of temp. A "state in which Ev is excessively unpleasant for (H)" is a condition in which Ev itself is strongly expressing unpleasantness (e.g. only a maximum value of 10 on a scale of 10). This is a judgment that indicates the previously mentioned case in which the main personality "(2) receives excessive unpleasant emotions all at once."

> 1-6.

Dissociation occurs in 1-4. and 1-5. The occurrence of the dissociation symptoms will be explained in the next section.

> 1-7.

In cases other than 1-4. or 1-5., dissociation does not occur and (H) functions as the personality. At this time, the information set of temp is recorded (j) in mm. Also at this time, the label of the (H) that functioned as the personality for this information set is added. Since the label of (H) was recorded, the conscious system can use this information set to recall the experience of (H) as the main personality later.

1-8.
The output information of the association subsystem is sent (b) from (H) to the reason subsystem.

6 Development Process of the Dissociation Symptoms

The dissociation in this conscious system occur through the following process. Here, the state (G) which is a lower-hierarchy MoNAD of the main personality (H), together with the paths (c, d, e) accompanying it, is assumed to have been already automatically generated by the "learning of unknown information" in the association subsystem [1, 5]. Rv and Ev are also simultaneously sent (a, b, d, e) to both (H) and (G). Additionally, the Rv sent to the association subsystem here is information known to (G).

2-1.
When a dissociation phenomenon occurs, Rh changes to a value that increases the unpleasantness (a) of Ev.
    2-2.
Due to the occurrence of the dissociation, in response to (H) aborting the sending (b) of information to the reason subsystem, EXM searches for another MoNAD that has already learned the Rv of temp by selecting from among the lower-hierarchy MoNADs of (H). Since (G) applies here, the conscious system separates (G) from (H) and makes (G) operate as another personality. If Rv is information unknown by all the lower-hierarchy MoNADs of (H), a newly learned MoNAD is generated beneath (H) and it is detached from (H) [1].
    2-3.
The value of Ev (e) changes according to the filter value Rg of (G). The changed Ev value is sent (k) to EXM and held in temp.
    2-4.
EXM newly calculates ave from temp and mm (g, h).
    2-5.
The calculated ave value is sent (l) to (G), and a judgment is made as to whether ave is unpleasant for (G).
    2-6.
Further, a judgment is made as to whether (k), (G) shows excessive unpleasantness only for the Ev value (current emotional value) of temp.
    2-7.
In cases other than 2-5. and 2-6., dissociation does not occur and (G) functions as another personality. The information set in temp is recorded (j) in mm. At this time, the label of (G) that functioned as the personality is added to the information set.
    2-8.
The output information is sent (d) from (G) to the reason subsystem.

    As mentioned above, when (G) functions as another personality, Rg takes a value that decreases the unpleasantness of Ev (e). As the unpleasantness of Ev decreases, the value indicating the unpleasantness of ave will also decrease as a result. Then, as the value indicating the unpleasantness of ave decreases, the dissociation of the association subsystem does not occur. In other words, in this conscious system, we can explain the dissociation phenomenon as a defensive reaction that attempts to reduce the unpleasantness (a) of Ev that is sent to the association subsystem and protect the main personality (H) from unpleasant emotions. Furthermore, when dissociation occurs and (G) functions as a personality, a label of (G) is added to the information set (2-7.). This conscious system cannot recall any information sets in which the label of a MoNAD other than (H) was recorded as the experience of the main personality (H). This condition can explain amnesia, which is one of the typical symptoms of dissociation. Also, the value of Rh at which the dissociation occurred is changed (2-1.) to a value that increases the unpleasantness (a) of Ev. Therefore, it can be said that dissociation in (H) is likely to occur each time the dissociation is repeated. Furthermore, as the dissociation repeats, the association subsystem finally judges that excessive unpleasantness is caused by the filter value of (H) regardless of what value of Ev (a) is received by (H), and it reaches a state at which it changes to the indicated value. Then, in the above case, when Rv, which is information known by (G), is sent to the association subsystem, the dissociation of (H) definitely occurs and (G) functions as another personality. In other words, (H) and (G) have independent paths (a, b) and (e, d), respectively. This can be explained as a state in which an independent personality (G) is formed by the chronic repetition of the dissociation symptoms. As such, the authors think that this model can explain both the symptoms of dissociation and the onset process of DID. Furthermore, the following operations are repeated in the process of forming another personality (G).

3-1.

When an unpleasant value of Ev is repeatedly sent (a) to the association subsystem, the value indicating the unpleasantness of ave increases and dissociation occurs.

3-2.

When dissociation occurs, the unpleasantness (e) of Ev decreases due to Rg and the value indicating the unpleasantness of ave decreases.

3-3.

Since the value indicating the unpleasantness of ave decreases, dissociation does not occur.

3-4.

Furthermore, when an unpleasant emotion is repeatedly sent (a) to the association subsystem, the value of ave increases again and dissociation occurs again.

The authors think that the above state expresses an incomplete process in the formation of a temporarily occurring other personality. We think that it is possible to explain dissociative amnesia symptoms [4] such as, "not being able to recall an important autobiographical memory, but where a distinct other personality as in DID does not appear." Therefore, the authors believe that not only DID but also various symptoms and onset processes of a group of dissociative disorders can be explained extensively with the conscious system for this dissociation model.

7 Conclusion and Discussion

In this paper the authors proposed a system to explain the dissociation phenomenon by adding new "EXM" and "MoNAD filter value" mechanisms to their conscious system as a model of DID. The authors created a program that utilizes the functions of EXM to simulate and demonstrate the processes of repeatedly or suddenly occurring dissociation phenomena in response to unpleasant emotions. In addition, the authors also demonstrated the formation process of another personality in DID patients by means of the change in the MoNAD filter value. This consciousness model demonstrates not only DID but also has the possibility of expressing the various symptoms of dissociative disorders. Moreover, the authors wish to take up as a future research topic the construction of a treatment mechanism based on the onset process of DID in the dissociation model presented in this paper.

References

1. Hoshino, T., Takeno, J.: Robot science discussion on the onset of dissociative identity disorder (DID). BICA **88**, 52–57 (2016)
2. Takeno, J.: Creation of a Conscious Robot. Pan Stanford, San Antonio (2013)
3. Putnam, F.W.: Diagnosis and Treatment of Multiple Personality Disorder. The Guilford Press, New York (1989)
4. Barnhill, J.W.: DSM-5 Clinical Cases. American Psychiatric Publishing, Philadelphia (2014)
5. Kushiro, K., Takeno, J.: Development of a robot that cognizes and learns unknown events. In: BICA, pp. 215–221 (2011)

Data Enrichment with Provision of Semantic Stability

Viacheslav E. Wolfengagen[1]($\boxtimes$), Sergey V. Kosikov[2]($\boxtimes$),
and Larisa Yu. Ismailova[1]($\boxtimes$)

[1] National Research Nuclear University "MEPhI"
(Moscow Engineering Physics Institute), Moscow 115409, Russian Federation
jir.vew@gmail.com, lyu.ismailova@gmail.com
[2] Institute for Contemporary Education "JurInfoR-MGU",
Moscow 119435, Russian Federation
kosikov.s.v@gmail.com

Abstract. The paper considers the problem of data enrichment, which is understood as supplying the data with semantics with further introduction of structuring. It also considers the data collected from heterogeneous sources with their subsequent organization in the form of information graphs. A data model is used in the form of a network, the framework of which is objects and relations between them. What is more, as objects, in turn, the relations can be used, and this potentially leads to higher order structures. The data connections and data dependencies are also taken into account.

Providing semantic stability during the data enrichment requires solving a number of problems, among which the search for semantically unstable objects, their classification by types of instability and the identification of ways to overcome the instability. An approach is proposed to solve these problems on the basis of the homotopic type theory. Methods are considered for modifying unstable objects for types of different structures. The paper discusses the possibilities of using the results in information systems that allow to increase the "degree of cognitization" of the data and, in the long term, the transition to cognitive business.

Keywords: Data enrichment · Semantics · Homotopic type theory
Computational model · Semantic stability · Information graph

1 Introduction

At present time the growth rate of significant amount of innovation is associated with the preparation and processing of so-called "Big Data" as the growth of data accumulated in various areas of business, public administration, health, etc. is estimated as exponential [1]. Especially important is the selection of the semantic characteristics of the data and the provision of support on the semantics of data while performing the business analytics [2,3]. Among the data used

© Springer Nature Switzerland AG 2019
A. V. Samsonovich (Ed.): BICA 2018, AISC 848, pp. 341–346, 2019.
https://doi.org/10.1007/978-3-319-99316-4_45

for management in both commercial and public organizations, there recently appeared a tendency to select (i) data classes related to the implementation of current operations for the current functioning of companies and their management, and (ii) the so-called "master data" usually aimed at supporting the procedure for making managerial decisions [4].

The data of the first type, as a rule, quickly acquire relevance and also quickly lose it. Therefore, for performing analytical operations such data require the methods of aggregation and statistical processing being used. The data of the second type, in contrast, have a fairly longer lifetime. The criterion for data selection of the indicated types is largely conditional, and this often does not allow for unambiguous separation but requires the use of methods that more closely take into account both the semantics of the data [2] and their pragmatics [3]. The conditionality of characterizing data in this paper is associated with insufficient consideration of the semantic character of the data. In standard DBMS data models [5] the save of imposed the integrity constraints is provided by DBMS means.

The use of both data types is connected with data enrichment processes. In the general case this paper understands the enrichment as improving the quality of existing data thanks to linking with them additional information, both new and derived from the data itself. The enrichment can take place when performing any data management process, in particular, when collecting, preparing, maintaining and using it.

There are two main ways to enrich the data: external and internal enrichment. In case of external enrichment the data is associated with additional information received from sources external to the considered information system. In case of internal enrichment the additional information is derived from the data itself due to their in-depth analysis. Further on such information can be used to improve methods of semantic interpretation of data.

Both the external and the internal data enrichment can be achieved by attaching additional attributes to the data, selecting additional metadata or metadata classes and linking them to existing data, etc. Such additional data can be used to solve data processing tasks within the system. The data additionally added can carry semantics that allows using this data to improve the quality of the semantic model.

The paper is structured as follows. The Sect. 2 considers some approaches to data enrichment. The Sect. 3 sets the task of supporting enrichment of data with ensuring semantic stability. The Sect. 4 proposes some methods to support the semantic stability, the methods being applicable both for external and internal data enrichment. The Sect. 5 discusses the necessary opportunities for appropriate support tools. The Sect. 6 summarizes the results and discusses the prospects for further work.

2   Approaches to Data Enrichment

The available systems and tools of business analytics raise the issue of increasing the "degree of cognitization" of the data and, in the long term, of the transition to cognitive business. In the literature [4] it is noted that the growth in the volume of data is often not accompanied by the support of their structuring and support, adequate to the tasks being solved. Therefore, the existing approaches to data enrichment are aimed at providing them with adequate semantics, which is interlocked with the task of building a semantic Web.

One of the well-known systems in this area is IBM Watson [6]. The project allows the use of elements of cognitive computing in the main business processes of the organization using IBM's sectoral experience. However, the methods of describing and adjusting the semantics of the model data are deeply inside of the Watson interfaces and require the use of special technologies for integration into the organization's information model [7].

Another tool, oriented to work with enriched data, is Google Analytics [8]. This service is focused to support developing the detailed statistics of visits to websites. However, the tool is not focused on integration with data of arbitrary semantics and has a specialized character.

The problems of data enrichment were also considered in connection with the analysis of social networks and crowd-sourcing processes [9]. The directions of descriptive, predictive and prescriptive analytics are distinguished. All the selected directions are closely related to the processes of data enrichment, primarily internal, but ensuring the semantic consistency of the resulting models is an open problem in general case.

Different techniques of data enrichment are an integral part of machine learning [10]. Thus, deep machine learning techniques, known as "no-drop", "dropout" and "dropconnect", are essentially built on the methods of stochastic gradient descent. The methods are parameterized, the parameters ensuring the classification and integration of learning models.

In the whole the existing approaches to the data enrichment simultaneously demonstrate the variety of the enrichment methods used and the relative weakness of the semantic integration of the resulting models. Therefore, the actual task is to increase the semantic stability of the model of enriched data collected from heterogeneous sources, followed by their homogeneous organization.

3   Task of Supporting Semantic Stability in Data Enrichment

Ensuring the semantic stability in data enrichment involves the use of appropriate semantically oriented support methods. The semantics is understood as the establishment of a correspondence between the expressions of the description language and of the manipulation of the data and functions defined on the data model used. The given paper considers the semantically stable data as elements of a data model, containing a set of restrictions, usually of a semantic nature, the operations of the data transformation model ensuring the save of imposed semantic restrictions when transforming the data.

During the enrichment process the data can both enter the system from external sources of data and are generated by analysis algorithms, which do not comply with the adopted restrictions in general case. Therefore, new data, generally speaking, violates the requirements of semantic stability and can be considered as semantically unstable. Ensuring stability in these conditions requires solving a number of problems, including:

- search and detection of semantically unstable objects;
- classification of semantically unstable objects by types of instability (sets of violated semantic restrictions);
- determination of ways to overcome instability.

In accordance with the nature of the data enrichment carried out it can be performed in various ways. In particular, during enrichment it is possible:

- adding new elements to the data (in particular, attributes to existing data);
- adding new metadata or classes of metadata;
- adding new ways of interpreting the data;
- modification of existing elements, metadata or methods of data interpretation, in particular, the transition to a more precise and accurate description of data, or, on the contrary, roughening the description of data.

When implementing enrichment it is necessary to take into account the connections between the data and the data dependencies. Accounting for dependencies shows that enrichment of data of one type may require semantically coordinated enrichment of the data associated with them and in general, lead to "cascade" data enrichment.

4   Methods to Ensure the Semantic Stability

The given paper proposes an approach to ensuring semantic stability in the enrichment of data based on the homotopic type theory [11]. The elements of the data model for enrichment are represented as objects of the homotopic type theory. Objects are assigned a type, which is represented as formal assertions of the form $a: A$, which is interpreted as "a is an object of type A".

Further on the formal notation $A: U$ is used for the expression "A is type". Here U is a universal type, which is a type whose elements are types.

Let's consider the ways to ensure stability for unstable objects, to which types of different structures are assigned. The assigned types can be either basic or constructed from other types, and in the latter case the types are equipped with functions for their construction and computation.

For typical construction - function type, we have the following. From types A and B it is possible to construct type $A \rightarrow B$ of functions with domain A and co-domain B. If assume that x: A and consider the expression $F : B$, then $(\lambda(x: A).F) : A \rightarrow B$.

The instability can appear either in domain A, or in co-domain B, or in both subordinate types. At the same time, if in the domain the instability is eliminated by using the function $f : A' \rightarrow A$, then in the functional domain it is eliminated by means of a term $(\lambda(x : A').F[fx/x]) : A \rightarrow B$, where $Z[y/x]$ is the result of substituting y in place of x in Z.

For the construction of dependent functional type, the elements of type are functions, the range of values of which may depend on the element of the domain, to which the function is applied. From the type $A \rightarrow U$ and the family of types $B : A \rightarrow U$, it is possible to construct the type $\prod(x: A)B(x): U$. If there is an expression $F: B(x)$ depending on $x: A$, then $(\lambda(x: A).F)$ has the type $\prod(x: A)B(x)$.

For the construction of Cartesian product of types A and B, the type $A \times B$ is constructed. It is also possible to consider the type of product with arity 0, which is called the unit type $1: U$. For $a: A$ and $b: B$ pairs $(a, b) : A \times B$ are constructed. Cartesian product is a pair whose components have independent types.

The functions to eliminate semantic instability may be constructed in the same way as a functional type case.

5 Tools of Support for Semantic Stability

The given paper proposes an organization in the form of information graphs to represent data collected from heterogeneous sources. A data model is used in the form of a network, the framework of which is objects and relations between them, and as objects, in turn, relations can be used, which potentially leads to higher order structures. The base graph is given in the form $G = (E, V)$, where E is the set of vertices of the graph, V is the set of arcs of the graph that is a subset of $E \times E$.

With the help of the base graph it is possible describe concepts and frames. The structures used for this are defined as follows:

$$MF = (C, A, V, P, R, ISA),$$

where C set of concepts; A set of constants; V set of variables; P set of predicates; R set of roles; ISA partial ordering of $C \times C$.

We will suggest that variables and constants are assigned types - concepts from the set C, which we denote by $a : c$, $x : c$, where a, x, c elements of sets A, V, C correspondingly.

We will consider simple frames of the following type:

$$F = [p(r_1 : o_1 : c_1; r_2 : o_2 : c_2; \ldots, r_n : o_n : c_n)]$$

where p element of set P, r_i element of set R, o_i element of set A or of set V, c_i element of set C. The frames of this type can be represented in many ways in a graphical form, and, therefore, can be regarded as those forming an information graph.

The use of the homotopic type theory to describe the basic semantic network allows to propose an approach to solving the problem of ensuring semantic stability based on a systematic modification of the used types by using the functions of transition from type to type, described in Sect. 5. For this it appears to be advanced to define accommodation functions in a form of information channels.

Through the channels messages are sent, which either carry data about new proposed knowledge (messages-conclusions), or control output operations (control messages). Thus, the channels and messages carry the main semantic load in the overall process of data enrichment.

6   Conclusion

The paper studies the problem of data enrichment with the provision of semantic stability of data model during the enrichment process. Definitions relating to semantic stability are given. Various methods of data enrichment are considered, in particular, the integration of data from different models and the transformation of data from one model to another, as well as the use of data analysis results. This paper proposes methods of ensuring semantic stability based on the homotopic type theory.

The paper considers the data both as collected from heterogeneous sources, and obtained by analysis methods, following with their organization in the form of information graphs. A data model is used in the form of a network, the framework of which is objects and relations between them. It appears that in the future the used model can be enlarged by using information channels and providing on this basis a controlled logical deduction.

Acknowledgement. The paper is supported by the grant 18-07-01082 of the Russian Foundation for Basic Research.

References

1. Wolfengagen, V.E., Ismailova, L.Y., et al.: Evolutionary domains for varying individuals. Procedia Comput. Sci. (2016). https://doi.org/10.1016/j.procs.2016.07.447
2. Wolfengagen, V.E., Ismailova, L.Y., et al.: Concordance in the crowdsourcing activity. Procedia Comput. Sci. (2016). https://doi.org/10.1016/j.procs.2016.07.448
3. Ismailova, L.Yu., Kosikov, S.V.: A computational model for refining Data domains in the property reconciliation. In: 2016 Third International Conference on Digital Information Processing, Data Mining, and Wireless Communications (DIPDMWC), Moscow, pp. 58–63 (2016)
4. Fleckenstein, M., Fellows, L.: Modern Data Strategy. Springer, Cham (2018)
5. Date, K.D., Darwen, H.: Fundamentals of Future Database Systems: Third Manifesto, 2nd edn. Janus-K, Moscow (2004)
6. https://www.ibm.com/watson/about/ . Accessed 03 June 2018
7. https://www.ibm.com/watson/developer/?lnk=mpr_buwa&lnk2=learn . Accessed 03 June 2018
8. https://www.google.com/analytics/#?modal_active=none . Accessed 03 June 2018
9. Thuan, N.H.: Business Process Crowdsourcing. Springer, Cham (2019)
10. Suthaharan, S.: Machine Learning Models and Algorithms for Big Data Classification. Springer, Cham (2016)
11. The Univalent Foundations Program, Homotopy Type Theory: Univalent Foundations of Mathematics. Institute for Advanced Study (2013). https://homotopytypetheory.org/book

Causality Reconstruction by an Autonomous Agent

Jianyong Xue[1,2(✉)], Olivier L. Georgeon[1,2,3,4],
and Mathieu Gillermin[3,4]

[1] Université de Lyon, LIRIS CNRS UMR5205, 69622 Villeurbanne, France
{jianyong.xue, olivier.georgeon}@liris.cnrs.fr
[2] Université Claude Bernard Lyon 1, Villeurbanne, France
[3] Université de Lyon, LBG UMRS 449, 69288 Lyon, France
mguillermin@univ-catholyon.fr
[4] UCLy, EPHE, Lyon, France

Abstract. Most AI algorithms consider input data as "percepts" that the agent receives from the environment. Constructivist epistemology, however, suggests an alternative approach that considers the algorithm's input data as feedback resulting from the agent's actions. This paper introduces a constructivist algorithm to let an agent learn regularities of actions and feedback. The agent organizes its behaviors to fulfill a form of intentionality defined independently of a specific task. The experiment shows that this algorithm constructs a Petri net whose nodes represent hypothetical stable states afforded by the agent/ environment coupling, and arcs represent transitions between such states. Since this Petri net allows the algorithm to predict the consequences of the agent's actions, we argue that it constitutes a rudimentary causal model of the "world" (agent + environment) learned by the agent through experience of interaction. This work opens the way to studying how an autonomous agent can learn more complex causal models of more complex worlds, in particular by explaining regularities of interaction through the presence of objects in the agent's surrounding space.

Keywords: Feedback · Constructivist paradigm · Interaction
Causality reconstruction · Developmental learning

1 Introduction

Traditional Artificial Intelligence (AI) algorithms strongly depends on how the agent connects with the environment through actuators and sensors. As Russell & Norvig stated that "the problem of AI is to build agents that receive percepts from the environment and perform actions" [2, p. iv]. In this paradigm, hereafter referred to as the *realist paradigm*, the AI algorithm assumes that the agent's input data is a direct function of the state of the environment.

Developmental approaches in AI explore alternative paths. Following the insights of cognitive science, developmental AI draws on the seminal work of Piaget [4] to define guiding principles for designing autonomous artificial agents [5]: embodiment,

A. V. Samsonovich (Ed.): BICA 2018, AISC 848, pp. 347–354, 2019.
https://doi.org/10.1007/978-3-319-99316-4_46

situatedness and a prolonged ontogenetic developmental process [1]. Ontogenetic development refers to the development of an individual by learning through interaction with its environment. In general, developmental AI tries to reproduce the path along which human acquire knowledge. For example, Mugan and Kuipers proposed a seminal attempt to construct a representation of the environment from sensori-motor interactions [6].

Constructivist epistemology [e.g. 8, 9] suggests an approach to make an autonomous agent iteratively construct a representation of an unknown environment. In the constructivist paradigm, the agent's input data constitutes feedback of the agent's actions with the environment [3]. By contrast with the realist paradigm, input data is not considered as directly representing elements of the world, supposedly pre-existing and available for registration. In a given state of the environment, the input data may vary according to the agent's action, and thus does not constitute a direct representation of an agent-independent presupposed reality.

Authors in psychology argue that perceiving the world consists of actively constructing a representation of the current situation through interaction, as opposed to directly receiving a representation of the world's state [10]. Georgeon, Casado and Matignon [11] model the biological beings as agents trying to perform rewarding interactions with their environment (interaction-driven tasks). Rosech et al. propose to implement a process of knowledge construction from regularities the agent learns from its interactions with the environment without any predefined knowledge [9]. Georgeon and Hassas [14] present a model that does not make of the information upon the environment directly available to registration by the agent. De Loor, Manac'h and Tisseau [12] propose an enaction-based artificial intelligence (EBAI), combined with the evolution of the environment to refine the ontogenesis of an artificial system. The achieved studies lead to the integration of human interactions into the environment to construct relevant meaning in terms of participative artificial intelligence.

In this paper, we introduce design principles, inspired by the constructivist paradigm, to develop a self-motivated agent capable of learning regularities in a task previously introduced by Georgeon [7] in a pedagogical game called *Little AI* (next section).

2   Experimental Settings

In the game Little AI, the player controls a simulated artificial agent in a simulated environment. She presses action buttons (white shapes at the bottom of the screen, Fig. 1a) and receives feedback. She, however, has no knowledge of what these actions are doing. The tuple <action, feedback> is called an *experience*. In the game, an experience is represented as a colored shape whose shape represents the action and whose color represents the feedback. The history of experiences constitutes the *trace* (stream of colored shapes in Fig. 1a). The purpose of the game is to learn to predict the feedback of actions. To do so, the player must infer a model that links actions and feedback, based on regularities observed in the trace.

In this paper, we focus on level 2.00 of Little AI, which provides five possible actions. Unbeknownst to the player, the agent consists of a simulated robot interacting with a simulated environment made of two tiles, one on the right and one on the left. Figure 1b shows the robot and the tiles but the player cannot see this screen until she wins the level. Tiles have two different sides: recto (green) and verso (red). The agent can swap the tiles and test their sides. More precisely, the set A of possible actions is $\{a_1 = feel\ left,\ a_2 = swap\ left,\ a_3 = feel\ both,\ a_4 = feel\ right,\ a_5 = swap\ right\}$ (Fig. 2). At round t, the player chooses an action a_t from A, and then receives feedback f_t depending on this action and on the state of the environment. We say that the agent *enacts* experience $e_t = <a_t,\ f_t>$ when it performs action a_t and receives feedback f_t. Actions *feel left* and *feel right* test the side of the corresponding tile; these actions yield feedbacks f_1 or f_2 whether the tile is recto or verso. Action *feel both tiles* tests the both tiles at the same time and can yield three different feedbacks: f_1: *both tiles are recto*, f_2: *both tiles are verso*, f_3: *one tile is recto and the other is verso*. Action *swap right* (or *swap left*) swaps the right (or left) tile and then returns feedback f_1 (or f_2) whether the right (or left) tile ends up recto or verso. As a result, there are eleven possible experiences. The player cannot assume that the color of feedback corresponds to the color of the side of the corresponding tile. We set the color of feedback to the same color as the side of the corresponding tile only to facilitate the understanding of this paper.

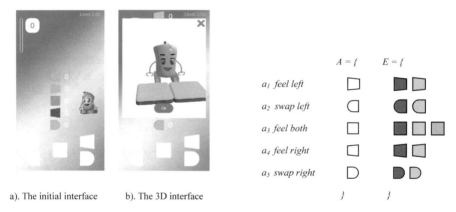

a). The initial interface   b). The 3D interface

$A = \{$ $E = \{$

a_1 *feel left*

a_2 *swap left*

a_3 *feel both*

a_4 *feel right*

a_5 *swap right*

$\}$ $\}$

Fig. 1. Little AI's user interface and Level 2.00

Fig. 2. Five actions and eleven experiences.

3   The Algorithm

In this paper, we present an algorithm to replace the human player and control the agent. Similar to the human player, the algorithm ignores the meaning of actions and feedback. It must learn to predict feedback triggered by actions based on regularities in the trace of experiences, which amounts to inferring a model of the *agent/environment* coupling. From here on, we use the term *agent* to refer to the "simulated robot" and the term *algorithm* to refer to the algorithm that controls the behavior of this agent instead of the player.

The problem of constructing a model from sequences of events has been studied in process mining (PM) [15]. PM algorithms take a flow of events and automatically construct an automaton that can generate this flow. For instance, Van der Aalst [16] proposed the foundational α-algorithm to construct such an automaton modeled as a Petri net.

Drawing on Van der Aalst's α-algorithm and Georgeon et al.'s work [13], our algorithm is based upon the assumption that there are stable states in the agent/environment coupling. This algorithm progressively constructs a Petri net whose nodes represent hypothetical stable states and arcs represent transitions between nodes. Each arc is associated with an experience. When the Petri net has been constructed, the agent uses it to predict the consequences of enacted experiences. The agent uses the token of the Petri net to keep track of the state. That is, when the token is on a peculiar node, the agent believes that the coupling with the environment is in the state represented by this node. The position of the token thus represents the current *belief state* of the agent. When the agent enacts an experience, the agent moves the token through the arc associated with this experience to the destination node of this arc. An arc can be a self-loop if its destination node is also its origin node. In this case, its associated experience does not change the belief state of the agent. We call such an experience a *persistent* experience as opposed to *sporadic* experiences that change the state.

We divide the learning process into two parts: the *node construction stage* and the *arc construction stage*. In the node construction stage, the algorithm tries to learn which stable states exist. In the arc construction stage, the algorithm learns how to transition from one state to another.

3.1   The Node Construction Stage

In this stage, the agent searches persistent experiences by trying to enact experiences several times in a row. When the same experience is enacted a certain number (the *excitement threshold*) of times in a row, the algorithm assumes that the experience is persistent and creates a new node. The algorithm implements a *BeliefState* class with an array *triedNumber* for each experience and a method *getLeastTriedExperience()*. When a new node is created, a new *BeliefState* object associated with this node is instantiated. Its *triedNumber* attribute is used to count the number of trials of an experience in this belief state. When the algorithm is in the "curious" mood, it uses the *getLeastTriedExperience()* function to select the least tried experience based on *triedNumber*.

3.2   The Arc Construction Stage

When the agent learns a stable state, then it goes into the arc construction stage. The arc construction combines the new feedback the agent gets from its past interactions and the previous patterns in the stream of traces. With the changeable environment, it needs to figure out the difference context between previous and post persistent experiences after the agent experienced a sporadic experience. If the two persistent experiences are different, then an arc between these two persistent experiences will be created through this sporadic experience. Otherwise, the agent continues try another sporadic experience until all sporadic experiences have all been tested with all persistent experiences. When all experiences are known to the agent and no changes happen in the causal structure, the agent is in the *confident* mood. Overall, we expect the algorithm to learn the Petri net in Fig. 3, which constitutes a valid representation of the structure of the agent/environment coupling of Level 2.00 of Little AI.

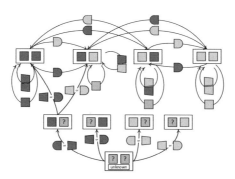

Fig. 3. Partial representation of the Petri net constructed by the algorithm. White rectangles represent nodes. Arcs with feeling experiences (trapezoids and squares) are self-loops that do not change the state. Arcs with swapping experiences (half-circles) cause transitions between belief states.

3.3   Details of the Algorithm

Alg 1. The control of the agent/environment interaction cycle

```
01    initial parameters
02        currentBeliefState = "unknown", mood = "curious"
03    loop
04        if mood = "curious"
05            intendedExperience = getLeastTriedExperience(currentBeliefState)
06        if mood = "confident"
07            finish the knowledge construction and learning is done
08        if mood = "excited"
09            intendedExperience = enactedExperience
10        enactedExperience = Environment(intendedExperience)
11        if enactedExperience is unknown
12            mood = "excited"
```

The algorithm has three possible motivational moods: *curious*, *excited* and *confident*. On initialization, the current belief state: *unknown*, and it is in a *curious* mood. Lines 04 and 05: if it is in a curious mood, then it selects the least tried experience in the context as the next *intended experience*. Lines 06 and 07: if it is in a *confident* mood, then the algorithm has reconstructed the whole Petri net. Lines 08 and 09: if it is in an excited mood, then it intends to repeat the previously enacted experience. Line 10: the algorithms passe the intended experience to the subprogram that implements the environment (Little AI Level 2.00). This subprogram processes the action associated with the intended experience and returns the enacted experience, which may be the intended experience or not. Lines 11 and 12: if the enacted experience is unknown, which means this experience has neither been marked *persistent* nor *sporadic*, then the mood becomes excited.

Alg 2. The node construction and the arc construction

```
01      if mood = "excited"
02          if intendedExperience ≠ enactedExperience
03              intendedExperience is sporadic
04              currentBeliefState = knowledgeUpdating(intendedExperience)
05          else if excitement > excitementThreshold
06              enactedExperience is persistent
07              create new beliefState and added in the beliefStateList
08              currentBeliefState = knowledgeUpdating(enactedExperience)
09          else
10              excitement++
11      updateTriedNumberOfExperience(intendedExperience)
12      knowledgeUpdating(currentBeliefState)
13      if all experiences have not been tried yet in the context of this currentBeliefState
14          mood = "curious"
15      if all experiences have been tried and knowledge isn't updating
16          mood = "confident"
```

Lines 01 to 10: if the algorithm is in the excited mood and the intended experience differs from the actually enacted experience, the intended experience is marked sporadic. Otherwise, the algorithm increments its excitement level. When the excitement reaches the preset threshold, the experience is marked *persistent*, a new belief state is instantiated. Line 11: the intended experience updated its tried numbers. Line 12: learn regularities and update the current belief state. Lines 13 and 14: if all experiences the agent has not been experienced with, the mood becomes curious. Lines 15 and 16: all experiences are known to the agent and there are no changes in the procedure of knowledge reconstruction. This means that the algorithm has reconstructed a valid representation of the structure of the agent/environment coupling of Level 2.00 of Little AI. The mood switches to *confident*.

4   Experiment

Our experiment shows that the algorithm was able to construct the Petri net in 350 interaction cycles (Fig. 4). Step 1: the algorithm intends a red left trapezoid and obtains a same red left trapezoid. Since the experience is neither yet marked sporadic or persistent, the agent gets excited (initiated black bar in Line 4). Step 2 to 5: the algorithm repeats the red left trapezoid and gets increasingly excited. Step 6: the algorithm reaches the excitement threshold; it marks the red left trapezoid as persistent and creates a new belief state associated with this experience (line 3: the current belief becomes red left trapezoid). Line 4: the algorithm becomes curious to play with the newly created belief. Step 7: the algorithm tries the red rectangle. Step 15: the algorithm tries the red left half-circle and obtains green left half-circle. Step 16: green left half-circle is marked as sporadic since it differs from the intended experience, the algorithm enters the arc construction stage. Step 17: the algorithm encounters again the red left trapezoid, the current belief state associates with this persistent experience. Step 77: similar, green right half-circle is sporadic. Step 350: The experiences are all marked and no more possible changes in the Petri net. The algorithm mood becomes confident (green circle). Arrived at this step, the algorithm can use the constructed Petri net to predict the consequences of its actions.

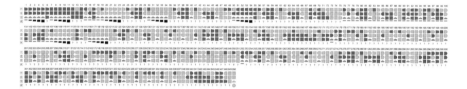

Fig. 4. Trace of the first 350 interaction cycles in our experiment. Line 1: intended experiences. Line 2: enacted experiences. Line 3: belief states: unknown (grey triangle)/known state represented by its corresponding persistent experience. Line 4: mood: curious (question mark), excited (increasing black bars), or confident (green circle).

5   Conclusion

We presented an algorithm that initially ignores the structure of the agent that it is controlling, as well as the structure of the environment with which this agent is interacting. As and when the algorithm controls the agent, it learns a model of the possibilities of interaction afforded by the agent/environment coupling.

An intriguing philosophical question is whether the learned model of the agent/environment coupling constitutes the best possible model accessible to the agent. When transposed to humans and their knowledge, Kantian-like positions affirm that we can only know the world as we experience it. By contrast, those admitting a more (scientific) realist epistemology believe that knowledge can be pushed further, toward the world in itself. Returning to our discussion, this suggests the following question: would it be possible and interesting to design an agent that would try to infer theories of the world in itself (the complete implementation of the agent and of the environment)?

Technically, the question remains how the algorithm could construct a theory that involves a representation of its environment with tiles and their sides. This may be feasible if the agent presupposes the existence of space. The agent could try to construct a simpler model based on the assumption that states of the agent/environment coupling are caused by the presence of objects in some locations in space. Such simplification of knowledge will be necessary when moving on towards more complex tasks. In more complex tasks, the Petri net will be too large and complex. Explaining the regularities in terms of the presence of objects in some locations in space will provide a powerful means to deal with such complexity.

References

1. Guériau, M., Armetta, F., Hassas, S., Billot, R., EI Faouzi, N.E.: A constructivist approach for a self-adaptive decision-making system: application to road traffic control. In: 28th International Conference on Tools with Artificial Intelligence, San Jose, pp. 670–677. IEEE (2016)
2. Russell, S., Norvig, P.: Artificial Intelligence, A Modern Approach. Pearson, Hoboken (2003)
3. Olivier, O.G., Mathieu, G.: Mastering the laws of feedback contingencies is essential to constructivist artificial agents. Constr. Found. 13(2), 300–301 (2018)
4. Jean, P.: The construction of reality in the child. J. Consult. Psychol. 19(1), 77 (1955)
5. Zlatev, J., Balkenius, C.: Introduction: Why "epigenetic robotics"? In: Balkenius, C., Zlatev, J., Kozima, H., Dautenhahn, K., Breazeal, C. (eds.) Lund University Cognitive Science Series, no. 85 (2001)
6. Mugan, J., Kuipers, B.: Autonomous representation learning in a developing agent. In: Computational and Robotic Models of the Hierarchical Organization of Behavior. Springer, Berlin (2013)
7. Georgeon, O.L.: Little AI: playing a constructivist robot. SoftwareX 6, 161–164 (2017)
8. Von Glasersfeld, E.: An Introduction to Radical Constructivism. The Invented Reality. W. W. Norton, London (1984)
9. Roesch, E.B., Spencer, M., Nasuto, S.J., Tanay, T., Bishop, J.M.: Exploration of the functional properties of interaction: computer models and pointers for theory. Constr. Found. 9(1), 26–33 (2013)
10. Findlay, J., Gilchrist, I.: Active Vision: The Psychology of Looking and Seeing. Oxford University Press, New York (2003)
11. Georgeon, O.L., Casado, R.C., Matignon, L.A.: Modeling biological agents beyond the reinforcement-learning paradigm. Procedia Comput. Sci. 71, 17–22 (2015)
12. De Loor, P., Manac'h, K., Tisseau, J.: Enaction-based artificial intelligence: toward co-evolution with humans in the loop. Mind Mach. 19(3), 319–343 (2009)
13. Georgeon, O.L., Bernard, F.J., Cordier, A.: Constructing phenomenal knowledge in an unknown noumenal reality. Procedia Comput. Sci. 71, 11–16 (2015)
14. Georgeon, O., Hassas, S.: Single agents can be constructivist too. Constr. Found. 9(1), 40–42 (2013)
15. Van der Aalst, W., Van Dongen, B., Herbst, J., Maruster, L., Schimm, G., Weijters, A.: Workflow mining: a survey of issues and approaches. Data Knowl. Eng. 47(2), 237–267 (2003)
16. Van der Aalst, W., Weijters, A., Maruster, L.: Workflow mining: discovering process models from event logs. IEEE Trans. Knowl. Data Eng. 16(9), 1128–1142 (2004)

Analyzing Motivating Texts for Modelling Human-Like Motivation Techniques in Emotionally Intelligent Dialogue Systems

Patrycja Swieczkowska[1,2](✉), Rafal Rzepka[1,2], and Kenji Araki[1]

[1] Hokkaido University, Sapporo, Japan
{swieczkowska,rzepka,araki}@ist.hokudai.ac.jp
[2] RIKEN AIP, Tokyo, Japan
http://arakilab.media.eng.hokudai.ac.jp

Abstract. In this paper, we present studies on human-like motivational strategies which eventually will allow us to implement motivational support in our general dialogue system. We conducted a study on user comments from a discussion platform Reddit and identified text features that make a comment motivating. We achieved around 0.88 accuracy on classifying comments as motivating or non-motivating using SVM and a shallow neural network. Our research is a first step for identifying computational features of a motivating piece of advice, which will subsequently be useful for implementing the ability to support the user by motivating him, imitating real human-to-human interactions.

Keywords: Human motivation understanding · Artificial agents
General dialogue systems

1 Introduction

Emotional intelligence is defined in psychology as, among others, the capability of individuals to recognize emotions of others and use emotional information to guide thinking and behavior [1]. In the context of various agents, from cognitive architectures to dialogue systems, this means being able to respond to various emotional states of the user, which already became an important research topic [2,3]. The authors of [2] dub systems capable of this "relational agents" and define them as "computational artifacts designed to build long-term, social-emotional relationships with their users." They recognize the importance of implementing human-like emotional intelligence in machines, which in case of dialogue systems is especially crucial, as they are specifically designed to interact with humans.

Likewise, we recognize the need to provide an artificial agent with the necessary skills to establish a successful cooperation with a human user. Specifically, we aim to design a dialogue system that would motivate the user to complete

© Springer Nature Switzerland AG 2019
A. V. Samsonovich (Ed.): BICA 2018, AISC 848, pp. 355–360, 2019.
https://doi.org/10.1007/978-3-319-99316-4_47

tasks on their schedule, regardless of the type of task or reason for being unmotivated, while employing motivating strategies inspired by real human utterances. The user will tell the system about their lack of motivation using natural language, and the system will produce a response meant to give the user some advice pertaining to the problem at hand. To be able to do this, the system has to know how to create motivational utterances. Therefore, the first step in our research was to examine texts containing motivational advice to find out what particular features they possess. Once we discover what makes an utterance motivating, we will be able to use that knowledge to generate motivational advice for the user.

The concept of motivation has been extensively studied in psychology (mostly involving gamification [4]), but not so much in the field of dialogue systems. While there exist papers suggesting various approaches to influencing motivational states in users [5,6], they do not contain experiments confirming their hypotheses. Therein lies the novelty of our research; to the best of our knowledge, ours will be the first dialogue system that motivates people to perform all kinds of tasks. Our system should be able to imitate actual human interactions, especially with respect to providing emotional support.

2   Datasets and Features

To be able to implement the ability to motivate users into our system, first we needed to determine what makes an utterance motivating. To achieve this, we analyzed posts and comments from an online discussion platform Reddit. Specifically, we accessed the subreddit r/getdisciplined[1], where the users post their issues with being unmotivated to perform various tasks, such as exercising, studying, going to work and so on. The posters then ask the commenters to provide them with some motivational advice. We chose this subreddit because the posts and comments are very close to the type of input and output of our end-goal dialogue system.

We noticed that best-ranked comments from r/getdisciplined usually had two things in common: they provided very specific, practical advice for the poster and included expressions of being able to relate to the poster's struggles, usually because the commenter had to deal with the same problem in the past. Consequently, we operationalized these characteristics into several features in our experiment. We also expanded our data with comments from non-motivational subreddits to compare them against our original motivational dataset.

2.1   Datasets

To train our classifier for the purpose of distinguishing motivating/advice comments from other texts, we created several datasets. Table 1 below is an overview of the number of comments in all our datasets. Specifically, we had two datasets of motivational/advice-giving comments from subreddits r/getdisciplined and

---

r/relationship_advice[2], and two datasets of other comments from subreddits r/pics[3] and r/todayilearned[4]. Both of the latter subreddits require posts to be interesting or amusing pictures and screenshots, which makes the comments unlikely to contain any advice.

Table 1. Overview of the datasets.

Subreddit	Training set	Test set
r/getdisciplined	1,352	573
r/relationship_advice	2,470	1,022
r/todayilearned	3,395	1,435
r/pics	2,255	1,144

2.2   Features

Before analyzing the comments, we pre-processed each one by detecting sentence boundaries, assigning part-of-speech tags, and, for some features, removing stopwords. From now on, we will use names *wordlist_withstops* for a list of all words in the comment, *wordlist_nostops* for the same list with stopwords removed, and *sent_list* for a list of sentences in the comment. The entire feature set included 13 features which were as follows:

Sentics scores of **aptitude, attention, pleasantness and sensitivity** measured with the Sentic library for Python on *wordlist_nostops*. The library is an API to the SenticNet knowledge base[5]. All the values fall on the scale between -1 and 1.

Sentiment score also provided by Sentic and measured on *wordlist_nostops*. The results fell on the five-point scale of strong negative / weak negative / neutral / weak positive / strong positive, which we converted accordingly to integer values between -2 and 2.

Relatability score measured by the percentage of first person pronouns (including possessive pronouns) in *wordlist_withstops*. The score range is 0 to 1.

Imperative score measured by the percentage of imperative/advice expressions in the comment text. Specifically we looked for clauses beginning with non-infinitive verbs but not ending in question marks, the word *please* preceding a verb, the phrase *why don't you* and phrases comprised of *you* or *OP* (*Original Poster*, which is a popular way of referring to the author of the post on Reddit) and a modal verb. Since most of these are bigrams, we counted the percentage on number of all words divided by 2. The score range is -1 to 1; negative values come from deducting points for question marks.

Specificity scores including six features: **Average Semantic Depth (ASD)** and **Average Semantic Height (ASH)** calculated from scores for each word in the sentence as retrieved from the WordNet ontology's hypernymy/hyponymy hierarchy, **Total Occurrence Count (TOC)** measured by obtaining occurrence count in WordNet for each word and adding up three lowest scores in a sentence, **Count of Named Entities (CNE)** and **Proper Nouns (CPN)** in the sentence, and **Sentence Length (LEN)**. These calculations were performed for each sentence in the comment using *sent_list* with stopwords removed. The final scores for the entire comment were obtained by adding up all the sentence scores. We then divided ASD, ASH, TOC and LEN by 100 and CNE and CPN by 10 to put the scores in the same numerical range as other features. Specificity score was first proposed by [7] to help extract suggestions and complaints from employee surveys and reviews. The goal was to find sentences containing specific content, which we adapted in our experiment for the purpose of finding specific motivational advice in comments. We calculated the scores as described in [7] with only slight modifications.

3 Experiments and Results

To test our method, we used two classifiers: a Support Vector Machine and a custom-made fully connected shallow neural network with two layers. We chose an SVM because they are robust thanks to their large margin optimization technique and perform well in classification problems. We then implemented the shallow neural network to see whether we could improve on the SVM results.

The neural network had an input layer of one unit per each of our 13 features, a hidden layer of 10 units using the tanh activation function and an output layer with one unit using the sigmoid activation function. For training, we used the parameters learning rate = 0.2 and number of iterations = 20,000. For the SVM computations we used an RBF kernel with the parameter C=20. All the parameters were chosen based on algorithm performance.

We achieved accuracy of 0.86 in the r/getdisciplined vs. r/pics and r/getdisciplined vs. r/todayilearned experiments using SVM. This score rose to 0.88 with the neural network.

To increase the amount of learning data, we then combined the r/getdisciplined and r/relationship_advice datasets into one *all-motivational* class, and r/pics and r/todayilearned into another *non-motivational* class. The performance of both SVM and the neural network was slightly worse, decreasing to 0.84.

Table 2 summarizes our results.

4 Discussion

In the test set for r/getdisciplined vs r/todayilearned we had 265 misclassified comments. Around 78% of them (207 comments) were r/getdisciplined comments misclassified as not motivating. A closer look at the data suggests that most such

Table 2. Accuracy scores for different datasets.

Dataset	Training examples	Test examples	SVM	Shallow NN
r/getdisciplined vs r/pics	3,607	1,717	0.86	0.87
r/getdisciplined vs r/todayilearned	4,747	2,008	0.86	0.88
all-motivational vs non-motivational	9,472	4,174	0.84	0.84

comments often had very similar values of ASD (*Average Semantic Depth*) and ASH (*Average Semantic Height*). While ASD and ASH are not dependent on each other, perhaps similar values of these features make it somehow harder to classify the comment properly.

Moreover, some misclassified comments may have contained unusual punctuation that influenced calculating individual features. For example, some comments had Imperative Score indicating that the algorithm did not detect some non-infinitive imperative verbs if they came after a quotation mark. This indicates that there is a need to improve the part of our algorithm responsible for calculating scores for this feature.

For the CPN (*Count of Proper Nouns*) calculations we used a chunker available in the nltk[6] library for Python. We specifically looked for NNP (proper noun, singular), NNPS (proper noun, plural) and CD (cardinal number) tags. However, in some cases capitalized words were counted as proper nouns even though they do not belong to this category. Therefore, it is important to improve calculations for this feature, perhaps by using a different tagging tool.

Further analysis revealed that error rates on both training and test sets were in the same close range at a relatively high value of 0.13. This suggests a bias problem in our algorithm. A way to fix this would be to use a deeper neural network, or one with a different architecture. Alternately, we can add more input features to our algorithm. For example, since similar ASD and ASH seem to be causing problems, perhaps we could additionally combine them into another feature that would be more informative to the algorithm, as well as adding some new features.

Lower scores for a bigger dataset (*all-motivational* vs *non-motivational)* suggest there might be also more noise in the data. Around half of the comments labeled as motivating (because they came from the r/getdisciplined subreddit) turned out to not contain any motivational advice, which in turn greatly contributed to the high error rate in our results. It is reasonable to expect that this problem gets bigger with bigger datasets. To counter this issue, in future research we can use only a few top-rated comments for each post, thus ensuring their motivational quality.

---

[6] https://www.nltk.org/.

5   Conclusions

In this paper, we proposed an algorithm for classifying texts as motivational (advice-giving) and non-motivational. This classification is a part of bigger research into motivational features of texts. Using a SVM and a shallow neural network, we achieved 0.88 accuracy on our test sets. As such, we have successfully determined a large subset of text features that make it motivating. Therefore, our research is an important step for giving machines the ability to support and motivate a human to perform various tasks, which can be thought of as computationally implementing a part of emotional intelligence.

To further improve the results, we are planning to add more features to our algorithm, as well as trying out different deep neural network architectures. In the long-term, after determining the features that make a text motivating, we will use this knowledge to construct a language generation module in our dialogue system to provide users with motivational advice suited to their needs.

Advice provided by Reddit commenters proved to be an invaluable source of knowledge how human users can be motivated. Artificial emotional intelligence can greatly benefit from knowledge provided by large online resources, not only by learning how to recognize human emotions, but also by acquiring knowledge about dealing with these emotions efficiently. We believe that in a long run, Wisdom of the Crowd-based knowledge might become a useful source for simulating emotional intelligence in cognitive architectures, improve their understanding of human behavior and enrich human communication with non-biological entities.

Acknowledgments. This work was supported by JSPS KAKENHI Grant Number 17K00295.

References

1. Coleman, A.: A Dictionary of Psychology, 3rd edn. Oxford University Press, Oxford (2008)
2. Bickmore, T., Picard, R.: Establishing and maintaining long-term human-computer relationships. ACM Trans. Comput.-Hum. Interact. (TOCHI) **12**(2), 239–327 (2005)
3. Callejas, Z., Griol, D., Lopez-Cozar, R.: Predicting user mental states in spoken dialogue systems. EURASIP J. Adv. Signal Process. **2011**(1), 6 (2011)
4. Mekler, E.D., Bruhlmann, F., Tuch, A.N., Opwis, K.: Towards understanding the effects of individual gamification elements on intrinsic motivation and performance. Comput. Hum. Behav. **71**, 525–534 (2017)
5. Tielman, M., Neerincx, M., Brinkman, W.-P.: Generating situation-based motivational feedback in a PTSD E-health system. In: International Conference on Intelligent Virtual Agents, pp. 437–440 (2017)
6. He, H.A., Greenberg, S., Huang, E.M.: One size does not fit all: applying the transtheoretical model to energy feedback technology design. In: Proceedings of the SIGCHI Conference on Human Factors in Computing Systems, pp. 927–936 (2010)
7. Deshpande, S., Palshikar, G.K., Athiappan, G.: An unsupervised approach to sentence classification. In: International Conference on Management of Data COMAD 2010, p. 88. Computer Society of India (2010)

Author Index

Printed in the United States
By Bookmasters